Animal Behavior

How and Why Animals Do the Things They Do

Volume 2: Function and Evolution

Ken Yasukawa, Editor

PRAEGER

AN IMPRINT OF ABC-CLIO, LLC
Santa Barbara, California • Denver, Colorado • Oxford, England

Copyright 2014 by Ken Yasukawa

All rights reserved. No part of this publication may be reproduced, stored in a retrieval system, or transmitted, in any form or by any means, electronic, mechanical, photocopying, recording, or otherwise, except for the inclusion of brief quotations in a review, without prior permission in writing from the publisher.

Library of Congress Cataloging-in-Publication Data

Animal behavior : how and why animals do the things they do / Ken Yasukawa, editor.
  pages cm
Includes bibliographical references and index.
  ISBN 978–0–313–39870–4 (hard copy : alk. paper) — ISBN 978–0–313–39871–1 (ebook) 1. Animal behavior. I. Yasukawa, Ken, 1949–
QL751.A6498   2013
591.5—dc23   2013012228

ISBN: 978–0–313–39870–4
EISBN: 978–0–313–39871–1

18 17 16 15 14   1 2 3 4 5

This book is also available on the World Wide Web as an eBook.
Visit www.abc-clio.com for details.

Praeger
An Imprint of ABC-CLIO, LLC

ABC-CLIO, LLC
130 Cremona Drive, P.O. Box 1911
Santa Barbara, California 93116-1911

This book is printed on acid-free paper ∞

Manufactured in the United States of America

Contents

1. Finding Food: Foraging Affects All Aspects of an Animal's Life   1
   Anna Dornhaus

2. Predation and Antipredator Behavior   23
   Theodore Stankowich

3. Animal Communication: How and Why Animals Communicate—or Do They?   51
   Zuleyma Tang-Martínez

4. Mating Systems and the Measurement of Sexual Selection   99
   Kenyon B. Mobley

5. The Evolution of Ornaments and Armaments   145
   Geoffrey E. Hill

6. Sexual Conflict: All's Not Fair in Love—It's War!   173
   Zenobia Lewis

7. A Nest of Vipers: Conflict and Cooperation in Families   195
   Michelle Pellissier Scott

8. Make Space Enough between You: Intraspecific Variation in Animal Spacing *Nancy G. Solomon and Brian Keane*	221
9. Let's Get Together: The Evolution of Social Behavior *Walter D. Koenig and Janis L. Dickinson*	257
10. Ecological and Evolutionary Feedbacks in the Evolution of Aggression *Renée A. Duckworth*	295
11. Altruism and Kinship *Lee Alan Dugatkin*	327
12. Evolutionary History of Behavior *Terry J. Ord*	339
Glossary	359
About the Editor and Contributors	377
About the Editorial Board	381
Index	383

1

Finding Food: Foraging Affects All Aspects of an Animal's Life

Anna Dornhaus

INTRODUCTION

Why do horses sleep only 3 hours a day, while lions may sleep 18 hours? Why do birds commonly sing at dawn rather than in the afternoon? Why are chimpanzees violent and patriarchal in their social life, while the closely related bonobos are peaceful and females dominate males? **Behavioral ecology** is the science that seeks to explain the diversity in lifestyle we find in animals, including the origins of our own human habits. Foraging plays a central role in this: all of the above questions have been answered by referring to the foraging needs of these respective species, demonstrating the central role that finding food plays in the evolution of animal behavior.

Foraging in animal behavior research refers to finding and collecting resources. Foraging for most animals is one of the most energetically costly and risky activities, particularly if the animal has to leave its home or shelter and cover great distances. Because of this, the requirements of foraging affect many other aspects of an animal's life. For example, plant matter has a much lower energy density than meat, which means that herbivores, such as horses (*Equus ferus*), may need to spend a lot more time collecting food than carnivores such as lions (*Panthera leo*), who can afford to spend a lot of time resting (Lesku et al., 2009). Some animals have so little reserves that they have to ration energetically expensive activities depending on the amount of food found each day—this is thought to be the case for small birds in cold climates, who can only afford to sing (to defend territory or attract mates) when they have survived the night and find that they still have fuel left over, leading to a dawn chorus (Hutchinson et al., 1993; Thomas, 1999). In the longer term, adaptations to the necessities of foraging change a species' overall life style. For example, bonobos (*Pan paniscus*) eat leaves and shoots of plants, which are common and easy to find. This allows groups of related females to stay together, which led to the evolution of strong female-female bonds. Chimpanzees (*Pan troglodytes*), on the other hand, mostly eat fruits and other foods that are patchy in distribution and require travel—which disadvantages females with infants, who travel slowly and thus have to search on their own if they want to avoid arriving at a resource already depleted by others in the group. This may prevent intense bonding and thus lead to the more patriarchal, aggressive societies we see in chimpanzees compared to bonobos (Wrangham, 1993). Even our unique human traits, such as large brains, cooperation, and trading, are thought to depend in large part on novel foraging behaviors by early hominids (Pennisi, 1999; Boehm, 2004).

Foraging is among the best-studied animal behaviors, due to the fact that foraging animals typically have to leave their shelter and move around, making them more visible and susceptible not only to predation but also to research (Kamil and Roitblat, 1985). The second advantage of foraging behavior, from a research perspective, is that success or failure in foraging is more easily defined and measured than in many other activities: a foraging animal collects a food of a certain caloric value in a certain amount of time. This led to foraging behavior being one of the first topics in which quantitative and predictive models were used intensively. The field of modeling foraging behavior came to be known as ***optimal foraging*** (Pyke et al., 1977; Stephens et al., 2007). The fact that measuring foraging success seemed so straightforward allowed models to actually predict optimal behavioral strategies in extraordinary detail. This led to some criticism, as some felt that such precise predictions were unrealistic or unhelpful (reviewed in Stearns & Schmid-Hempel, 1987; Raine et al., 2006). However, all fields of studying animal behavior, and perhaps of organismal biology, have benefited from the idea that the "target" of selection, or the optimal behavior, is not necessarily obvious to a casual observer and can be derived quantitatively and objectively within given constraints. Much of biology is now emulating similar approaches, trying to clarify precisely which factors will drive evolution of organismal traits and in what directions. These can be nonintuitive, particularly in cases where subtle differences in assumptions can actually lead to differences in the direction of selection (Kacelnik, 1993).

One behavior thus studied is the phenomenon of "partial loads" in ***central-place foragers***, particularly bees (Raine et al., 2006). Honey bees visiting a plentiful nectar source will often not carry as much back to their hive as they could, despite the fact that the same forager will make multiple subsequent trips to the same source to collect more nectar. This seems counterintuitive and indeed inefficient: *why not carry a full crop with every trip?* Mathematical models have helped us to understand the conditions under which partial, instead of full, loads may be advantageous to bees. First, if a flower offers nectar at decreasing returns, that is, the last bit of nectar is harder to extract, then it may be worth leaving before a full load has been collected in order to visit more other sources (Charnov, 1976). Second, it may be that if heavy loads impose high transport costs, and if a bee's lifespan is affected by how hard it has to work, the total nectar collected in the life of a foraging bee is in fact higher if only partial loads are collected (Schmid-Hempel et al., 1985; Wolf & Schmid-Hempel, 1989). Third, bees can recruit their nestmates to a rich food source; it may be worth limiting the amount of time the first forager spends there in favor of recruiting more bees to the resource before it is discovered by competitors, thus increasing the success of the group as a whole at the expense of the success of that first individual (Núñez, 1982; Varju & Núñez, 1991). Finally, since information about new, superior sources could arrive at any time at the hive, a forager may prefer to return to the hive frequently to check for such information in order to prevent wasting time at an inferior resource (Dornhaus et al., 2006). Mathematical models have allowed researchers to derive clear, quantitative predictions not only about how early a forager should leave a resource but also about which factors should affect this decision and how—for example, according to the third hypothesis, bees should collect smaller loads if the resource is of high quality, whereas the fourth hypothesis predicts that they should collect larger loads in this case. Empirical tests have revealed that the former is the case, supporting the idea that honey bee foragers sacrifice their own foraging success in order to improve that of the hive as a whole (Varju & Núñez, 1993). Without mathematical descriptions of the factors involved, behavioral scientists would not have been able to distinguish between these hypotheses. Moreover, the models pointed to what had to be investigated empirically in order to understand this aspect of foraging behavior.

In the example above, models helped explain a trait that had already been observed in bees, namely partial loads. But *what if the animals do not perform the behavior the model predicts?* Animals may not be optimally adapted, and any studies trying to identify the adaptive function of a trait or behavior are sometimes criticized as being ***adaptationist***, that is, naïvely assuming that evolution will be able to lead to perfection (Gould & Lewontin, 1979). However, this is misunderstanding the aim of such models. First, without quantitative models, we would not be able to determine whether a particular behavior in fact is the one that, given certain constraints, is most likely to allow the animal to reach its goal. Second, models help in defining the problem (the constraints and the goal) more sharply (Stephens et al., 2007). *What is the "goal" of any animal behavior?* The behaviors that are most likely to persist in a population are those that allow individuals to contribute maximally to the genetic makeup of the next generation, largely by maximizing the number of offspring produced. This is what we call ***evolution***. There are many components to this—finding and selecting a mate, avoiding predators and diseases, finding shelter or otherwise dealing with the elements, and in many cases establishing a place in a social community (Davies et al., 2012). But for many animals, the amount of food collected is a limiting factor in how many offspring can be produced. This means that behaviors that allow an animal to collect more food often have a direct effect on ***fitness***, enabling researchers to employ the shorthand of using "food collected" or "food collected per time" as so-called ***fitness proxies***, that is, as the currencies that animal behavior is adapted to maximize (Ydenberg & Hurd, 1998). Indeed animals perform behaviors that we can only understand as strategies to increase foraging success by just a few percent points; the existence of such behaviors shows that evolution can lead to subtle adaptations and optimization. It is important, however, to recognize the limitations of such fitness proxies—primarily, that other aspects of the animal's life are ignored for the purpose of a particular study.

If an animal appears to behave in a way very different from the predictions of a model, this may mean that the animal has not yet evolved the optimal behavior. However, in many cases, it is likely to mean that we have not yet properly understood which problem the animal is trying to solve: we may not understand the ***constraints*** under which it is operating, or we may not know the ***currency*** being maximized—for example, perhaps the animal needs to avoid predators or attempts to find mates while foraging (McArthur et al., 2012). Researchers now also study how the computational mechanisms that enable animals to choose behaviors themselves evolve (McNamara & Houston, 2009; Fawcett et al., 2012a, 2012b). For example, this allows us to understand predictable mistakes that result from applying a rule of thumb that provides the best results when averaging over the situations the animal commonly finds itself in but that may not appear optimal in a particular situation, or laboratory experiment (Kamil & Roitblat, 1985; Stephens et al., 2004; Seth, 2007; Houston, 2009). Interestingly, behaviors that are identified with such mental states as optimism, disappointment, and impulsivity may be interpreted as resulting from such rules of thumb—more on this below.

THE FORAGING TRIP

A honey bee leaves the hive to search for flowers that may provide nectar. The bee starts this trip because she is old enough, has a genetic predisposition to nectar foraging, and has sensed some commotion near the hive entrance, a likely indication that some of her nestmates are already foraging successfully. She (all honey bee workers are female) flies to a stand of cherry trees about a mile from the hive, using her knowledge of local landmarks to find them. Although she foraged successfully from these trees a few days ago, their bloom has faded and the bee starts searching elsewhere. Relying on her innate preferences, she picks some blue wildflowers, but they either do not contain nectar or it is too deep for her to reach. After each unsuccessful attempt, she travels a bit further until she can taste the sweet sugary nectar in some yellow flowers on a meadow. After first locating nectar, she changes her travel path to a more meandering one, testing every yellow flower but bypassing flowers of other colors or shapes. Many of the yellow flowers are empty, but some contain nectar—never more than a tiny drop, often less than a few microliters. After couple hundred flower visits and having collected about a third of her body weight (although she could carry more than two times as much), the bee returns to her hive. There, she passes the collected nectar on to hive-workers; but since she had to queue to complete this task, she decides not to communicate her discovery but to return immediately to the meadow for another foraging trip.

Foraging behavior has many facets. In an animal as well studied as the honey bee, we know many of the decisions made and the factors that will affect them. In bees, early naturalists as well as modern behavioral scientists have studied the factors that prompt bees to start and stop foraging trips, choose or reject certain flowers, navigate away from and back to their nest, and communicate with their nestmates to affect their foraging behavior in turn. I will discuss each of these aspects of foraging separately below.

TO FORAGE OR NOT TO FORAGE

At each trip or bout of foraging activity, an animal has to make several decisions, starting with whether to attempt to collect food at all, or when to do so. This, as any other decision, depends on the benefits and costs of foraging compared to another activity the animal could engage in instead. Benefits of foraging of course depend on the prospects for success, and many animals innately prefer to forage when cues such as weather indicate that resources will be plentiful. For example, desert ants are sensitive to overheating and will forage only under cool or moist conditions (Cerda et al., 1998). On the other hand, ants can also learn to associate temperature with food (Kleineidam et al., 2007). Many animals learn cues associated with resources, good foraging conditions, or even sophisticated combinations; for example, bees learn not only which areas are most profitable but also at what times of day which locations are most likely to provide food. They can even learn how likely it is that resources will change in profitability, that is, learn when it is worth learning (Dornhaus & Franks, 2008).

Whether a particular animal attends to environmental cues in its decision to forage should depend on how much these cues predict foraging success. For example, the availability of edible leaf material will not change in the short term for a caterpillar sitting on a plant, but the availability of nectar in flowers changes quickly and drastically with the time of day and weather, making it useful for bees to attend to such cues. Whether a particular cue provokes an innate response or is only used after the animal has learned to associate it with foraging success depends on how predictable its effects are over the long term (Dunlap & Stephens, 2009). The life cycle and **circadian rhythm** of specialist cactus bees, which forage on the flowers of a particular cactus, are adapted to the timing of nectar availability in this plant (McIntosh, 2005). The timing of foraging in generalist honey bee foragers, on the other hand, is flexible and is shaped by the bees' experience with the nectar availability of plants in the current season and site. This difference between **specialist** and **generalist foragers** has wide-ranging consequences for species resilience under environmental change.

The benefits of foraging depend not only on the expected returns but also on how the food is used. *Is food necessary to satisfy the immediate energy needs of the animal, is it stored for future need, or is it fed directly to offspring or other individuals?* If food collected is used primarily for **provisioning**, that is, stored or provided to other individuals, then foragers may attempt to minimize energetic expenditure of foraging because any food used up for the foraging activity itself will not be available to provision. When just foraging for themselves, however, animals may behave so as to maximize daily gain without regard for how much is spent attaining it (Ydenberg et al., 1994). For example, the sooty shearwater (*Puffinus griseus*), a seabird, may travel 1,500 km from its nest to catch fish in productive waters to maximize its own energy gain, but when feeding chicks will often perform short trips, which lead to less expensive travel but are less productive (Weimerskirch, 1998).

Do animals ever have "enough food"? The idea that foraging behavior may not so much evolve to maximize food collected as ensure that a certain minimum food requirement is met has been called ***satisficing*** in foraging research and has been somewhat controversial (Nonacs & Dill, 1993; Ydenberg & Hurd, 1998). It seems obvious from observation that many animals spend a lot of time in inactivity when they could be collecting more food—remember the sleeping lions mentioned earlier in this chapter. However, evolutionary considerations suggest that only if additional food can be converted into offspring or stored as a buffer against lean times, it should be collected regardless of current energy requirements. Indeed, many cases of inactivity are explained by the fact that either another factor limits reproduction in the long term or that current energetic costs of foraging exceed its benefits (Herbers, 1981). For example, there may be limits to the speed with which food can be digested or processed (Burd & Howard, 2005). Alternatively, if there is temporal variation in food availability and food cannot be stored, how many offspring can be raised may be limited by the minimum amount of food that can always be collected; in brief periods of food abundance, an animal would then appear to forage less than expected compared to food availability simply because not enough offspring are present to consume this food. Overall, however, the fact remains that animals vary in the degree to which they display inactivity (Careau et al., 2008) or sleep (Lesku et al., 2009), and that the function of neither is completely understood (Cirelli & Tononi, 2008).

FINDING FOOD

Once an animal has decided to forage, it has to actually locate its prey or food source. The pattern in which animals search is thought to resemble a ***random walk***, in which each new step is in a new, random direction. "Correlated" random walks imply that the new direction is affected by the previous direction, usually such that the animal is more likely to keep a straight heading rather than choosing a particular different direction. Even within these assumptions of a seemingly mindless search algorithm, a search path can be optimized for the presumed density and clumping of food sources (Nathan et al., 2008; Sims et al., 2008). However, many animals usually forage in a known area and may return to previously profitable locations or travel along known paths (called ***traplines***). For example, bumble bees tend to visit several known profitable flowers in a fixed sequence (Thomson, 1996). These bees are doing multiple foraging trips in a row, and the traplining strategy evens out the time intervals after each flower is revisited. It turns out that this is the optimal strategy for resources that slowly but at a constant rate renew themselves, such as flowers that typically produce nectar continuously (Ohashi et al., 2007). If a food item has been located by a forager, the forager will often concentrate further search in the same area, a phenomenon loosely termed "win-stay, lose-shift" (implying that if you find something, stay close; if not, shift to another area), a strategy that is optimal if food distribution is patchy in nature (Hodges, 1985).

WHAT NOT TO EAT

Once a forager comes near a prey item, it has to both detect it as a stimulus and decide to attack it. Detection will depend on the foragers' physiological sensitivity to the color, pattern, smell, or sounds produced by the prey but also on the foragers' experience and current state of mind. Detecting a stimulus from a noisy scene is often hard; much research in computer science is devoted to trying to replicate the skills of even simple animals in visually recognizing particular items under varying and often patchy lighting, with a noisy background, and from different angles (Smeulders et al., 2000; Galleguillos & Belongie, 2010). In addition, although some food items, such as flowers for bees, actively signal their presence (because flowers are pollinated in exchange for the nectar they offer), others, such as prey animals to be eaten, will be trying to avoid detection. Also, foragers will likely encounter many objects that are either harmful or a waste of time but that resemble the sought prey items. For example, orchids have no nectar but mimic other flowers to fool bees (Dafni, 1984; Gigord et al., 2002), yellow or white crab spiders will camouflage on flowers of the same color (Chittka, 2001), and noxious butterflies may look similar to tasty ones to a bird (Skelhorn & Rowe, 2010; Stoddard, 2012). Foragers thus need to be discriminating and will often evolve particular sensory sensitivity to the most useful stimulus associated with their preferred food (Sabbah et al., 2010). In addition, foragers will often reject items that appear too similar to unwanted items, even to the point of rejecting many suitable ones; this is called the ***peak shift*** phenomenon in ***signal detection theory*** (Chittka et al., 2009; Lynn et al., 2005).

If a forager is searching for a particular prey type, it tends to be able to detect items of this type more easily than other potential prey; the forager is said to have a ***search image*** (Dukas & Ellner, 1993; Dukas, 2004). Why this is the case is not entirely clear, but it may be that even if foragers can recognize several prey types they have difficulty holding more than one in their working memory, where it is readily accessible for comparison with the scene encountered. For example, bees are known to show "flower constancy," where profitable flowers are ignored in favor of visiting a series of flowers of the same type, sometimes for an extended period. If a bee switches to a different flower type, it appears to take longer to recognize it. This may reflect the neural costs of accessing information that is not currently in the working memory of the bee (Waser et al., 1996). Several other explanations for flower constancy have also been proposed, however, from the difficulty of learning to handle multiple flower types to avoiding competition with other bees. Indeed, specialization on particular food sources is thought to confer several advantages—ability to specialize on finding, handling, and digesting food as well as savings in cognitive complexity of the forager (Bernays, 2001).

Once a food item has been identified by the forager, the forager is still faced with choices: sometimes among several foods, sometimes between choosing to attack a particular food item now or to ignore it in favor of searching for a better one (Stephens et al., 2007). These choices are referred to as ***diet selection*** and may be based simply on how long it takes to handle particular food items or on a multitude of factors affecting item quality, including specific nutrients required by the forager (Sih & Christensen, 2001). Animals use a variety of direct and indirect ways to assess the quality of a resource. One of the interesting aspects of this is that the quality assessment appears to be strongly influenced by recent experience: animals exhibit behavior that appears to show disappointment or even optimism and pessimism (Morgan et al., 2012; Nettle & Bateson, 2012). Together, these phenomena have been termed ***contrast effects*** because they appear when animals experience a contrast between what they have experienced in the past and what they are experiencing now. It is not entirely clear why they occur—after all, if a particular food item is worth more than the effort to handle it, the animal should be using it regardless of whether it is better or worse than expected. However, this is not how animals act. Similarly unexpected, apparently nonoptimal behavior has been shown in other contexts, such as when animals choose a small reward now over a much bigger one later ("impulsivity" [Stevens et al., 2005]), when risky alternatives are preferred or rejected compared to stable ones with the same average yield ("risk sensitivity" [Caraco et al., 1990; Cartar & Abrahams, 1996]), or when irrelevant options bias choice ("irrationality" [Shafir et al., 2002]). New research is now uncovering the actual cognitive mechanisms, or ***decision rules***, that animals use and how they have evolved. Often, particular decision rules evolve because they usually, across environments or stimuli typically encountered by a particular type of animal, function well; this does not mean that they provide optimal solutions in all contexts (Houston et al., 2007; Fawcett et al., 2012b). Contrast effects, impulsivity, risk sensitivity, and irrationality may be examples of such rules—for example, animals adapted to catching flighty insects may be more impulsive than those using stable resources such as plants (Stevens et al., 2005). In addition, previous experience allows the animal to assess its own state as well as that of its environment, allowing better prediction of what is likely to be experienced in the near future (Schuck-Paim et al., 2004). This provides an intriguing angle to view human emotions: *are our irrational-seeming reactions in some contexts the result of evolved decision rules* (Bateson et al., 2011; Nettle & Bateson, 2012)? Many of the insights on diet selection, or how animals judge whether to use a particular resource, also apply to contexts other than foraging, for example when selecting mates; very similar models are also used for humans. For example, models in **sequential choice theory** calculate whether a prey item of a particular quality and taking a particular time to handle and digest should be chosen depending on what other items the forager is likely to encounter later (Stephens et al., 2007); the same models can be used to predict whether an animal should choose a mating partner of a particular quality (given the distribution of potential mating partners) (Castellano et al., 2012) and can also be used to calculate whether a person should take a job offered based on what he or she estimates will be offered in the future and other problems of this nature.

Choices while foraging are also affected by other aspects of an animal's life, such as the risks of encountering predators, mates, or competitors (McArthur et al., 2012; Wajnberg, 2012). Animals are almost always vigilant to avoid a predator attack and often switch between looking for food and surveying the area for possible threats. If the probability of being attacked is higher in particular patches, animals demand "hazardous duty pay" (Brown & Kotler, 2004); in other words, such patches have to be particularly profitable to be visited by a forager. Another aspect is that some areas may be safer, for example if they provide more cover, and will be preferred in the presence of predators; this can lead to changes in overall food use, as in deer who avoid open aspen stands when wolves are present in the area (Fortin et al., 2005). Similarly, animals often avoid competition with individuals of the same or other species. This may be achieved, for example, by specializing on a subset of the suitable foods or by restricting activity to particular times. Dominant individuals or species will then often forage during the most profitable times and on the most profitable sources, while others may be displaced to suboptimal ones (Alanara et al., 2001). For example, in desert ants foraging on cacti for nectar (from **extrafloral nectaries**; see below), the less dominant species forage later in the day, at higher temperatures, putting them at higher risk of dehydration and overheating (Morris et al., 2005).

When a forager ends its foraging bout or leaves a patch of resources, it may then return to its nest to unload and start a new foraging bout or turn to other activities. If it does this after every trip, it is central-place foraging (Orians & Pearson, 1979); other animals have no fixed home base and instead roam widely within a home range or over large distances (Hays et al., 2006). When to stop foraging in a patch is a decision that depends largely on how the rate of acquiring food from the patch changes over time and how long the forager has to travel between the food patch and its home base. If the food source provides ***decreasing marginal returns***, that is, it becomes harder and harder to extract food, the forager will eventually leave it to find a new, fresh resource, possibly well before it has collected all it can (see discussion on partial loads above). How foragers navigate the route back to their nest or among foraging patches is another interesting research area that is discussed in Volume 1, Chapter 12.

EXTRACTING REWARD

So far, we have mostly discussed how to find and recognize food and how to decide whether to attack it—whether it is a prey animal or a plant or anything else. However, actually capturing the prey and extracting the food reward can be a challenge in itself and may be a stereotyped behavior or exhibit a high degree of innovation; it may also involve sophisticated behavioral sequences, whether innate or learned. Consider the emerald cockroach wasp (*Ampulex compressa* [Fouad et al., 1996; Libersat, 2003]). The wasp needs live but subdued insect prey to feed to her developing offspring. The female will hunt for cockroaches and, once they are found, sting into the brain of the roach precisely in the suboesophagal ganglion. This, apparently, turns the roach into the wasp's puppet zombie! The roach appears to lose all will of its own and is guided—live and able to walk—to its doom in the form of the wasp's nest and her larvae (Libersat et al., 2009). This foraging strategy on the part of the wasp is not only surprising and unique, it is also fairly complex but very likely completely genetically determined. Many other animals, on the other hand, have been shown to learn, and even pass on to others in a type of ***culture***, particular foraging strategies useful in a local area (Volume 1, Chapter 4). For example, dolphins off the coast of western Australia have been shown to harvest cup-shaped sponges to use as a "glove" over their sensitive snouts when digging on the sea floor (Krutzen et al., 2005); great tits (*Parus major*) famously learned to open milk bottles with aluminum caps left on British doorsteps (Fisher & Hinde, 1949; Lefebvre, 1995); and apes such as chimpanzees and orangutans have been shown to master a variety of tool-using techniques in foraging, from cracking nuts with wooden hammers and anvils to fashioning spears for hunting monkeys (Whiten et al., 1999; van Schaik et al., 2003). *Why do some animals have such sophisticated, flexible foraging* *strategies*? It was long thought that carnivores, hunting for live animal prey, needed to be particularly clever and perhaps had evolved more cognitive and sensory abilities as a result. However, the evidence for this is not clear (Garamszegi et al., 2002). On the other hand, it seems that innovative foraging behavior enables some animals to invade new habitats more easily (Sol et al., 2002).

SOCIAL FORAGING

Many animals do not just rely on their own skills and experience to find and identify palatable and profitable food sources (Galef & Giraldeau, 2001; Laland, 2004; see also Volume 1, Chapter 4). For example, many birds will closely watch others in the flock searching for seeds on the ground; if another bird appears to have found something, they will then quickly join it to exploit that same resource. This strategy has been termed "scrounging" (as opposed to "producing," which refers to birds searching for food themselves) (Giraldeau et al., 1994). Bees may also watch conspecifics choose flowers and instead of attempting to use the exact same flower may use this **public information** to pick other flowers of that same color (instead of using **private information**, or sampling many flower types themselves to determine which species provide the most nectar) (Leadbeater & Chittka, 2007). Rats are known to learn the scent of palatable food from the smell at the mouth of other rats, enabling them to quickly spread information about novel food types in their group (Galef, 2009). Cliff swallows, and other birds that nest in large groups, will watch for other birds returning with a mouthful of insects; if they spot one, they will follow it out to the presumably profitable hunting grounds (Brown, 1988). If information is exchanged at a nest site or colony as in this case, the colony is referred to as **information center**. In all these examples, individuals are using **cues** gained simply by observing behavior of other foragers; the producers of that information do not behave in such a way as to improve the information exchange. However, many cooperative group-living animals also purposefully exchange information in the form of **signals** (Seeley, 1998): honey bees signal the direction and distance of flower patches in the "waggle dance" (Gruter & Farina, 2009), some ants lay pheromone trails to lead nestmates to food sources (Hölldobler & Wilson, 1990), other ant species **teach** (facilitate learning by others at a cost to themselves) their nestmates the route to food by leading them in a tandem run to it (Richardson et al., 2007; Hoppitt et al., 2008). Highly cooperative groups may also improve their foraging efficiency in other ways, such as by employing division of labor (Powell & Franks, 2006; Burns & Dyer, 2008); sometimes this even occurs in non-cooperating groups (Giraldeau & Lefebvre, 1986).

Interestingly, both the explicit signals and the cues generated by foragers may also be exploited by animals from other species (Coolen et al., 2003). In addition, many organisms rely partially or completely on other species to aid them in food acquisition—such interactions are termed mutualisms (Sachs & Simms, 2006; Aanen & Hoekstra, 2007). For example, plants provide all food for bees and many other pollinators (nectar and pollen) and gain pollination services in return. Plants also provide food in return for protection from herbivores, such as to many ants, which feed on extrafloral nectaries. Corals are colonies of polyps (cnidarian animals) that rely on their algal symbionts to generate nutrients by photosynthesis. Such algal symbionts are also present in several other animals and even fungi: lichens are the result of such a fungal-algae cooperation. Cleaner fish rely on larger fish, called clients, to come to their "cleaning station," where the cleaner fish eats both parasites on the skin as well as sometimes the skin of the client—a very interesting interaction with subtle choices made by both clients and cleaners about whom to pick and whom to pick on (Barbu et al., 2011). Such mutualisms are quite common, and in fact all of us carry many microorganisms, which are relying on us for food and shelter. Some of them help us digest complex foods, some crowd out harmful bacteria (e.g., Li et al., 2008). And each animal cell contains mitochondria, which are often called the power plants of the cell but used to be free-living bacteria. These were engulfed by an early eukaryotic cell, starting a cooperation based on mutual exchange of nutrients and initiating the evolution of complex life on earth (Dyall et al., 2004).

INDIVIDUAL TASTES

We sometimes like to think that each human is particularly unique because we all come with our own personalities, preferences, and behavioral idiosyncrasies. This is so; but it turns out that this is also true for probably all other animals (Dall et al., 2012; Sih et al., 2004). Some individual fish are bolder than others of the same species, seeming to taunt predators (Ward et al., 2004); some spiders are more aggressive to both their mates and enemies than others (Pruitt et al., 2008); some bumble bees will only forage on purple flowers while their own sisters prefer other flower types (Raine & Chittka, 2007). These "animal personalities" at first seemed strange to scientists: after all, there must be one optimal strategy, and *why would any individuals stray from it*? However, we now understand that **personalities**, stable behavioral differences among individuals of the same species, can evolve for many reasons. For example, herbivores need to not only ingest enough of all necessary nutrients but also avoid accumulating toxins (practically all plants produce toxins to deter herbivory). This can lead to individual preferences or "tastes" being affected by individual experience, if grazing animals need to learn which foods are palatable or develop resistances to particular toxins (Provenza et al., 2003). Individuals may also differ in their body size or growth rate, leading to differences in risk taking and thus food and food-patch preferences (Stamps, 2007); this can lead to a linkage between multiple behavioral traits, called a ***behavioral syndrome*** (Sih et al., 2004). In addition, different individuals may specialize on particular foods to avoid competition with others of the same species (Dall et al., 2012).

FOOD AND LIFESTYLE

Food types differ enormously in their spatial and temporal distribution. Some, such as plant leaves (e.g., grass) are available continuously over wide areas; others are hard to find and extinguish quickly, such as flowers that may only offer pollen within an hour of sunrise; and yet others must be subdued before they can be consumed, such as dangerous prey animals. In the first case, the main concern of a forager will be selecting the right type of food, such as a plant with few toxins; in the second case, a forager will focus on how to collect information about food availability as efficiently as possible; and in the third, a forager may need to acquire sophisticated handling or hunting strategies. Although all of these foragers have to locate, select, and handle food, their resulting foraging strategies will be different. They will be likely to have sophisticated and subtle adaptations in those parts of the foraging process that make the most difference to foraging success.

Such differences in priorities can also occur within species. For example, the reproductive success of males in many species is limited more by how many mating partners they can acquire than by how much food they can collect. Conversely, the reproductive success of females is rarely limited by the number of mating partners because males typically can provide more sperm in a single mating than a female can produce offspring. This has the effect that female distribution in space is often determined by resource availability, whereas male distribution in space may mostly depend on where females are. The different types of overlap between male and female ranges can lead to different mating systems, such as pair bonds, promiscuity, or harems defended by a single male (Davies & Lundberg, 1984).

EXPLAINING DIVERSITY IN ANIMAL BEHAVIOR

Honey bee foragers are parts of a highly coordinated collective foraging system and visit hundreds of individual flowers to which they may travel half a million times their body length (this equals 1,000 km, or 625 miles, for humans) in a single trip but return home after each excursion to share their food (Seeley, 1985). Chimpanzees may or may not share food but will copy each other's sophisticated techniques for extracting it. Birds may not search for food at all but follow others to take it from them when they have located any. Snakes and spiders often gain enough nutrients in a single meal that they can afford to be inactive for weeks or months afterwards; hummingbirds and bats must find food every day or risk starvation; and sloths must spend practically all their waking time eating to be able to extract enough nutrients from their hard-to-digest food. These are examples of the amazing diversity of lifestyles and diets in animals. Behavioral ecology, as a science, attempts to explain why this diversity exists, that is, which factors led to the evolution of different foraging strategies in different animals (Davies et al., 2012). Experiments in which the behavior of the animal or its food source is manipulated can show which foraging strategy performs best under particular conditions. Comparative studies, in which the foraging strategies of several species of animal are compared and related to other factors such as habitat type, body size, and so forth, can show which of these factors appear to have affected the evolution of particular foraging behaviors historically. Modeling studies can give precise predictions of how subtle adaptations may affect foraging success. The insights gained from this research have direct and practical applications, for example in the management of grazing animals (Provenza et al., 2003) or pollination of crops (Weinberg & Plowright, 2006). But we have also learned that the beautiful variety seen in nature of animal lifestyles, cognition, and social behavior can only be understood if we understand the foraging habits of these species.

ACKNOWLEDGMENTS

I would like thank the NSF (grants IOS-1045239 and IOS-0841756) for support.

REFERENCES AND SUGGESTED READING

Aanen, D. K. & R. F. Hoekstra. (2007). The evolution of obligate mutualism: If you can't beat 'em, join 'em. *Trends in Ecology and Evolution*, 22, 506-509.

Alanara, A., M. D. Burns, & N. B. Metcalfe. (2001). Intraspecific resource partitioning in brown trout: The temporal distribution of foraging is determined by social rank. *Journal of Animal Ecology*, 70, 980–986.

Barbu, L., C. Guinand, R. Bergmuller, N. Alvarez, & R. Bshary. (2011). Cleaning wrasse species vary with respect to dependency on the mutualism and behavioural adaptations in interactions. *Animal Behaviour*, 82, 1067–1074.

Bateson, M., B. Brilot, & D. Nettle. (2011). Anxiety: An evolutionary approach. *Canadian Journal of Psychiatry—Revue Canadienne De Psychiatrie*, 56, 707–715.

Bernays, E. A. (2001). Neural limitations in phytophagous insects: Implications for diet breadth and evolution of host affiliation. *Annual Review of Entomology*, 46, 703–727.

Boehm, C. (2004). What makes humans economically distinctive? A three-species evolutionary comparison and historical analysis. *Journal of Bioeconomics*, 6, 109–135.

Brown, C. R. (1988). Enhanced foraging efficiency through information centers: A benefit of coloniality in cliff swallows. *Ecology*, 69, 602–613.

Brown, J. S. & B. P. Kotler. (2004). Hazardous duty pay and the foraging cost of predation. *Ecology Letters*, 7, 999–1014.

Burd, M. & J. J. Howard. (2005). Central-place foraging continues beyond the nest entrance: The underground performance of leaf-cutting ants. *Animal Behaviour*, 70, 737–744.

Burns, J. G. & A. G. Dyer. (2008). Diversity of speed-accuracy strategies benefits social insects. *Current Biology*, 18, R953–R954.

Caraco, T., W. U. Blanckenhorn, G. M. Gregory, J. A. Newman, G. M. Recer, & S. M. Zwicker. (1990). Risk-sensitivity: Ambient temperature affects foraging choice. *Animal Behaviour*, 39, 338–345.

Careau, V., D. Thomas, M. M. Humphries, & D. Reale. (2008). Energy metabolism and animal personality. *Oikos*, 117, 641–653.

Cartar, R. V. & M. V. Abrahams. (1996). Risk-sensitive foraging in a patch departure context: A test with worker bumble bees. *American Zoologist*, 36, 447–458.

Castellano, S., G. Cadeddu, & P. Cermelli. (2012). Computational mate choice: Theory and empirical evidence. *Behavioural Processes*, 90, 261–277.

Cerda, X., J. Retana, & S. Cros. (1998). Critical thermal limits in Mediterranean ant species: Trade-off between mortality risk and foraging performance. *Functional Ecology*, 12, 45–55.

Charnov, E. (1976). Optimal foraging: The marginal value theorem. *Theoretical Population Biology*, 9, 129–136.

Chittka, L. (2001). Camouflage of predatory crab spiders on flowers and the colour perception of bees. *Entomologia Generalis*, 25, 181–187.

Chittka, L., P. Skorupski, & N. E. Raine. (2009). Speed-accuracy tradeoffs in animal decision making. *Trends in Ecology and Evolution*, 24, 400–407.

Cirelli, C. & G. Tononi. (2008). Is sleep essential? *PLoS Biology*, 6, e216. doi: 10.1371/journal.pbio.0060216.

Coolen, I., Y. van Bergen, R. L. Day, & K. N. Laland. (2003). Species difference in adaptive use of public information in sticklebacks. *Proceedings of the Royal Society of London, B*, 270, 2413–2419.

Dafni, A. (1984). Mimicry and deception in pollination. *Annual Review of Ecology and Systematics*, 15, 259–278.

Dall, S. R. X., A. M. Bell, D. I. Bolnick, & F. L. W. Ratnieks. (2012). An evolutionary ecology of individual differences. *Ecology Letters*, 15, 1189–1198.
Davies, N. B., J. R. Krebs, & S. A. West. (2012). *An Introduction to Behavioural Ecology*. Fourth Edition. Hoboken, NJ: Wiley-Blackwell.
Davies, N. B. & A. Lundberg. (1984). Food distribution and a variable mating system in the dunnock, *Prunella modularis*. *Journal of Animal Ecology*, 53, 895–912.
Dornhaus, A., E. J. Collins, F. X. Dechaume-Moncharmont, A. I. Houston, N. R. Franks, & J. M. McNamara. (2006). Paying for information: Partial loads in central place foragers. *Behavioral Ecology and Sociobiology*, 61, 151–161.
Dornhaus, A. & N. R. Franks. (2008). Individual and collective cognition in ants and other insects (hymenoptera: Formicidae). *Myrmecological News*, 11, 215–226.
Dukas, R. (2004). Causes and consequences of limited attention. *Brain Behavior and Evolution*, 63, 197–210.
Dukas, R. & S. Ellner. (1993). Information-processing and prey detection. *Ecology*, 74, 1337–1346.
Dunlap, A. S. & D. W. Stephens. (2009). Components of change in the evolution of learning and unlearned preference. *Proceedings of the Royal Society of London, B*, 276, 3201–3208.
Dyall, S. D., M. T. Brown, & P. J. Johnson. (2004). Ancient invasions: From endosymbionts to organelles. *Science*, 304, 253–257.
Fawcett, T., S. Hamblin, & L.-A. Giraldeau. (2012a). We can study how mechanisms evolve without knowing the rules of chess or the workings of the brain. *Behavioral Ecology*, 24, 14–15.
Fawcett, T. W., S. Hamblin, & L.-A. Giraldeau. (2012b). Exposing the behavioral gambit: The evolution of learning and decision rules. *Behavioral Ecology*, 24, 2–11.
Fisher, J. & R. A. Hinde. (1949). The opening of milk bottles by birds. *British Birds*, 42, 347–357.
Fortin, D., H. L. Beyer, M. S. Boyce, D. W. Smith, T. Duchesne, & J. S. Mao. (2005). Wolves influence elk movements: Behavior shapes a trophic cascade in Yellowstone National Park. *Ecology*, 86, 1320–1330.
Fouad, K., F. Libersat, & W. Rathmayer. (1996). Neuromodulation of the escape behavior of the cockroach *Periplaneta americana* by the venom of the parasitic wasp *Ampulex compressa*. *Journal of Comparative Physiology A: Sensory Neural and Behavioral Physiology*, 178, 91–100.
Galef, B. G. (2009). Strategies for social learning: Testing predictions from formal theory. *Advances in the Study of Behavior*, 39, 117–151.
Galef, B. G. & L.-A. Giraldeau. (2001). Social influences on foraging in vertebrates: Causal mechanisms and adaptive functions. *Animal Behaviour*, 61, 3–15.
Galleguillos, C. & S. Belongie. (2010). Context based object categorization: A critical survey. *Computer Vision and Image Understanding*, 114, 712–722.
Garamszegi, L. Z., A. P. Møller, & J. Erritzoe. (2002). Coevolving avian eye size and brain size in relation to prey capture and nocturnality. *Proceedings of the Royal Society of London, B*, 269, 961–967.

Gigord, L. D. B., M. R. Macnair, M. Stritesky, & A. Smithson. (2002). The potential for floral mimicry in rewardless orchids: An experimental study. *Proceedings of the Royal Society of London, B*, 269, 1389–1395.

Giraldeau, L.-A. & L. Lefebvre. (1986). Exchangeable producer and scrounger roles in a captive flock of feral pigeons: A case for the skill pool effect. *Animal Behaviour*, 34, 797–803.

Giraldeau, L.-A., C. Soos, & G. Beauchamp. (1994). A test of the producer-scrounger foraging game in captive flocks of spice finches, *Lonchura punctulata*. *Behavioral Ecology and Sociobiology*, 34, 251–256.

Gould, S. J. & R. C. Lewontin. (1979). The spandrels of San Marco and the Panglossian paradigm: A critique of the adaptationist programme. *Proceedings of the Royal Society of London, B*, 205, 581–598.

Gruter, C. & W. M. Farina. (2009). The honeybee waggle dance: Can we follow the steps? *Trends in Ecology and Evolution*, 24, 242–247.

Hays, G. C., V. J. Hobson, J. D. Metcalfe, D. Righton, & D. W. Sims. (2006). Flexible foraging movements of leatherback turtles across the North Atlantic Ocean. *Ecology*, 87, 2647–2656.

Herbers, J. M. (1981). Time resources and laziness in animals. *Oecologia*, 49, 252–262.

Hodges, C. M. (1985). Bumble bee foraging: Energetic consequences of using a threshold departure rule. *Ecology*, 66, 188–197.

Hölldobler, B. & E. O. Wilson. (1990). *The Ants*. Cambridge, MA: Harvard University Press.

Hoppitt, W. J. E., G. R. Brown, R. Kendal, L. Rendell, A. Thornton, M. M. Webster, & K. N. Laland.. (2008). Lessons from animal teaching. *Trends in Ecology and Evolution*, 23, 486–493.

Houston, A. I. (2009). Flying in the face of nature. *Behavioural Processes*, 80, 295–305.

Houston, A. I., J. M. McNamara, & M. D. Steer. (2007). Do we expect natural selection to produce rational behaviour? *Philosophical Transactions of the Royal Society of London, B*, 362, 1531–1543.

Hutchinson, J. M. C., J. M. McNamara, & I. C. Cuthill. (1993). Song, sexual selection, starvation and strategic handicaps. *Animal Behaviour*, 45, 1153–1177.

Kacelnik, A. (1993). Leaf-cutting ants tease optimal foraging theorists. *Trends in Ecology and Evolution*, 8, 346–348.

Kamil, A. C. & H. L. Roitblat. (1985). The ecology of foraging behavior: Implications for animal learning and memory. *Annual Review of Psychology*, 36, 141–169.

Kleineidam, C. J., M. Ruchty, Z. A. Casero-Montes, & F. Roces. (2007). Thermal radiation as a learned orientation cue in leaf-cutting ants (*Atta vollenweideri*). *Journal of Insect Physiology*, 53, 478–487.

Krutzen, M., J. Mann, M. R. Heithaus, R. C. Connor, L. Bejder, & W. B. Sherwin. (2005). Cultural transmission of tool use in bottlenose dolphins. *Proceedings of the National Academy of Sciences, USA*, 102, 8939–8943.

Laland, K. N. (2004). Social learning strategies. *Learning and Behavior*, 32, 4–14.

Leadbeater, E. & L. Chittka. (2007). The dynamics of social learning in an insect model, the bumblebee (*Bombus terrestris*). *Behavioral Ecology and Sociobiology*, 61, 1789–1796.

Lefebvre, L. (1995). The opening of milk bottles by birds: Evidence for accelerating learning rates, but against the wave-of-advance model of cultural transmission. *Behavioural Processes*, 34, 43–53.

Lesku, J. A., T. C. Roth II, N. C. Rattenborg, C. J. Amlaner, & S. L. Lima. (2009). History and future of comparative analyses in sleep research. *Neuroscience and Biobehavioral Reviews*, 33, 1024–1036.

Li, M., B. Wang, M. Zhang, M. Rantalainen, S. Wang, H. Zhou, Y. Zhang, J. Shen, X. Pang, M. Zhang, H. Wei, Y. Chen, H. Lu, J. Zuo, M. Su, Y. Qiu, W. Jia, C. Xiao, L. M. Smith, S. Yang, E. Holmes, H. Tang, G. Zhao, J. K. Nicholson, L. Li, & L. Zhao. (2008). Symbiotic gut microbes modulate human metabolic phenotypes. *Proceedings of the National Academy of Sciences, USA*, 105, 2117–2122.

Libersat, F. (2003). Wasp uses venom cocktail to manipulate the behavior of its cockroach prey. *Journal of Comparative Physiology A: Neuroethology Sensory Neural and Behavioral Physiology*, 189, 497–508.

Libersat, F., A. Delago, & R. Gal. (2009). Manipulation of host behavior by parasitic insects and insect parasites. *Annual Review of Entomology*, 54, 189–207.

Lynn, S. K., J. Cnaani, & D. R. Papaj. (2005). Peak shift discrimination learning as a mechanism of signal evolution. *Evolution*, 59, 1300–1305.

McArthur, C., P. Orlando, P. B. Banks, J. S. Brown. (2012). The foraging tightrope between predation risk and plant toxins: A matter of concentration. *Functional Ecology*, 26, 74–83.

McIntosh, M. E. (2005). Pollination of two species of ferocactus: Interactions between cactus-specialist bees and their host plants. *Functional Ecology*, 19, 727–734.

McNamara, J. M. & A. I. Houston. (2009). Integrating function and mechanism. *Trends in Ecology and Evolution*, 24, 670–675.

Morgan, K. V., T. A. Hurly, M. Bateson, L. Asher, & S. D. Healy. (2012). Context-dependent decisions among options varying in a single dimension. *Behavioural Processes*, 89, 115–120.

Morris, W. F., W. G. Wilson, J. L. Bronstein, & J. H. Ness. (2005). Environmental forcing and the competitive dynamics of a guild of cactus-tending ant mutualists. *Ecology*, 86, 3190–3199.

Nathan, R., W. M. Getz, E. Revilla, M. Holyoak, R. Kadmon, D. Saltz, & P. E. Smouse. (2008). A movement ecology paradigm for unifying organismal movement research. *Proceedings of the National Academy of Sciences, USA*, 105, 19052–19059.

Nettle, D. & M. Bateson. (2012). The evolutionary origins of mood and its disorders. *Current Biology*, 22, R712–R721.

Nonacs, P. & L. Dill. (1993). Is satisficing an alternative to optimal foraging theory? *Oikos*, 67, 371–375.

Núñez, J. A. (1982). Honeybee foraging strategies at a food source in relation to its distance from the hive and the rate of sugar flow. *Journal of Apicultural Research*, 21, 139–150.

Ohashi, K., J. D. Thomson, & D. D'Souza. (2007). Trapline foraging by bumble bees: IV. Optimization of route geometry in the absence of competition. *Behavioral Ecology*, 18, 1–11.

Orians, G. H. & N. E. Pearson. (1979). On the theory of central place foraging. In D. J. Horn, R. D. Mitchell, & G. R. Stairs (eds.), *Analyses of Ecological Systems* (pp. 154–177). Columbus, OH: Ohio State University Press.

Pennisi, E. (1999). Human evolution: Did cooked tubers spur evolution of big brains? *Science*, 283, 2004–2005.

Powell, S. & N. R. Franks. (2006). Ecology and the evolution of worker morphological diversity: A comparative analysis with *Eciton* army ants. *Functional Ecology*, 20, 1105–1114.

Provenza, F. D., J. J. Villalba, L. E. Dziba, S. B. Atwood, & R. E. Banner. (2003). Linking herbivore experience, varied diets, and plant biochemical diversity. *Small Ruminant Research*, 49, 257–274.

Pruitt, J. N., S. E. Riechert, & T. C. Jones. (2008). Behavioural syndromes and their fitness consequences in a socially polymorphic spider, *Anelosimus studiosus*. *Animal Behaviour*, 76, 871–879.

Pyke, G. H., H. R. Pulliam, & E. L. Charnov. (1977). Optimal foraging: A selective review of theory and tests. *Quarterly Review of Biology*, 52, 137–154.

Raine, N. E. & L. Chittka. (2007). Flower constancy and memory dynamics in bumblebees (hymenoptera: Apidae: Bombus). *Entomologia Generalis*, 29, 179–199.

Raine, N. E., T. C. Ings, A. Dornhaus, N. Saleh, & L. Chittka. (2006). Adaptation, genetic drift, pleiotropy, and history in the evolution of bee foraging behavior. *Advances in the Study of Behavior*, 36, 305–354.

Richardson, T. O., P. A. Sleeman, J. M. McNamara, A. I. Houston, & N. R. Franks. (2007). Teaching with evaluation in ants. *Current Biology*, 17, 1520–1526.

Sabbah, S., R. L. Laria, S. M. Gray, C. W. Hawryshyn. (2010). Functional diversity in the color vision of cichlid fishes. *BMC Biology*, 8, 133. doi: 10.1186/1741-7007-8-133

Sachs, J. L. & E. L. Simms. (2006). Pathways to mutualism breakdown. *Trends in Ecology and Evolution*, 21, 585–592.

Schmid-Hempel, P., A. Kacelnik, & A. I. Houston. (1985). Honeybees maximize efficiency by not filling their crop. *Behavioral Ecology and Sociobiology*, 17, 61–66.

Schuck-Paim, C., L. Pompilio, & A. Kacelnik. (2004). State-dependent decisions cause apparent violations of rationality in animal choice. *PLoS Biology*, 2, e402. doi: 10.1371/journal.pbio.0020402.

Seeley, T. D. (1985). *Honeybee Ecology: A Study of Adaptation in Social Life*. Princeton, NJ: Princeton University Press.

Seeley, T. D. (1998). Thoughts on information and integration in honey bee colonies. *Apidologie*, 29, 67–80.

Seth, A. K. (2007). The ecology of action selection: Insights from artificial life. *Philosophical Transactions of the Royal Society of London, B*, 362, 1545–1558.

Shafir, S., T. A. Waite, & B. H. Smith. (2002). Context-dependent violations of rational choice in honeybees (*Apis mellifera*) and gray jays (*Perisoreus canadensis*). *Behavioral Ecology and Sociobiology*, 51, 180–187.

Sih, A., A. Bell, & J. C. Johnson. (2004). Behavioral syndromes: An ecological and evolutionary overview. *Trends in Ecology and Evolution*, 19, 372–377.

Sih, A. & B. Christensen. (2001). Optimal diet theory: When does it work, and when and why does it fail? *Animal Behaviour*, 61, 379–390.

Sims, D. W., E. J. Southall, N. E. Humphries, G. C. Hays, C. J. A. Bradshaw, J. W. Pitchford, A. James, M. Z. Ahmed, A. S. Brierley, M. A. Hindell, D. Morritt, M. K. Musyl, D. Righton, E. L. C. Shepard, V. J. Wearmouth, R. P. Wilson, M. J. Witt, & J. D. Metcalfe. (2008). Scaling laws of marine predator search behaviour. *Nature*, 451, 1098–1102.

Skelhorn, J. & C. Rowe. (2010). Birds learn to use distastefulness as a signal of toxicity. *Proceedings of the Royal Society of London, B*, 277, 1729–1734.

Smeulders, A. W. M., M. Worring, S. Santini, A. Gupta, & R. Jain. (2000). Content-based image retrieval at the end of the early years. *IEEE Transactions on Pattern Analysis and Machine Intelligence*, 22, 1349–1380.

Sol, D., S. Timmermans, & L. Lefebvre. (2002). Behavioural flexibility and invasion success in birds. *Animal Behaviour*, 63, 495–502.

Stamps, J. A. (2007). Growth-mortality tradeoffs and "personality traits" in animals. *Ecology Letters*, 10, 355–363.

Stearns, S. C. & P. Schmid-Hempel. (1987). Evolutionary insights should not be wasted. *Oikos*, 49, 118–125.

Stephens, D., J. S. Brown, & R. C. Ydenberg. (2007). *Foraging: Behavior and Ecology*. Chicago: University of Chicago Press.

Stephens, D. W., B. Kerr, & E. Fernández-Juricic. (2004). Impulsiveness without discounting: The ecological rationality hypothesis. *Proceedings of the Royal Society of London, B*, 271, 2459–2465.

Stevens, J. R., E. V. Hallinan, & M. D. Hauser. (2005). The ecology and evolution of patience in two new world monkeys. *Biology Letters*, 1, 223–226.

Stoddard, M. C. (2012). Mimicry and masquerade from the avian visual perspective. *Current Zoology*, 58, 630–648.

Thomas, R. J. (1999). Two tests of a stochastic dynamic programming model of daily singing routines in birds. *Animal Behaviour*, 57, 277–284.

Thomson, J. D. (1996). Trapline foraging by bumblebees: I. Persistence of flight-path geometry. *Behavioral Ecology*, 7, 158–164.

van Schaik, C. P., M. Ancrenaz, G. Borgen, B. Galdikas, C. D. Knott, I. Singleton, A. Suzuki, S. S. Utami, & M. Merrill. (2003). Orangutan cultures and the evolution of material culture. *Science*, 299, 102–105.

Varju, D. & J. Núñez. (1991). What do foraging honeybees optimize? *Journal of Comparative Physiology A: Sensory, Neural, and Behavioral Physiology*, 169, 729–736.

Varju, D. & J. Núñez. (1993). Energy balance versus information exchange in foraging honeybees. *Journal of Comparative Physiology A: Sensory, Neural, and Behavioral Physiology*, 172, 257–261.

Wajnberg, E. (2012). Multi-objective behavioural mechanisms are adopted by foraging animals to achieve several optimality goals simultaneously. *Journal of Animal Ecology*, 81, 503–511.

Ward, A. J. W., P. Thomas, P. J. B. Hart, & J. Krause. (2004). Correlates of boldness in three-spined sticklebacks (*Gasterosteus aculeatus*). *Behavioral Ecology and Sociobiology*, 55, 561–568.

Waser, N. M., L. Chittka, M. V. Price, N. Williams, & J. Ollerton. (1996). Generalization in pollination systems, and why it matters. *Ecology*, 77, 1043–1060.

Weimerskirch, H. (1998). How can a pelagic seabird provision its chick when relying on a distant food resource? Cyclic attendance at the colony, foraging decision and body condition in sooty shearwaters. *Journal of Animal Ecology*, 67, 99–109.

Weinberg, D. & C. M. S. Plowright. (2006). Pollen collection by bumblebees (*Bombus impatiens*): The effects of resource manipulation, foraging experience and colony size. *Journal of Apicultural Research*, 45, 22–27.

Whiten, A., J. Goodall, W. C. McGrew, T. Nishida, V. Reynolds, Y. Sugiyama, C. E. G. Tutin, R. W. Wrangham, & C. Boesch. (1999). Cultures in chimpanzees. *Nature*, 399, 682–685.

Wolf, T. J. & P. Schmid-Hempel. (1989). Extra loads and foraging life span in honeybee workers. *Journal of Animal Ecology*, 58, 943–954.

Wrangham, R. W. (1993). The evolution of sexuality in chimpanzees and bonobos. *Human Nature: An Interdisciplinary Biosocial Perspective*, 4, 47–79.

Ydenberg, R. & P. Hurd. (1998). Simple models of feeding with time and energy constraints. *Behavioral Ecology*, 9, 49–53.

Ydenberg, R. C., C. V. J. Welham, R. Schmid-Hempel, P. Schmid-Hempel, & G. Beauchamp. (1994). Time and energy constraints and the relationships between currencies in foraging theory. *Behavioral Ecology*, 5, 28–34.

2
Predation and Antipredator Behavior

Theodore Stankowich

INTRODUCTION: THE DANGER LURKING

It is dusk on the coast of northern California, and a five-year-old adult female black-tailed deer quietly drops her head in a clearing of short coastal grass to feed on small, soft, green shoots. The wind is blowing gently from the west off the ocean, and the air is calm and quiet. As the dark of night approaches, she gradually diverts more attention to her surroundings, aware of the increasing danger.

Meanwhile a lone cougar picks up a faint scent of deer and quietly begins to make his way through the brush towards its sources. He steps quietly between two rows of bushes that line a grassy clearing when the deer comes into view about 40 meters away—he immediately freezes. The cougar stares at the deer with interest, but he's a bit too far away to attack just yet. He needs to get closer—much closer. He tries to survey the clearing for a better spot to launch an attack—it has to be downwind of her so she won't be able to smell him coming but close enough to capture her quickly. When he spots a large clump of brush about 10 meters from her, he starts to creep his way quietly through the brush. Suddenly, a crackle of dead sticks and leaves under his paw pierces the night air and catches the attention of the deer, causing her to raise her head to investigate. He freezes.

She sniffs the air for any signs of danger. Staring into the bushes, bits and pieces of a lurking figure are visible—a patch of tawny fur, a tail tip, then a pair of eyes stare in her direction from a gap in the leaves. It's unclear what kind of animal it is. If it's a cougar it's far enough away that she would be easily able to flee an attack. A coyote, on the other hand, could be one of many in the area, making escape a difficult proposition. Becoming visibly agitated, she expels air through her nose loudly, stomping her right foreleg on the earth. Her tail flags back and forth. The heads of four other deer in the vicinity pop up from the vegetation—two adult females and their fawns. Cuing in to what this deer is staring at, they join in, staring into the bushes, snorting, and footstamping. She takes a few steps toward the bushes to try to get a better look, stomping and snorting frequently.

The noise and movement of the deer have made it clear to the cougar that he has been spotted and has lost a possible meal. With his quarry on high alert, there is no way to use stealth to get close enough to have any chance of a successful attack. The shadowy figure moves away and disappears deep into the brush.

After several minutes, the collective snorts become softer and the deer begins to move off to forage in a nearby wooded area that might offer more concealment. She has survived another evening but will probably avoid this area for a while.

This vignette of a dangerous encounter between a black-tailed deer (*Odocoileus hemionus hemionus*) and a cougar (*Puma concolor*), while somewhat anthropomorphic and dramatized, illustrates the dynamic interplay of behavioral decisions that both predators and prey must make when they encounter each other. Very few encounters result in attacks and captures by predators, so a large volume of literature has examined the roles of behavioral decision making and morphological adaptations that help prey reduce the probability of being captured by predators and help predators increase the probability of successfully acquiring a meal.

Several behavioral scientists have developed frameworks for conceptualizing the diversity and functions of antipredator behaviors. Steve Lima and Lawrence Dill (1990) developed an attack sequence for predator-prey encounters that compartmentalizes behavioral decisions that each party must make at each step, resulting in a probability of death given an encounter situation. A modified form of this flow chart appears in Figure 2.1a. Briefly, they divide the interaction into four main steps: encounter, detection, attack, and **capture/escape**. Because there are many possible nonlethal outcomes of the interaction, often only a small proportion of all encounters lead to prey death. This is a useful framework to build from because it nicely illustrates the different opportunities each party has to either halt the sequence from leading to death (in the case of the prey) or push the sequence forward toward prey death (in the case of the predator), and there are many behavioral decisions and morphological adaptations that help both parties in their cause. To this end, R. Brian Langerhans (2007) presented a different type framework for studying

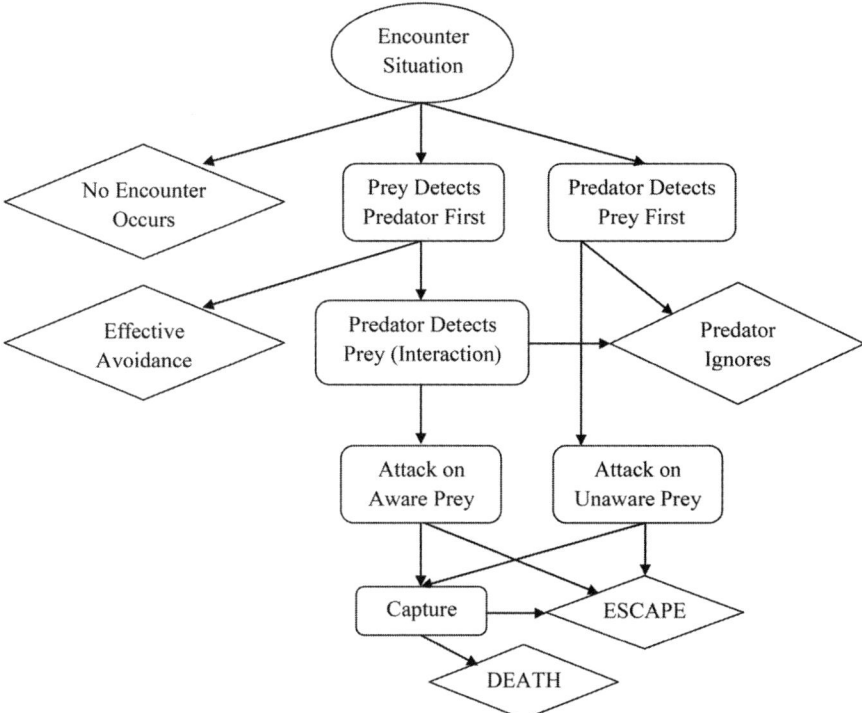

Figure 2.1a. Frameworks for understanding predator-prey interactions. Redrawn and modified substantially from Lima and Dill (1990), this flow chart shows the possible outcomes of an "encounter situation" between a prey and its predator, emphasizing that only a proportion of all encounters lead to the death of the prey through consumption or fatal injury. (Copyright 1990 Canadian Science Publishing. Used by permission.)

such antipredator traits in terms of how they affect prey fitness (Table 2.1). His brief review of antipredator adaptations examines the myriad of ways animals seek to avoid encountering predators, avoid a predator's detection, and deter attacks and consumption by predators.

In this chapter, I modify the Lima and Dill (1990) sequence, adding several new intermediate steps, and integrate Langerhans's (2007) trait-based approach to develop a comprehensive framework that demonstrates the opportunities for natural selection to shape both prey and predator traits that enhance fitness over the course of the predator-prey encounter sequence. In Figure 2.1b, ***recognition*** has been added to the ***detection phase*** to emphasize the fact that often an animal may detect the presence of another animal

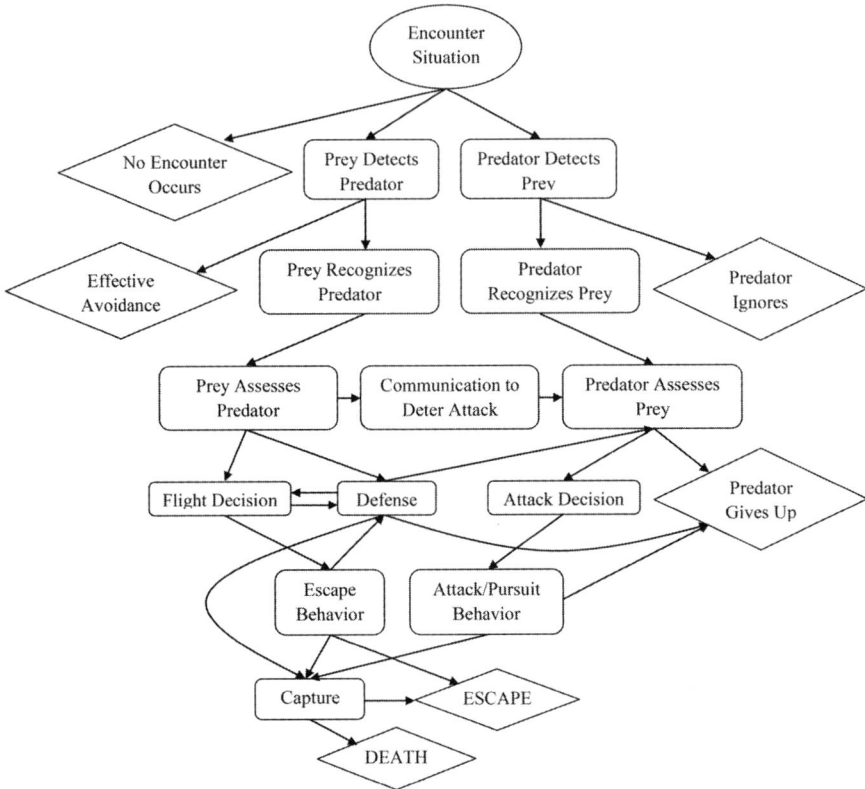

Figure 2.1b. Framework used in this chapter to demonstrate the myriad of ways that natural selection can shape assessment, behavioral decisions, and morphology in predators and prey.

in their area but still lack enough information to determine if it a dangerous or innocuous species. I have also added an ***assessment/communication phase*** before the ***attack/flight phase*** to account for incidences where the prey judges how much of a risk this particular individual poses to its safety, when the predator weighs the costs and benefits of attack and the likelihood of success, and when there is overt exchange of signals or cues between the parties that allow them to assess each other more rapidly. Finally, I have added the option of ***defense*** to the final stage to capture the cases where prey have formidable behavioral or morphological defenses with which to repel predators during physical combat or to deter them prior.

Using this general framework, my goal in this chapter is to give a brief overview of the many directions that predator-prey interactions can take given a

Table 2.1. Redrawn and edited from Langerhans (2007), this is a framework for understanding how antipredator traits affect direct fitness and survivorship, organized by the chronological stage of the predatory encounter in which they are employed. Some prey traits listed in the table are not discussed in this chapter. ASF: avoiding a predator's sensory field; DSF: avoiding detection within a predator's sensory field; ATD: attack deterrence; CPD: capture deterrence; CND: consumption deterrence.

Prey trait	How the antipredator trait directly enhances individual survivorship and fitness	
	Avoidance	Escape
Activity level	ASF	
Crypsis	DSF	
Development time	ASF	
Use predator-free habitat	ASF	
Active defense		CPD, CND
Aposematism		ATD
Attack diversion		CPD
Autotomy		CPD
Chemical defense		ATD, CPD, CND
Death feigning		ATD
Deimatic behavior		ATD, CPD
Mimicry		ATD
Protective morphologies		ATD, CPD, CND
Rapid retreat, protean behavior		CPD
Grouping	ASF, DSF	ATD, CPD, CND
Use protective habitat	ASF, DSF	ATD, CPD, CND
Vigilance	ASF, DSF	ATD, CPD, CND

single interaction. From the perspective of each party, the opponent is not a static, predictable automaton, but instead each party alters its behavior based on its opponent's behavior, the environmental situation, its own physical status, and its past experience with its opponent's species. New and important model systems of predator-prey study will be highlighted throughout the chapter to illustrate the past, present, and future of this exceptionally vibrant field of study. Finally, it should become clear by the end of this chapter that future studies of predator-prey interaction should focus on the behavioral decisions of *both* parties either simultaneously in natural interactions when possible or in dual studies of each party responding to dynamic controllable models of their opponent.

ENCOUNTER

In order for the attack sequence to even begin and for there to be any possibility of death for an animal, the predator and prey must come near enough to each other for the possibility of detection to occur. From the perspective of prey, they behave to minimize the probability that they will **encounter** potential predators, while hungry predators will behave to maximize this possibility. Strategies vary from selecting permanent home ranges or habitat types with more prey or fewer predators, to coordinated hunting or foraging trips by predators, to, as is detailed below, daily or seasonal migrational patterns.

Many species move around in their habitat to change the probability of coming within a detectable distance of predators or prey, and these cyclic migrations may occur daily or seasonally. Eduardo Arraut and colleagues (2010) showed that male Amazonian manatees (*Trichechus inunguis*) spend the summer high-water season in deeper **várzea** lakes to take advantage of abundant foraging opportunities but migrate to rivers when the water level in lakes falls during the low-water season. Low water levels in lakes increase the risk of predation by jaguars (*Panthera onca*), caimans, and humans, and rivers offer fewer predators but also fewer foraging opportunities. **Diel** (daily) **vertical migration (DVM)** in pelagic environments is a common pattern of migration that has been associated with increased foraging opportunities and decreased probability of predation. Many studies have shown that zooplankton move to deeper waters during the daylight hours and return to shallower depths at night in order to avoid predatory fish that rely on vision to detect prey: capture rates in shallow illuminated waters are far greater than in darker environments (Zaret & Suffern, 1976). DVM, however, does occur in **mesopredatory** fish (midlevel predators) as well. Mark Scheuerell and Daniel Schindler (2003) found that juvenile sockeye salmon (*Oncorhynchus nerka*) showed similar DVM patterns, but there was a lack of DVM in their zooplankton prey. Zooplankton in the Alaskan lakes they studied remained at the well-lit surface of the lake throughout the day, while the salmon dove to greater depths during the day (Figure 2.2), maintaining a constant-light environment, thereby allowing them to reduce the odds of visual detection by larger **piscivorous** (fish-eating) fish, in which DVM were not detected.

Simply changing activity rates and schedules may affect the likelihood that predators and prey will encounter each other in a meaningful way. For

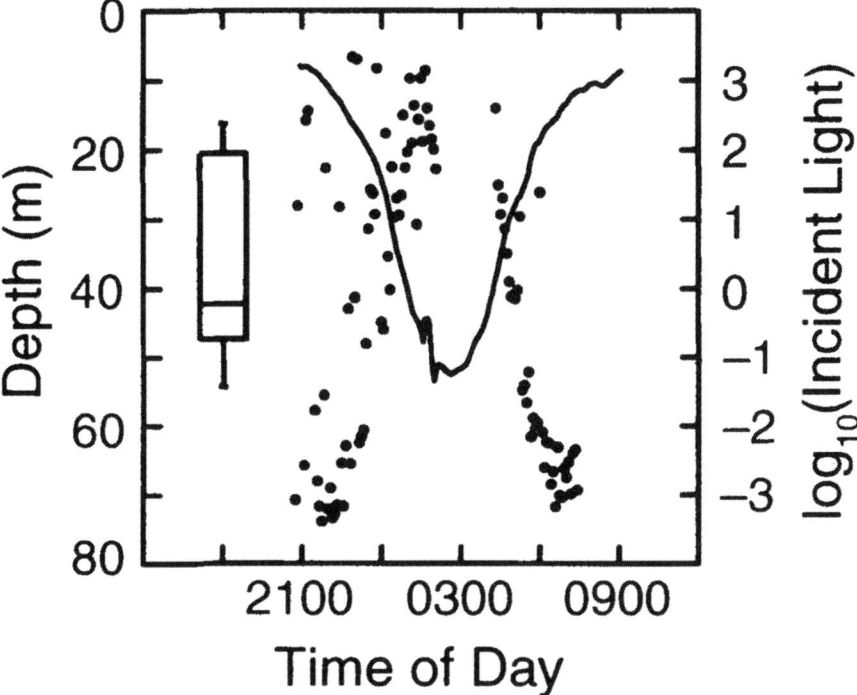

Figure 2.2. Box plot indicates the depth distribution of predatory fish of Nerka Lake (from Scheurell & Schnidler, 2003). Black dots show the time series of depth of juvenile sockeye salmon (*Oncorhynchus nerka*). The solid black line shows the log of light intensity at the surface of the water. Salmon exhibit DVM patterns, rising to the surface during the night when incident light is minimal to maintain a constant low-light environment when in the presence of potential predators, which stay near the surface. (Copyright 2003 Ecological Society of America)

terrestrial mammals and birds, **nocturnality** and **diurnality** offer different advantages and disadvantages. For prey, most large mammalian carnivores are nocturnal hunters, so being active during the day minimizes the likelihood of encountering them. However, most birds of prey are diurnal, and if you are small enough to be captured and killed by a large predatory bird, the daytime offers tremendous danger. Shifts in activity can be paired with migration patterns to add further control of one's susceptibility to predation or likelihood of encountering prey (e.g., DVM). Other physiological and foraging considerations (e.g., thermoregulation, availability of food) contribute to the decision of when to be active, and many species hedge their bets by staying close to sources of refuge that can offer protection should a predator enter the area.

Burrows, dense vegetation, shells, rocks, and so forth all offer outstanding protection from predators and reduce a prey animal's detectability.

Finally, living or hunting in groups may affect an individual's probability of encountering a predator or prey animal. For prey, if aggregations of prey are no more detectable than singletons, then predators may suffer increased search times between encounters with prey (Krause & Ruxton, 2002). Often groups of predators can increase their odds of detecting faint cues of distant prey animals. If one member of the group picks up on the scent of potential prey, others that did not detect the scent may benefit by following that trail and, in general, increasing the probability of encountering prey.

DETECTION AND RECOGNITION

Once prey and predator have encountered each other in time and space so that they are potentially detectable to each other, actual detection of the other party is not a foregone conclusion. There are multiple morphological (e.g., camouflage body shape and coloration) and behavioral (e.g., freezing) adaptations used to minimize the likelihood of detection, and there are sensory modalities both prey and predators use (e.g., binocular versus lateral vision, olfactory cues, electroreception, echolocation, vibrations) to maximize the likelihood of detecting the other party. From the above discussion of sockeye salmon (*Oncorhynchus nerka*) maintaining a constant-light environment to minimize detection by larger predators, it is clear that the distinction between adaptations that influence either the encounter or detection phase are "fuzzy" at best, but it is helpful to make a distinction between being in the same place at the same time so detection of another organism is even possible and actually detecting that other organism.

Detection of Other Organisms

Animals use every sensory modality to detect other organisms in their environment. While they are typically more useful for long-distance detection of organisms in the surrounding area, **olfactory cues** from urine, scent markings, and wind-blown odors of hidden organisms are helpful for detecting the presence of potential predators or prey in the immediate vicinity as well. For terrestrial organisms, the most commonly relied-on modality for detection of predators and prey is vision, and visual systems tend to evolve to suit both predator detection and foraging needs (e.g., predators often form **search images**, giving them the heightened ability to detect a prey item; see Chapter 1). Esteban Fernández-Juricic and his collaborators study the components of the avian visual system that aid in detection and how predatory species and prey species differ in their visual abilities. Typically predatory species have high acuity and larger binocular fields while prey have wider lateral areas to detect predators from all around them and lower visual acuity. Recent research shows that, for example, white-crowned sparrows (*Zonotrichia leucophrys*) and California towhees (*Pipilo crissalis*) are both prey species that have wider lateral visual fields and faster head-movement rates (a proxy for scanning behavior) than predatory species to enhance detection ability (Fernández-Juricic et al., 2011). However, these prey species show wider binocular fields than their aerial avian predators, which may be related to their ability to detect prey (e.g., seeds and small insects) at close distances from the beak. Further, studying visual morphology in concert with behavioral responses to predators and conspecifics allows these ornithologists to understand (1) trade-offs between opposing sensory needs (distant predator detection from all directions versus binocular vision for detecting and manipulating food on the ground) and (2) how animals can behaviorally compensate for differences in visual ability due to morphology (towhees have lower spatial resolution on the periphery of the retina than white-crowned sparrows and, therefore, have faster sideways head movements to compensate). Similarly, visual-field configuration and scanning behavior in birds of prey is matched to prey-searching strategies and environmental obstructions: red-tailed hawks (*Buteo jamaicensis*), which scan for prey in open habitats from high perches, have smaller binocular areas, a lower degree of eye movement, and make fewer head movements than Cooper's hawks (*Accipiter cooperi*), which ambush prey through brush in forested areas (O'Rourke, Hall, et al., 2010; O'Rourke, Pitlik, et al., 2010). Finally, increased prey group size can enhance detection speed (Krause & Ruxton, 2002; Pulliam, 1973; Treherne & Foster, 1980). This is most evident in species that respond behaviorally immediately upon detection of a predator, and individuals in these larger groups typically have reduced individual rates of vigilance (Cresswell, 1994).

Camouflage: Avoiding Detection

While both predator and prey have an interest in detecting the other as quickly as possible, both parties also seek to avoid being detected by the other, and there are multiple behavioral and morphological adaptations that have evolved to make organisms less detectable to other species. There are multiple forms of camouflage (Endler, 1981; Ruxton et al., 2004). **Background matching** is simply having a coloration that blends in with the background habitat (e.g., vegetation, soil) and is both common and effective. **Masquerade**, a variant of background matching, occurs when the organism is detectable but resembles an inedible object, such as a leaf swaying in the wind or a stick. This strategy is particularly common in terrestrial and aquatic arthropods that have exoskeletons that can be easily altered in shape and color by natural selection. ***Disruptive coloration*** is a phenomenon similar to background matching where animals use bold blocks of color or projections of the body to break up their body outline against the background environment. Cuttlefish provide a striking example of an animal capable of changing its coloration to either match the background or be disruptive. Another camouflage strategy particularly common in vertebrates is ***countershading***, where the animal's back is darker in color than its belly. Because light usually comes from above the animal, the darker shadow beneath the animal is compensated for by the lighter coloration, making the animal appear less tubular in shape if viewed from the side. This strategy is particularly effective in aquatic animals because, when viewed from above, their darker back helps blend in with the dark depths of the water, and when viewed from below, the lighter belly helps blend in with the light coming from above the water's surface. ***Transparency***, the dominant form of camouflage in invertebrates that live in the open water where there are no sources of refuge and no background surfaces to blend into, can be found in some jellyfish, comb jellies, and larval fish.

One very specialized alternative to traditional camouflage is called ***aggressive mimicry***, where the predator mimics some signal that either is benign to the prey so they do not avoid it or is attractive to the prey (i.e., a lure) so the prey approaches the predator. Deep-sea anglerfish live in dark ocean environments and use a bioluminescent fleshy appendage that hangs in front of their head to attract unsuspecting fish toward their mouth. Therefore, with aggressive mimicry, the predator avoids detection by diverting the target's attention towards some attractive object, allowing it to close the distance between itself and the prey without ever moving.

In addition to these morphological adaptations, many species have behavioral tactics to help avoid detection. Probably the most basic example is ***stillness***. Motion against the background environment is typically very easy to detect. Many predators remain motionless for long periods to avoid detection by their prey, and many species will remain motionless when they are not required to feed or move through their environment in order to minimize the chances of predators detecting their movements. Further, as an alternative to fleeing, many species will freeze if they have been moving when they detect a predator or prey (David, 2005). Young white-tailed deer (*Odocoileus virginianus*), for example, will freeze more often than adults, and freezing is more common in vegetative cover (Smith, 1991). Finally, masquerading species often accentuate their ruse by mimicking the natural movements of the objects they resemble. Mantids and leaf-like insects often sway their bodies from side to side to resemble foliage moving naturally in the wind.

The vast majority of both predator and prey species adopt some form of the above coloration to avoid detection, but some have evolved morphologies that make them more detectable. Often, greater detectability results from coloration that makes them more attractive to potential mates; many others are boldly colored to advertise their defensive abilities to predators, a phenomenon known as ***aposematism***. I will discuss this topic in detail later.

Recognition

Once detection has occurred, each party must also recognize the other as potential prey, irrelevant, a conspecific, or a potential threat. Recognition after detection is not a foregone conclusion: I have personally witnessed a black-tailed deer stare at a model leopard (extinct in the study site for more than 600,000 years) from 10 meters away without concern or recognition of danger as she chewed her food (Stankowich & Coss, 2007). There is a great literature on neural mechanisms governing recognition of objects as dangerous or potentially edible. Both processes are the product of inherited recognition patterns that evoke adaptive alert responses and plastic learning capabilities that allow for changes in local predator composition and temporal or spatial variation in prey availability. Richard Coss (Coss, 1999, 2010; Coss & Ramakrishnan, 2000; Stankowich & Coss, 2007) and Eberhard Curio (Curio, 1975, 1993) have studied the evolution and persistence of predator recognition in a variety of mammals (ground squirrels, black-tailed deer, and bonnet macaques [*Macaca radiata*]) and in birds (pied flycatchers [*Ficedula hypoleuca*]), respectively. In some prey species, perceptual capabilities are integrated with higher-order inferences of predator hunting strategies in different environments (Coss, 2010). Predators and prey use several different sensory modalities to recognize unique biologically important patterns and features of their opponent above background noise. Olfactory cues in the form of alarm odors of depredated conspecifics (e.g., ***Schreckstoff***) can be a potent cue for many aquatic species (Kats & Dill, 1998). Many terrestrial species are able to winnow out extraneous sounds and remain sensitive to relevant sounds of predators and prey, and countless studies across the animal kingdom have shown that common biological patterns (e.g., rosettes or spots, forward-facing eyes of predators) are potent recognition cues (Coss & Goldthwaite, 1995; Coss & Ramakrishnan, 2000; Coss et al., 2005; Curio, 1975). Recognition of some cues can be heritable so that recognition of evolutionarily constant predators or prey happens innately, or selection may also favor the ability to learn to recognize morphological characteristics in particular settings when encounters occur in unpredictable settings (see "Discussion" section for remarks on the role of experience and learning in predator-prey interactions). For sit-and-wait predators, prey may learn to avoid particular locations where threatening encounters or attacks have occurred previously, much like the deer in the introductory vignette. It is noteworthy, however, that sometimes recognition can be difficult, even to the point that animals must assess the organism for some time before deciding whether it is a potential predator or potential prey (Cooper & Stankowich, 2010).

ASSESSMENT AND COMMUNICATION

Assessment

Once one or both parties have recognized the other as either a potential threat or a potential meal, there may be a phase of assessment where the organisms gather information regarding the likelihood of attack or capture. Recently, there has been a growing interest in the factors that influence risk assessment during encounters with predators. Dan Blumstein and I (Stankowich & Blumstein, 2005) analyzed more than 100 previous studies of flight responses by prey to determine what types of factors influence how prey assess risk and to what degree they perceive fear. Animals perceive greater threat when predators behave as if they intend to attack or are targeting the subject, when they are more familiar with the predator due to experience (naïve individuals may be relatively fearless), when they are farther from refuge, and when they lack cryptic or defensive morphologies like armor or chemical defenses (Figure 2.3). Black-tailed deer (*Odocoileus hemionus columbianus*), like many other species, are particularly attentive to the behavior of the approaching predator, perceiving greater risk when the predator approaches more directly and at greater speeds (Stankowich & Coss, 2006), but they are not attuned to fine-scale changes in a predator's gaze or posture during approach that might further inform their decision to flee. A similar suite of factors (e.g., size, handling time, probability of success, hunger state) likely influences predators during their assessment of potential prey. David Scheel (1993), for example, found that lions (*Panthera leo*) spent longer durations stalking larger groups of prey and prey groups with more individuals in the head-up position than they did with smaller prey groups or those with fewer vigilant individuals, suggesting that lions assess more vigilant groups for longer because they are more difficult to attack. Ultimately, many models of assessment during foraging are applicable to predators assessing prey (see "Attack and Flight" section below and Chapter 1 in this Volume; Brown et al., 2011). In general, theory suggests that animals take in much more

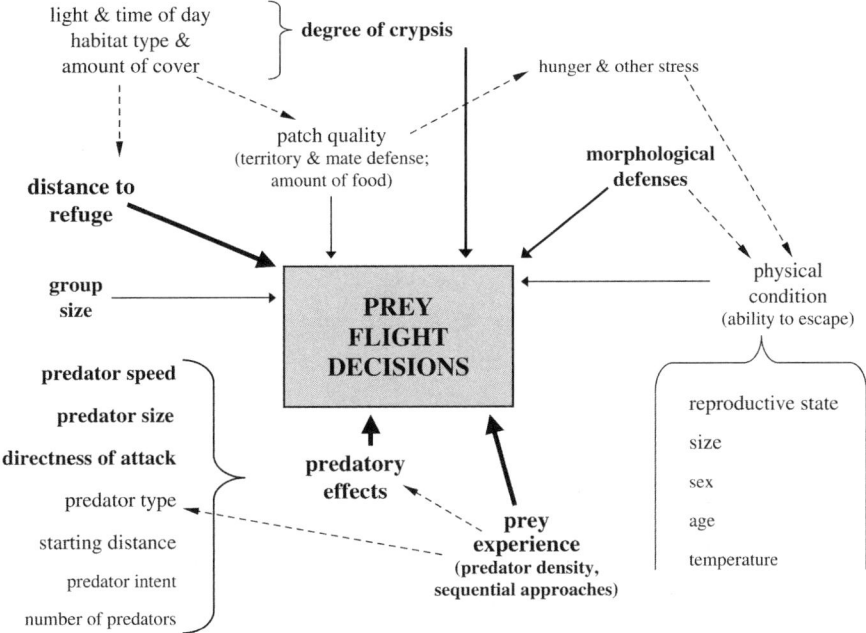

Figure 2.3. The diagram summarizes the results of a review and meta-analysis of the many factors that potentially influence flight-initiation distance in animals (from Stankowich & Blumstein, 2005). Thicker lines and bolder, larger fonts indicate that greater statistical and theoretical support exists for the influence of the factor on flight decisions. Among others, threatening predator behavior and longer distances significantly and consistently increase perceptions of fear in animals, while others like group size and physical condition had weaker or less consistent effects. (Copyright 2005 Royal Society and Highwire Press)

information than they are capable of dealing with and likely weigh and deemphasize certain factors during assessment: natural selection should favor individuals that pay attention to only one or a few factors to streamline decision making, something very important during fleeting interactions with potential predators and potential prey (Bernays & Wcislo, 1994).

Communicating with Predators

In many predator-prey interactions, the prey assesses that it would be successful in escaping to safety or defending itself should the predator choose to attack. However, any such escalation raises the risk of injury to the prey animal and, at the very least, prolonged fleeing behavior or combative interactions are certainly energetically costly. Natural selection, therefore, has favored individuals that can successfully communicate to a potential predator that any such attack would be fruitless and would only cause similar unnecessary risk and energy expenditure on the part of the predator.

When predators hunt using stealth, often simply advertising to the predator that it has been detected is enough to deter the predator from further pursuit. In the introductory vignette, the snorting and footstamping of the deer, in addition to potentially warning other deer in the area of a potential threat, is also detectable by the predator and is available as a signal that its cover has been blown. Often these behaviors become more overt and even risky advertisements of prey quality (Stankowich, 2010), whereby prey behave or harass the predator in a way signaling that any attack would likely be unsuccessful given the escape or defensive ability of the prey animal. Don Owings and Matt Rowe studied the **tail flagging** and **snake-harassment behavior** of California ground squirrels (*Spermophilus beecheyi*) for more than 30 years, showing that squirrels may sequentially approach and retreat from snakes while kicking substrate at them, flagging their tails back and forth, and even biting in order to deter further attack by the snake and indicate the defensive quality of the squirrel (Owings & Coss, 1977; Owings & Morton, 1997; Rowe & Owings, 1990, 1996; Swaisgood, Owings, et al., 1999; Swaisgood, Rowe, et al., 1999; Swaisgood et al., 2003). Recently, Aaron Rundus and colleagues (2007) demonstrated that these signals are tuned to the sensory modalities of the species of snake with which the squirrels are confronted. California ground squirrels will flag their tails when accosted by both rattlesnakes (*Crotalus oreganus*) and gopher snakes (*Pituophis melanoleucus*) but will increase blood flow to their tails to raise the temperature in the tail to emit radiant heat when flagging towards a rattlesnake, which can detect infrared radiation using pit organs, but not towards a gopher snake, which lacks pit organs (Figure 2.4; Rundus et al., 2007). Further, rattlesnakes were more likely shift to defensive postures than predatory postures when presented with a robotic squirrel with a heated flagging tail compared to one with an unheated flagging tail.

Aposematism

Visual communication in the form of body coloration is also a common way to advertise to and warn potential predators that they should not attempt an attack. Bold, contrasting color patterns are very often found on species that bear some sort of defensive weapon or trait that would make them an unpleasant meal for predators. Aposematism, or warning coloration, is found in a variety of animal groups but most commonly in insects, marine gastropods, and amphibians. Several teams of researchers have examined the roles of predator

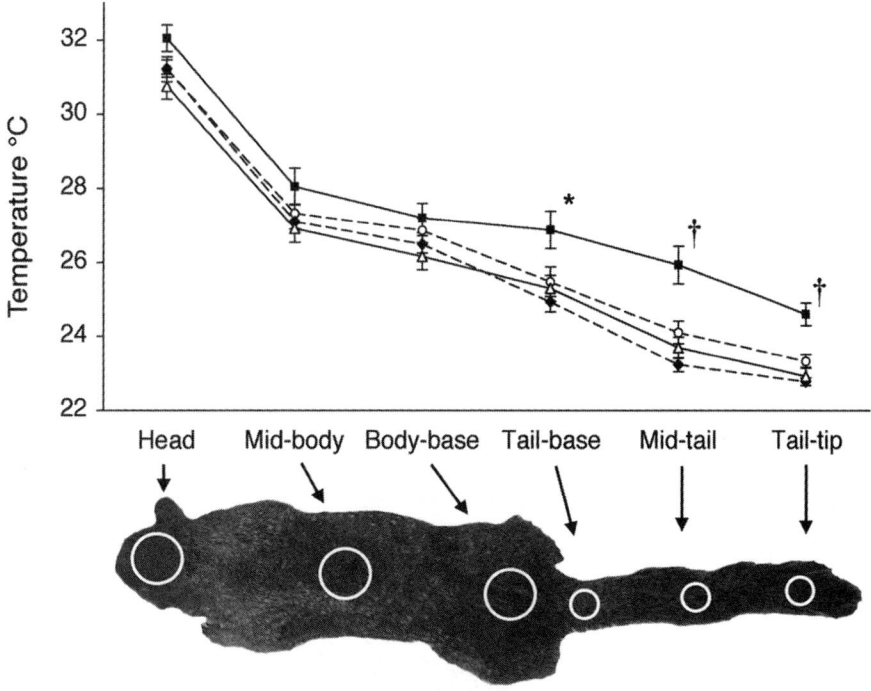

Figure 2.4. Mean body surface temperature in each condition across six body regions in California ground squirrels (from Rundus et al., 2007). Testing conditions are coded as: ■ = rattlesnake, △ = gopher snake, ♦ = conspecific, ○ = baseline. Squirrels showed increased emission of infrared radiation over baseline and control trials when exposed to rattlesnakes, which can detect infrared radiation, but not when exposed to gopher snakes, which cannot. This result suggests that squirrels assess risk and adjust their response both behaviorally and physiologically to the type of predator with which they are confronted. * $P < 0.01$ compared with conspecific and gopher snake; † $P < 0.01$ compared with baseline, conspecific, and gopher snake. (Copyright 2007 National Academy of Sciences, USA)

and sexual selection on coloration and toxicity in dendrobatid poison dart frogs. Poison dart frogs are brightly colored species (reds, blues, and yellows, often contrasted against black) with variably toxic secretions in their skin, and bright coloration has been shown to be an honest signal of toxicity, in at least some species (Maan & Cummings, 2011). Brice Noonan and Aaron Comeault (2009) placed clay frog models in the wild that varied in coloration: black body with yellow dorsum to represent the conspicuous local species, blue-bodied with black-and-yellow mottled dorsum to represent a novel conspicuous phenotype from a different location, and brown to represent a cryptic phenotype. By examining the markings in the clay of the attacked models, they were able to show that birds attacked the novel conspicuous models more often than both the familiar local conspicuous and cryptic models (Figure 2.5), suggesting that avian predators learn to associate specific color patterns with toxicity rather than simply avoiding all boldly colored frogs. This suggests that predator selection has a purifying effect on coloration in these populations so that most individuals have similar patterns. There is evidence, however, of trade-offs

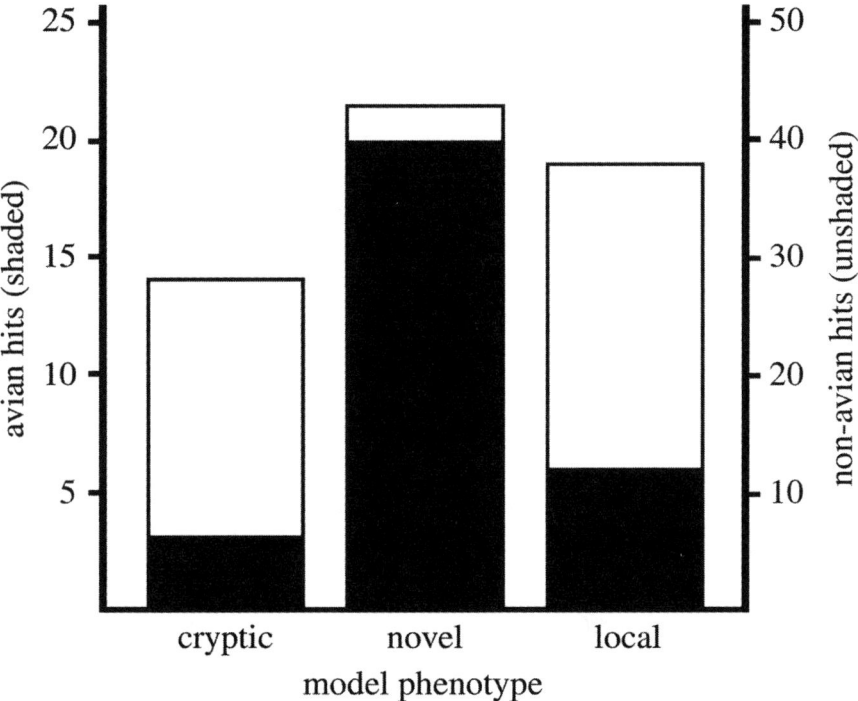

Figure 2.5. Bars indicate the frequency of cryptic brown frog models, novel blue frog models, and local yellow frog models attacked by visual (black) and nonvisual (white) predators (modified from Noonan & Comeault, 2009). While both novel and local frog models were conspicuously colored, birds preferentially attacked the novel blue models and avoided local yellow models, suggesting that lack of bird predation on yellow frogs is due to a learned association between yellow and toxic prey rather than a general avoidance of brightly colored prey. (Copyright 2007 Royal Society and Highwire Press)

between investing energy in toxicity versus conspicuousness, where degree of toxicity and conspicuousness can become decoupled, and investing in alternative strategies (either more conspicuous or more toxic) increases diversity in the warning signal and provides some level of protection for nontoxic mimic species (Darst et al., 2006). Michael Speed, Graeme Ruxton, and their colleagues have developed several models of the evolution of aposematism (Blount et al., 2009; Broom et al., 2006; Speed & Ruxton, 2005, 2007), one of which suggests that the great variation in the relationship between conspicuousness and toxicity in dendrobatid frogs can likely be linked to differences in ecological conditions where high costs to conspicuousness might favor heavier investment in toxicity as a defensive strategy, favoring further decoupling of the two traits (Speed & Ruxton, 2007).

Finally, some mammals display bold, conspicuous color patterns (e.g., skunks, polecats, honey badgers [*Mellivora capensis*]), which may also signal warnings to predators. A recent study by Stankowich, Caro, and Matt Cox found that bold, salient, black-and-white coloration in mammalian carnivores was found in stockier species, possessing noxious anal-gland secretions or pugnacious defensive dispositions, and living in more open habitats where they might be more exposed to predators and less able to rely on vegetation or water as a refuge (Stankowich et al., 2011). Contrasting coloration in other types of mammals (e.g., ungulates and whales), however, likely has camouflage or social communicative functions (Caro, 2005a; Caro, 2009; Caro et al., 2011; Caro & Stankowich, 2010).

ATTACK AND FLIGHT

Decision to Flee

The studies of risk assessment and decisions to flee are significantly intertwined: researchers typically study both using flight-initiation distance, the distance between the predator and prey at which the prey flees, as the primary metric. Ron Ydenberg and Larry Dill were the first to develop a theoretical economic model predicting when animals would flee from an approaching predator based on the costs of fleeing and the costs of remaining. When the costs of remaining exceeded the costs of fleeing, escape is predicted (Ydenberg & Dill, 1986). Hundreds of studies have examined the factors influencing the decision to flee (see "Assessment" above; Stankowich & Blumstein, 2005). While the original economic model is intuitive and accurately predicts most flight decisions, more recent models have focused on optimizing prey fitness as opposed to simply looking for a break-even point in the costs of remaining versus fleeing. William Cooper and William Frederick have developed general optimality models that maximize prey fitness and allow benefits to increase during the encounter (Cooper & Frederick, 2007). They have applied and refined these models to examine the effect of variable lethality and other factors on optimal flight decisions. When lizards lose their tails during nonlethal attacks by predators (i.e., tail **autotomy**), they suffer a reduction in locomotor performance and are thus more vulnerable to predators in future encounters (i.e., predatory attacks increase in lethality), and they may also suffer reduced fitness due to reduced social status and foraging ability. The net result of these reductions in initial fitness and increase in lethality is that optimal flight distance increases (Cooper & Frederick, 2010). The ability of these optimality models to account for changes in both initial fitness and performance during predatory encounters makes them straightforward to modify to examine changes in other factors (e.g., sex, age, health). Clearly, a wide variety of potential factors are weighed during an animal's decision of when to flee from an approaching predator (Figure 2.3).

Decision to Attack

There is a great deal of theory but few reliable empirical data on how predators select which type of prey to hunt and which individuals to attack (Lima, 1998). Granted, the great majority of acts of predation occur either (1) on prey that cannot defend themselves or flee, and thus the decision to attack is simply a matter of selecting the most energetically beneficial individual(s) to attack; or (2) with predators that will attack any individual prey item that it recognizes in its vicinity. In both cases, the decision of whether or not to attack is a foregone conclusion, and there are many empirical studies of prey selection on immobile or defenseless prey (see Chapter 1 for discussion). On the other hand there are many predator-prey interactions in nature where a predator must assess a potential prey item and decide whether to attack it, and if so, when and how. Very few empirical studies, however, are available that explore this decision in predators, probably because predators are difficult to study in the wild and witnessing natural acts of predation is rare. John Quinn and Will Cresswell have had some success in testing attack decisions of sparrowhawks (*Accipiter nisus*) in natural settings. Using intensive observational sampling of sparrowhawks hunting redshanks (*Tringa tetanus*), they found that sparrowhawks were more likely to attack redshanks in larger groups that were closer to cover and were spaced farther apart within those groups (Quinn & Cresswell, 2004). A subsequent study showed that the targeted redshanks were more widely spaced apart than their nearest neighbors that were not targeted (Quinn & Cresswell, 2006). These findings suggest that predators that hunt prey using costly attack styles and low probability of success pay close attention to the behavior of potential targets, target prey that appear more vulnerable, and decide to attack only when the circumstances are favorable.

PURSUIT, ESCAPE, AND DEFENSE

Pursuit and Capture

The means by which predators have evolved to pursue and capture prey are too numerous to cover here in any comprehensive way. Many predators pursue their prey over some distance, similar to a cheetah (*Acinonyx jubatus*) chasing down a gazelle at high speeds. Some predators pursue prey so as to keep the moving target in front of them, which results in the predator following a curved path (assuming a straight trajectory for the prey) toward the prey. In contrast, some predators move to intercept the prey at some point in the distance. Dragonflies, for example, aim their pursuit at some point in front of the prey, keeping it in a fixed position on its retina (Olberg et al., 2000). As opposed to long chases, many predators are stealth hunters, moving to within very short distances under cover before attacking. Large predatory cats and jumping spiders in the genus *Portia* among many others follow this strategy; most of these stealth hunters prefer sites that offer good cover, and some take advantage of environmental disturbances to use as a smokescreen to further cover their approach. Finally, some predators use traps to capture prey without an active pursuit. As Frodo Baggins can attest, the most well-known examples of predatory traps are spider webs. Orb-weaving spiders create the familiar circular webs between objects in their environment, and the sticky webs ensnare flying insects that do not detect them until it is too late. While most prey captured in webs are small (<2 mm), Samuel Venner and Jérôme Casas (2005) showed that spiders cannot survive without catching larger (>10 mm) prey that are rare and difficult to ensnare in the web.

Defense

Once a predator has attacked, an organism has two choices—attempt to flee to safety or attempt to hold its ground and defend itself physically—but sometimes these options bleed together. We can classify prey defenses into several different anthropomorphic categories: armor, combat weapons, and chemical weapons (Caro, 2005b; Emlen, 2008; Stankowich, 2012). Defensive armor is one of the most common morphological adaptations that prey have evolved for protecting themselves against attacking predators. Armor may take the form of spines or quills, as in sticklebacks, hedgehogs, and porcupines or of thick protective plates, as in turtles, shelled gastropods and arthropods, armadillos, and pangolins. When an organism only has one or a few spines, they often function to simply make it harder to swallow the animal (Hoogland et al., 1956). When spines or quills cover the entire body, on the other hand, they also make the prey harder to handle without risking injury, and in some cases may even cushion the animal if dropped or knocked off of tree branches or rocks (Vincent & Owers, 1986). Combat weapons include horns or antlers; large teeth, fangs, or tusks; or long claws, paired with physical strength. Chemical weapons include sprayed noxious or malodorous chemicals, venom, and secreted toxins, among others. **Mobbing** by prey is also an effective defense against single predators and is found in a wide variety of animal groups (e.g., muskoxen [*Ovibos moschatus*] protecting young offspring, bees protecting their hive, shorebirds protecting their nests). While there is a broad array of morphological and behavioral adaptations used in defense against predators, some may have evolved primarily in response to natural selection by predators, while others may be secondarily co-opted for use against predators but evolved primarily for some other function (e.g., combat with conspecifics or feeding). In mammals, weaponry that evolved primarily for defensive purposes tends to be found in intermediate-sized species that live in more exposed environments with fewer opportunities for refuge, whereas weaponry that evolved primarily for within-species combat (e.g., sexual weaponry in males) tends to be found in larger species where size influences the outcomes of competitive interactions (Stankowich, 2012; Stankowich & Caro, 2009).

Escape

If an animal is too small to defend itself against a predator or when its defensive tactics fail, it is forced to attempt an escape to safety. The specific escape strategy depends upon the medium and habitat but may include running, swimming, or flying away; dropping out of a tree; jumping out of the way; burrowing; or diving into water. When prolonged escape is used, species often use fast, erratic turns (i.e., ***protean behavior***) that make it difficult for heavier predators to adjust. If a group of animals scatters and flees erratically, it may confuse the predator, making it difficult to follow any one individual. In fact, if individuals all have different innate escape patterns, then it would be impossible for predators to learn the pattern to the escape path; therefore, the production and maintenance of such variation in escape pattern via genetic recombination may actually promote the evolution of sexual reproduction over asexual reproduction (the "red tooth hypothesis"; French, 2010). Paolo Domenici and colleagues (2008) directed puffs of air directly towards the heads of cockroaches (*Periplaneta americana*) to elicit an escape response and measured the angle of the escape trajectory: cockroaches turned their bodies and fled in four preferred trajectories (90°, 120°, 150°, and 180°). These four paths probably result in variation in escape response sufficient to keep the predator guessing. Further, protean behaviors, if performed at a distance, may also help communicate agility to the pursuing predator: Thomson's gazelles (*Eudorcas thomsonii*) often bound up and down as they flee (termed **stotting**) to advertise their quality to a predator and encourage it to stop the pursuit.

Species differences in escape strategies typically depend on a wide range of ecological and morphological factors. Steve Lima, for example, was able to group 43 emberizine finch species into five types based on sociality and vegetation density (Lima, 1993). Species may also vary their escape strategy based on the type of predator they are faced with. Redshanks suffer the lowest capture rates by flying away from sparrowhawks and merlins (*Falco columbarius*), but when faced with peregrine falcons (*F. peregrinus*), diving into creeks and crouching are far more effective modes of escape (Cresswell, 1996).

DISCUSSION

Using the attack sequence provides a useful framework for reviewing the many morphological and behavioral adaptations of predators and prey. Detailed coverage of all of these adaptations would fill several volumes (but see Caro, 2005b, for a good start), and it is clear from even this cursory overview that a huge array of factors besides predation and the need for food drive the evolution of these adaptations. Some recent studies have focused on the persistence of antipredator adaptations in the absence of predators (i.e., relaxation in selection). Some costly morphological traits may be lost rapidly if they are coded for by only one or a few genes, while many behavioral traits can take hundreds of thousands of years to degrade (Lahti et al., 2009). Predator recognition, for example, is highly variable in its tendency to degrade in the absence of predators. Moose (*Alces alces*) showed reduced recognition of wolves (*Canis lupus*) after only 50 to 130 years of absence (Berger et al., 2001), and tammar wallabies (*Macropus eugenii*) suffered a breakdown in visual predator recognition after 130 years of isolation from mammals (Blumstein et al., 2004). At longer time scales, Columbian black-tailed deer showed loss of recognition of spotted cats after 600,000 years (Stankowich & Coss, 2007), and California ground squirrels retained snake recognition after 300,000 years of allopatry but lost it after 3 to 5 million years (Coss, 1991; Goldthwaite et al., 1990).

A few central factors influence predator-prey evolution up and down the attack sequence. Group size has powerful effects on both predator hunting behavior and prey antipredator responses (reviewed at length in Krause &

Ruxton, 2002). Larger prey group sizes result in increased vigilance ability (*many-eyes effect*), greater speed of information transfer between individuals, dilution of risk, and the ability to confuse or mob predators. Spatial position within groups influences predation risk but also influences feeding rates by predators in foraging groups. Large prey groups also make for potentially larger meals for predators. Larger predator groups can allow for capture of prey that are larger than they would otherwise be able to bring down if hunting alone and improve the likelihood of prey detection. There are many benefits and costs to grouping behavior for all organisms, and their effects on predation risk and ability to locate prey are strong influences on optimal group size.

Lifetime experience also significantly influences predator-prey interactions up and down the attack sequence. For many organisms that are more developmentally plastic, exposure to certain predator cues at an early age can influence both morphological development and adult behavior, including predator recognition and risk assessment. Maud Ferrari, Doug Chivers, Grant Brown, and their colleagues have, in recent years, made great strides in understanding the effect of exposure to predator cues and conspecific alarm odors in early developmental environments on learning ability and predator recognition. They tested the ability of fathead minnows (*Pimephales promelas*) to generalize between a known predator species and other species that varied in evolutionary relatedness. They found that after teaching minnows to fear lake trout (*Salvenilus namaycush*) by pairing the lake trout odor with conspecific alarm odors, the minnows showed strong antipredator responses to the lake trout, brook trout (*S. fontinalis*, a member of the same genus as lake trout), and rainbow trout (*Oncorhynchus mykiss*, a member of a different genus but the same family), but they did not recognize odors of predatory northern pike (*Esox Lucius*, a member of a different order) or nonpredatory white suckers (*Catostomus commersoni*, a member of a different superorder) (Figure 2.6: Ferrari et al., 2007). In a similar study they found nearly identical generalization patterns in rainbow trout responding to predators of varying degrees of relatedness (Brown et al., 2011).

Clearly, the evolution of adaptive predatory and antipredatory behaviors and morphologies is influenced by many environmental factors and sources of selection, including the need to locate and attract potential mates. In fact, reproductive and defensive strategies often evolve to complement each other insomuch as some entire mating systems may be disadvantageous in light of morphological constraints based on predator avoidance or the need to hunt. In this respect, broad syntheses of **sexual** and **natural selection** are needed to look for general patterns and correlations between reproductive, feeding, and antipredator adaptations. As this chapter demonstrates, there many ways

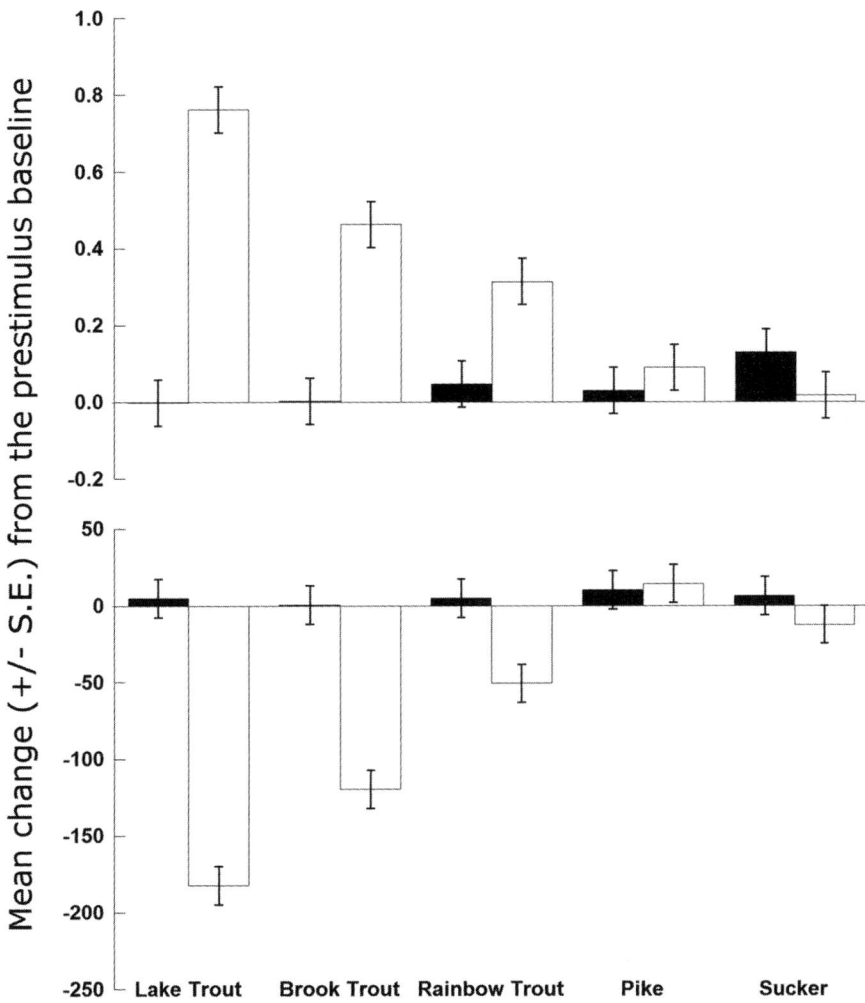

Figure 2.6. The mean change in shoaling index (top) and line crosses (bottom) for fathead minnows conditioned to fear lake trout odor paired with either water (black bars) or conspecific alarm cues (white bars) (redrawn from Ferrari et al., 2007). Minnows were able to generalize their conditioned fear of lake trout to brook trout (same genus as lake trout) and rainbow trout (different genus but same family as lake trout) but not to northern pike (different order from lake trout) or nonpredatory white suckers (different superorder from lake trout). Lifetime experience with predatory stimuli has significant effects on future antipredator responses. (Copyright 2007 Royal Society and Highwire Press)

that organisms can adapt to new threats or new foods in their environment, and these adaptations complement or trade off with other fitness-enhancing traits already present in the species' repertoire.

ACKNOWLEDGMENTS

Many thanks to E. Fernández-Juricic, M. Ferrari, and R. Coss for advice on particular references, access to original data or images, and comments on portions of this chapter.

REFERENCES AND SUGGESTED READING

Arraut, E. M., M. Marmontel, J. E. Mantovani, E. M. L. M. Novo, D. W. Macdonald, & R. E. Kenward. (2010). The lesser of two evils: Seasonal migrations of Amazonian manatees in the Western Amazon. *Journal of Zoology*, 280, 247–256.

Berger, J., J. E. Swenson, & I. L. Persson. (2001). Recolonizing carnivores and naive prey: Conservation lessons from Pleistocene extinctions. *Science*, 291, 1036–1039.

Bernays, E. A. & W. T. Wcislo. (1994). Sensory capabilities, information processing, and resource specialization. *Quarterly Review of Biology*, 69, 187–204.

Blount, J. D., M. P. Speed, G. D. Ruxton, & P. A. Stephens. (2009). Warning displays may function as honest signals of toxicity. *Proceedings of the Royal Society of London, B*, 276, 871–877.

Blumstein, D. T., J. C. Daniel, & B. P. Springett. (2004). A test of the multi-predator hypothesis: Rapid loss of antipredator behavior after 130 years of isolation. *Ethology*, 110, 919–934.

Broom, M., M. P. Speed, & G. D. Ruxton. (2006). Evolutionarily stable defence and signalling of that defence. *Journal of Theoretical Biology*, 242, 32–43.

Brown, G. E., M. C. O. Ferrari, P. H. Malka, S. Russo, M. Tressider, & D. P. Chivers. (2011). Generalization of predators and nonpredators by juvenile rainbow trout: Learning what is and is not a threat. *Animal Behaviour*, 81, 1249–1256.

Caro, T. (2005a). The adaptive significance of coloration in mammals. *BioScience*, 55, 125–136.

Caro, T. (2005b). *Antipredator Defenses in Birds and Mammals*. Chicago: University of Chicago Press.

Caro, T. (2009). Contrasting coloration in terrestrial mammals. *Philosophical Transactions of the Royal Society of London, B*, 364, 537–548.

Caro, T., K. Beeman, T. Stankowich, & H. Whitehead. (2011). The functional significance of colouration in cetaceans. *Evolutionary Ecology*, 25, 1231–1245.

Caro, T. & T. Stankowich. (2010). The function of contrasting pelage markings in artiodactyls. *Behavioral Ecology*, 21, 78–84.

Cooper, W. E. & W. G. Frederick. (2007). Optimal flight initiation distance. *Journal of Theoretical Biology*, 244, 59–67.
Cooper, W. E. & W. G. Frederick. (2010). Predator lethality, optimal escape behavior, and autotomy. *Behavioral Ecology*, 21, 91–96.
Cooper, W. E. & T. Stankowich. (2010). Prey or predator? Body size of an approaching animal affects decisions to attack or escape. *Behavioral Ecology*, 21, 1278–1284.
Coss, R. G. (1991). Context and animal behavior III: The relationship between early development and evolutionary persistence of ground squirrel antisnake behavior. *Ecological Psychology*, 3, 277–315.
Coss, R. G. (1999). Effects of relaxed selection on the evolution of behavior. In S. A. Foster & J. A. Endler (eds.), *Geographic Variation in Behavior: Perspectives on Evolutionary Mechanisms* (pp. 180–208). New York: Oxford University Press.
Coss, R. G. (2010). Predator avoidance: Mechanisms. In M. D. Breed & J. Moore (eds.), *Encyclopedia of Animal Behavior*, Vol. 2 (pp. 757–764). Oxford, UK: Academic Press.
Coss, R. G. & R. O. Goldthwaite. (1995). The persistence of old designs for perceptions. In N. S. Thompson (ed.), *Perspectives in Ethology*, Vol. 11 (pp. 83–148). New York: Plenum Press.
Coss, R. G. & U. Ramakrishnan. (2000). Perceptual aspects of leopard recognition by wild bonnet macaques (*Macaca radiata*). *Behaviour*, 137, 315–335.
Coss, R. G., U. Ramakrishnan, & J. Schank. (2005). Recognition of partially concealed leopards by wild bonnet macaques (*Macaca radiata*). The role of the spotted coat. *Behavioural Processes*, 68, 145–163.
Cresswell, W. (1994). Flocking is an effective anti-predation strategy in redshanks, *Tringa tetanus*. *Animal Behaviour*, 47, 433–442.
Cresswell, W. (1996). Surprise as a winter hunting strategy in sparrowhawks *Accipiter nisus*, peregrines *Falco peregrinus* and merlins *F. columbarius*. *Ibis*, 138, 684–692.
Curio, E. (1975). The functional organization of anti-predator behaviour in pied flycatcher: A study of avian visual perception. *Animal Behaviour*, 23, 1–115.
Curio, E. (1993). Proximate and developmental aspects of antipredator behavior. *Advances in the Study of Behavior*, 22, 135–238.
Darst, C. R., M. E. Cummings, & D. C. Cannatella. (2006). A mechanism for diversity in warning signals: Conspicuousness versus toxicity in poison frogs. *Proceedings of the National Academy of Sciences, USA*, 103, 5852–5857.
David, E. (2005). Die hard: A blend of freezing and fleeing as a dynamic defense—implications for the control of defensive behavior. *Neuroscience and Biobehavioral Reviews*, 29, 1181–1191.
Domenici, P., D. Booth, J. M. Blagburn, & J. P. Bacon. (2008). Cockroaches keep predators guessing by using preferred escape trajectories. *Current Biology*, 18, 1792–1796.
Emlen, D. J. (2008). The evolution of animal weapons. *Annual Review of Ecology, Evolution, and Systematics*, 39, 387–413.

Endler, J. A. (1981). An overview of the relationships between mimicry and crypsis. *Biological Journal of the Linnean Society*, 16, 25–31.

Fernández-Juricic, E., M. D. Gall, T. Dolan, C. O'Rourke, S. Thomas, & J. R. Lynch. (2011). Visual systems and vigilance behaviour of two ground-foraging avian prey species: White-crowned sparrows and California towhees. *Animal Behaviour*, 81, 705–713.

Ferrari, M. C. O., A. Gonzalo, F. Messier, & D. P. Chivers. (2007). Generalization of learned predator recognition: an experimental test and framework for future studies. *Proceedings of the Royal Society of London, B*, 274, 1853–1859.

French, R. M. (2010). The red tooth hypothesis: A computational model of predator-prey relations, protean escape behavior and sexual reproduction. *Journal of Theoretical Biology*, 262, 165–176.

Goldthwaite, R. O., R. G. Coss, & D. H. Owings. (1990). Evolutionary dissipation of an antisnake system: Differential behavior by California and Arctic ground squirrels in above- and below-ground contexts. *Behaviour*, 112, 246–269.

Hoogland, R., D. Morris, & N. Tinbergen. (1956). The spines of sticklebacks (*Gasterosteus* and *Pygosteus*) as means of defence against predators (*Perca* and *Esox*). *Behaviour*, 10, 205–236.

Kats, L. B.& L. M. Dill. (1998). The scent of death: Chemosensory assessment of predation risk by prey animals. [Review]. *Ecoscience*, 5, 361–394.

Krause, J. & G. D. Ruxton. (2002). *Living in Groups*. Oxford, UK: Oxford University Press.

Lahti, D. C., N. A. Johnson, B. C. Ajie, S. P. Otto, A. P. Hendry, D. T. Blumstein, et al. (2009). Relaxed selection in the wild. *Trends in Ecology and Evolution*, 24, 487–496.

Langerhans, R. B. (2007). Evolutionary consequences of predation: Avoidance, escape, reproduction, and diversification. In A. M. T. Elewa (ed.), *Predation in Organisms: A Distinct Phenomenon* (pp. 177–220). Berlin: Springer.

Lima, S. L. (1993). Ecological and evolutionary perspectives on escape from predatory attack: A survey of North American birds. *Wilson Bulletin*, 105, 1–47.

Lima, S. L. (1998). Stress and decision making under the risk of predation: Recent developments from behavioral, reproductive, and ecological perspectives. *Advances in the Study of Behavior*, 27, 215–290.

Lima, S. L. & L. M. Dill. (1990). Behavioral decisions made under the risk of predation: A review and prospectus. *Canadian Journal of Zoology*, 68, 619–640.

Maan, M. E. & M. E. Cummings. (2012). Poison frog colors are honest signals of toxicity, particularly for bird predators. *American Naturalist*, 179, E1–E14. http://www.jstor.org/stable/10.1086/663197.

Noonan, B. P. & A. A. Comeault. (2009). The role of predator selection on polymorphic aposematic poison frogs. *Biology Letters*, 5, 51–54.

Olberg, R. M., A. H. Worthington, & K. R. Venator. (2000). Prey pursuit and interception in dragonflies. *Journal of Comparative Physiology A: Neuroethology, Sensory, Neural, and Behavioral Physiology*, 186, 155-162.

O'Rourke, C. T., M. I. Hall, T. Pitlik, & E. Fernández-Juricic. (2010). Hawk eyes I: Diurnal raptors differ in visual fields and degree of eye movement. *PLoS ONE*, 5, e12802.
O'Rourke, C. T., T. Pitlik, M. Hoover, & E. Fernández-Juricic. (2010). Hawk eyes II: Diurnal raptors differ in head movement strategies when scanning from perches. *PLoS ONE*, 5, e12169.
Owings, D. H. & R. G. Coss. (1977). Snake mobbing by California ground squirrels: Adaptive variation and ontogeny. *Behaviour*, 62, 50–69.
Owings, D. H. & E. S. Morton. (1997). The role of information in communication: An assessment/management approach. In D. H. Owings, M. D. Beecher, & N. Thompson (eds.), *Perspectives in Ethology*, Vol. 12 (pp. 359–390). New York: Plenum Press.
Pulliam, H. R. (1973). On the advantages of flocking. *Journal of Theoretical Biology*, 38, 419–422.
Quinn, J. L. & W. Cresswell. (2004). Predator hunting behaviour and prey vulnerability. *Journal of Animal Ecology*, 73, 143–154.
Quinn, J. L. & W. Cresswell. (2006). Testing domains of danger in the selfish herd: Sparrowhawks target widely spaced redshanks in flocks. *Proceedings of the Royal Society of London, B*, 273, 2521–2526.
Rowe, M. P. & D. H. Owings. (1990). Probing, assessment, and management during interactions between ground squirrels and rattlesnakes. Part 1: Risks related to rattlesnake size and body temperature. *Ethology*, 86, 237–249.
Rowe, M. P. & D. H. Owings. (1996). Probing, assessment and management during interactions between ground squirrels (Rodentia: Sciuridae) and rattlesnakes (Squamata: Viperidae). 2: Cues afforded by rattlesnake rattling. *Ethology*, 102, 856–874.
Rundus, A. S., D. H. Owings, S. S. Joshi, E. Chinn, & N. Giannini. (2007). Ground squirrels use an infrared signal to deter rattlesnake predation. *Proceedings of the National Academy of Sciences, USA*, 104, 14372–14376.
Ruxton, G. D., T. N. Sherratt, & M. P. Speed. (2004). *Avoiding Attack: The Evolutionary Ecology of Crypsis, Warning Signals, and Mimicry*. New York: Oxford University Press.
Scheel, D. (1993). Watching for lions in the grass: The usefulness of scanning and its effects during hunts. *Animal Behaviour*, 46, 695–704.
Scheuerell, M. D. & D. E. Schindler. (2003). Diel vertical migration by juvenile sockeye salmon: Empirical evidence for the antipredation window. *Ecology*, 84, 1713–1720.
Smith, W. P. (1991). Ontogeny and adaptiveness of tail-flagging behavior in white-tailed deer. *American Naturalist*, 138, 190–200.
Speed, M. P. & G. D. Ruxton. (2005). Aposematism: What should our starting point be? *Proceedings of the Royal Society of London, B*, 272, 431–438.
Speed, M. P. & G. D. Ruxton. (2007). How bright and how nasty: Explaining diversity in warning signal strength. *Evolution*, 61, 623–635.

Stankowich, T. (2010). Risk-taking in self-defense. In M. D. Breed & J. Moore (eds.), *Encyclopedia of Animal Behavior*, Vol. 3 (pp. 79–86). Oxford, UK: Academic Press.

Stankowich, T. (2012). Armed and dangerous: Predicting the presence and function of defensive weaponry in mammals. *Adaptive Behavior*, 20, 34–45.

Stankowich, T. & D. T. Blumstein. (2005). Fear in animals: A meta-analysis and review of risk assessment. *Proceedings of the Royal Society of London, B*, 272, 2627–2634.

Stankowich, T. & T. Caro. (2009). Evolution of weaponry in female bovids. *Proceedings of the Royal Society of London, B*, 276, 4329–4334.

Stankowich, T., T. Caro, & M. Cox. (2011). Bold coloration and the evolution of aposematism in terrestrial carnivores. *Evolution*, 65, 3090–3099.

Stankowich, T. & R. G. Coss. (2006). Effects of predator behavior and proximity on risk assessment by Columbian black-tailed deer. *Behavioral Ecology*, 17, 246–254.

Stankowich, T. & R. G. Coss. (2007). The re-emergence of felid camouflage with the decay of predator recognition in deer under relaxed selection. *Proceedings of the Royal Society of London, B*, 274, 175–182.

Swaisgood, R. R., D. H. Owings, & M. P. Rowe. (1999). Conflict and assessment in a predator-prey system: Ground squirrels versus rattlesnakes. *Animal Behaviour*, 57, 1033–1044.

Swaisgood, R. R., M. P. Rowe, & D. H. Owings. (1999). Assessment of rattlesnake dangerousness by California ground squirrels: Exploitation of cues from rattling sounds. *Animal Behaviour*, 57, 1301–1310.

Swaisgood, R. R., M. P. Rowe, & D. H. Owings. (2003). Antipredator responses of California ground squirrels to rattlesnakes and rattling sounds: The roles of sex, reproductive parity, and offspring age in assessment and decision-making rules. *Behavioral Ecology and Sociobiology*, 55, 22–31.

Treherne, J. E. & W. A. Foster. (1980). The effects of group size on predator avoidance in a marine insect. *Animal Behaviour*, 28, 1119–1122.

Venner, S. & J. Casas. (2005). Spider webs designed for rare but life-saving catches. *Proceedings of the Royal Society of London, B*, 272, 1587–1592.

Vincent, J. F. V. & P. Owers. (1986). Mechanical design of hedgehog spines and porcupine quills. *Journal of Zoology*, 210, 55–75.

Ydenberg, R. C. & L. M. Dill. (1986). The economics of fleeing from predators. *Advances in the Study of Behavior*, 16, 229–249.

Zaret, T. M. & J. S. Suffern. (1976). Vertical migration in zooplankton as a predator avoidance mechanism. *Limnology and Oceanography*, 21, 804–813.

3

Animal Communication: How and Why Animals Communicate—or Do They?

Zuleyma Tang-Martínez

INTRODUCTION AND TERMS

As a child growing up in Venezuela, I loved to climb trees and then, from that vantage point, watch iguanas (which are very large lizards) scurrying around in neighboring trees and interacting with one another. One behavior, which involved the iguanas bobbing their heads rapidly up and down, was particularly noticeable, and I often wondered what this head bobbing could possibly mean. Years later, when I became an animal behaviorist, I learned that head bobbing is one of the most common behaviors that many lizards use to communicate with one another. Each species has its own pattern of head bobs, and they use these behaviors during courtship or aggressive interactions.

In the first part of this chapter, I will address only ***intraspecific communication***—communication that occurs between or among members of the same species—initially focusing specifically on more traditional views of communication. Later I also will discuss the possibility that communication can occur between members of different species—***interspecific communication***. Also later in the chapter, I will attend to recent debates and different perspectives (e.g., Dawkins & Krebs, 1978; Owren et al., 2010; Rendall et al., 2009) that question whether communication is honest and whether it is best described as the exchange of information or as a manipulative interaction; related to this issue is the question of whether animal signals can be said to encode information or are best understood by focusing on their effects or functions, independently of whether signals encode accurate information (see "Is Communication Honest?" and "Do Signals Really Convey Information?" below). As you will see, throughout the history of the study of communication, many debates have arisen, and as we work through these debates and consider different perspectives, new insights can be gained. Science, at its core, is about debating issues and testing alternative hypotheses; the study of communication is an excellent example of this approach.

At this point you might ask, *what is **communication***? and *how do we know that animals can communicate with one another*? In animal behavior we say that communication has occurred when the behavior of one individual (A) affects the behavior of another individual (B). Additionally, communication is usually assumed to result in "reduction of uncertainty" (e.g., Seyfarth et al., 2010). Some definitions (Slater, 1983) additionally require that animal A benefits (in terms of survival or reproductive success) from the response of animal B. Generally, the effect of communication is one of probabilities. For example, using an analogy in humans, if individual A is speaking and individual B does something (e.g., putting his finger to his lips) and individual A immediately stops talking, we can say that communication occurred. Likewise, in the lizards, if one individual head bobs and another individual that was approaching stops, then we can say that communication has occurred. In both cases the behavior of one individual affected the probability that the second individual would perform a certain behavior. In other words, the behavior of A decreased the probability that B would continue speaking, or in the case of the lizard, that B would continue approaching. However, the effects can also be an increase in the probability of a behavior occurring in B. For example, head bobbing by A could result in B suddenly beginning to also head bob—an increase in the frequency of head bobbing by B.

From a traditional perspective, all communication systems have a ***sender***, a ***signal***, and a ***receiver***. In the lizard example, individual A was the sender and individual B was the receiver. The head bob was the signal. Consider that you are listening to the radio. In this case, the broadcasting station is the sender, the sound coming from the radio is the signal, and you are the receiver. But what comes from the radio is not just random noise; rather, it contains information, so we say that all signals have ***information***, sometimes also called a ***message content*** (but see "But Do Signals Really Convey Information?" below for an alternative perspective). Now imagine that you and a friend are listening to the radio when a tornado warning is broadcast. Your friend is fascinated by storms, so she runs to the window to see if she can see the tornado approaching. You are frightened, so you immediately head down to the safety of your basement. Both you and your friend are receivers, but you have responded differently to the same message. We can say that your **interpretation** of the message is different.

The same thing can happen in animal communication. Consider a male bird singing in its territory. The ***information content*** of the song (signal) may be that he is a male in reproductive condition, he is looking for a mate, and he already has a territory that he will defend against any intruders. A female bird (receiver) listens to the song and interprets it as an invitation. She approaches the male as a potential mating partner. A male bird listening to the same song, and receiving the same messages, interprets the song as a warning and flies away to avoid getting into a fight with the territorial male. Thus, any time that we study communication we must be aware that the signal, information content, and interpretation are all different and that different receivers can have different interpretations—and therefore different responses to the same signal and message (e.g., Marler, 1961; Smith, 1977; but also see discussions in Ruxton & Schaefer, 2011; Searcy & Nowicki, 2005).

W. John Smith (1977), who first proposed the concept of "messages," also distinguished among different types of messages. Behavioral messages are those that provide information on what the sender is most likely to do next (what behavior, the probability that it will occur, and how intensely the behavior may be performed). Nonbehavioral messages provide information about the identity of the sender (i.e., they are "identifiers") or about location (of the sender or other relevant item such as food). This framework is the one that has dominated the study of animal communication, but as alluded to earlier, it has recently been challenged. There are other interesting aspects of communications systems. For example, some signals may be directed specifically at a particular individual. A baby bird that makes shrill noises when it is hungry is directing those calls specifically to its parent. On the other hand, a honeybee (*Apis mellifera*) that detects a danger (e.g., a predator) will produce an alarm odor (***alarm pheromone***) that alerts *all* bees in the vicinity that there is a danger (Boch & Shearer, 1971; Wager & Breed, 2000). In the latter case, when the sender produces a signal that is propagated throughout the environment without having one specific intended receiver, we refer to the signal as a ***broadcast signal***. Note also that any time there is a broadcast signal, and even in some cases where the signal is intended only for one receiver, ***eavesdropping*** can happen. For example, in the case of the baby bird's feeding calls, a nest predator may eavesdrop on the calls and use them to locate and eat the baby bird. Individuals in social species may give calls to indicate to group members that they have found food; other individuals who are not group members can eavesdrop and also take advantage of the food. Another well-known example involves the túngara frog (*Physalaemus pustulosus*). In this species males vocalize to attract females. However, there are frog-eating bats (*Trachops cirrhosus*) that can use the males' mating calls to locate the calling males and then eat them (Ryan et al., 1982). Other examples (e.g., Templeton & Green, 2007) include birds of one species, such as nuthatches (*Sitta canadensis*) eavesdropping and responding to the alarm calls of a different species, the black-capped chickadee (*Poecile atricapillus*). And, for an even more complex example involving several species of birds, see Rob Magrath and colleagues (2009). In fact, in some cases, birds and mammals may even respond to one another's alarm calls; hornbill birds (*Ceratogymna elata*), for example, respond to alarm calls of Diana monkeys (*Cercopithecus diana*) (Rainey et al., 2004). Obviously, broadcast signals and eavesdropping have recently become important and very active areas for study in animal communication. Some of these examples are also discussed in more detail in the section "Interspecific Communication" (below).

In addition to *signals*, animal behaviorists frequently also use the word **display**. Technically, a display is a **stereotyped** behavior that has evolved to function specifically as a signal (or a series of signals) in communication. Some displays are considered "derived" behaviors because, presumably, they were originally used in a different context (i.e., not in communication) but over time became modified specifically for use in communication. This issue will be discussed in more detail below, under "Ritualization and Origins of Displays." However, for the purposes of this chapter, I consider *signal* and *display* as interchangeable terms unless indicated otherwise.

MODALITIES OF COMMUNICATION

All animal communication involves the use of senses. Specifically, signals may be visual (e.g., postures, displays, coloration), acoustic (e.g., songs, calls), chemical (e.g., glandular secretions, saliva, urine), tactile (e.g., touching, stroking, plucking), electrical (e.g., electric pulses), or vibrational (e.g., waves transmitted through substrates such as the ground, plant stems, or water). I suspect the first four modalities did not surprise you because they are senses that humans are aware of and use routinely. On the other hand, you may not have realized the importance of electrical communication in some species. Weakly electric fishes, for example, communicate with a series of electric pulses produced by specialized organs (Hagedorn, 1986). These pulses can encode information by varying characteristics such as their pattern, repetition rate, or frequency. Likewise, many species are now known to communicate by sending waves through substrates. Sometimes it is difficult to distinguish these types of signals from sounds (after all, sounds are also waves that travel through the air). Certain insects (e.g., tree hoppers) communicate by producing waves that travel through the stems of plants, and they use different types of waves depending on what is being communicated (Cocroft, 2005; Cocroft & Rodriguez, 2005). Some spiders use parts of their bodies to generate waves that travel through the ground and are perceived by receivers (e.g., Gibson & Uetz, 2008; Sivalingehm et al., 2010), and some underground rodents either butt their heads against the roofs of their burrows (Heth et al., 1987; Rado et al., 1998), or footdrum on the ground (Narins et al., 1997; Randall & Lewis, 1997), generating waves that travel through the ground to other conspecifics. The term **seismic communication** is often used to describe signals that are propagated through the ground. In all cases of vibrational communication, humans cannot hear the "sounds" that are generated unless they use specialized equipment. However, using such equipment, for example in your backyard, can result in a cacophony of amazing sounds: whines, grunts, knocks, whistles, chirps, and so on—sort of the insect equivalent of "heavy metal"!

Visual Communication

Visual signals are used by many different species of animals and can be either **dynamic** or **static**. A static signal, such as the color pattern of the peacock's tail, is one that does not change. A dynamic signal can change while the display is taking place. For example, some fish have dark stripes on their sides, but they can make the stripes very dark or so light they are almost invisible (and everything in between) during an aggressive encounter (e.g., Baylis, 1974). The darkness of the stripes communicates motivation—such as whether the fish at that instant in time is feeling more aggressive or more scared. Visual signals may be **morphological**, consisting of special structures (e.g., crests or elongated tails in birds) or of colors or patches of colors (the black chest bib in house sparrows (*Passer domesticus*), or **behavioral**, consisting of movements, postures, or displays produced by the animal (e.g., head bobbing of lizards—but note here that this use of *behavioral* differs somewhat from that of Smith [1977], who uses this term specifically with regard to messages or information content of signals). Frequently, the morphological and behavioral signals are combined in one display. Many lizards have **dewlaps**—a fold of skin that can be extended into a semicircular structure—under their chins. Dewlaps can be very colorful and are often displayed during head bobbing, thereby combining a morphological signal with a behavioral one. In some species there is evidence that females can recognize males of their own species by the color of the dewlap and/or the pattern of head bobbing and push-ups, and that they may use these signals to avoid mating with the wrong species (e.g., Hunsaker, 1962; Jenssen, 1970). Some animals (e.g., some birds and insects) even produce and perceive ultraviolet signals that humans cannot see.

A special type of visual signal involves structures that are produced by the sender and then used as part of a display. The best know example of this is found in bowerbirds (Borgia, 1986). In this group, males build elaborate structures—called ***bowers***—out of twigs and other vegetation and then decorate them with colorful fruit, flower petals, string, shiny pieces of paper, and many other types of objects. Once the bower is completed the male courts passing females by bowing, posturing, and calling near the entrance to the bower. Females apparently choose males based on their bowers, as well as on the male's behaviors (Borgia, 1986).

Acoustic Communication

Acoustic signals are also widespread throughout the animal kingdom, being most commonly associated with birds, mammals, anuran amphibians (frogs and toads), and insects. Many sounds, such as the chirping of crickets, the songs of birds, the calling of frogs, and the roar or growling of mammals, are audible to the human ear. Most humans can only hear sounds that fall into the range of 20 to 20,000 hertz (Hz). However, in some cases animals produce acoustic signals that are above or below this audible range and cannot be heard by humans without the use of specialized equipment. For example, many rodents produce very high-frequency (***ultrasonic***—inaudible to us) calls that are important in parental care and sexual behavior (e.g., Ehret, 2005). In another interesting case, Katy Payne, studying elephants in a zoo, noticed behavior in elephants that suggested they were making sounds that she could not hear. She went on to demonstrate that elephants communicate over long distances by using very low-frequency (***infrasonic***) rumbling sounds that also are inaudible to humans (Payne et al., 1986; Payne, 1998). In fact, by some estimates, most of the sounds that elephants make are at such low frequencies that they cannot be heard by our ears. Because some of these sounds are also transmitted through the ground, they are sometimes classified as seismic signals rather than as true sounds.

In some cases, animals may use objects to generate sounds. A well-know example occurs in chimpanzees (*Pan troglodytes*). A variety of sounds used during communication are produced by chimps slapping or stomping on the ground or vigorously shaking or dragging tree branches (Muller & Mitani, 2005).

Acoustic signals tend to be extremely complex, and many different characteristics of the sound may be important in communication. In addition to loudness, duration, and frequency (expressed in hertz), other parameters of the sound, such the repetition rate, the number of pulses per second, the time intervals between pulses, and the patterns of the different sounds that make up the signal, can be critically relevant in animal communication. Different species may pay attention to different acoustic parameters. Not surprisingly, the study of the characteristics of sounds that are most important is an active and fertile area of research (especially in frogs and birds) in animal communication.

Chemical Communication

Chemical communication, which includes both olfactory and gustatory signals, is nearly ubiquitous, occurring throughout the animal kingdom, with the possible exception of among birds (but see below). Because olfaction and taste can be so difficult to distinguish, this chapter concentrates on olfaction and the use of odors or scents as signals. Frequently, odors are produced by specialized glands, but urine, feces, and saliva can also contain relevant information. An interesting characteristic of chemical signals is that, in many cases, they can be deposited and broadcast throughout the environment even when the sender is no longer present. One example you are probably familiar with is urination in dogs (*Canis familiaris*). Dogs will sniff and then repeatedly "mark" certain trees—or the proverbial fire hydrants—and then move on. It appears that each dog may be leaving a sort of "calling card" that keeps everyone informed of who has been there and, possibly, how long ago (Wyatt, 2003). An alternative explanation for this common dog behavior is that each dog may be "erasing" the previous dog's scent mark by urinating on top of it (Wyatt, 2003). Also in dogs, everyone is likely familiar with the **sex pheromone** produced by female dogs when they are in heat. This odor is also broadcast and is highly attractive to any male dogs in the general vicinity (Beach & Gilmore, 1949; Dunbar, 1977). Sex pheromones are important not only in mammals but also in many insects, crustaceans, fishes, salamanders, reptiles, and other taxa.

There often is a great deal of misunderstanding about the definition of **pheromone**, and many people think that all pheromones function in sexual behavior, but this is not correct. A pheromone is any chemical substance that is secreted to the outside by an individual of one species and has an effect on the behavior or physiology of a conspecific—that is, another individual of the same species. While it is true that many pheromones are used in sexual contexts, they can also function in other contexts, including in communicating alarm (as in the case of the honey bees mentioned earlier), social status, and individual or group identity; in eliciting aggregations; or in creating scent trails. Moreover, the definition also makes it clear that chemical cues that act between different species are not pheromones.

Bark beetles of the genus *Ips* are a good example of species that produce pheromones used specifically for facilitating aggregations of conspecifics (Young et al., 1973; Bakke, 1978; Byers, 1983). *Ips* attacks on pine trees can result in terrible infestations that may kill hundreds or thousands of living trees. When the first beetle finds a suitable tree (usually an older tree or one that has been weakened by fire or disease), it immediately begins producing an ***aggregation pheromone*** that attracts large numbers of conspecific beetles. These beetles produce additional pheromone, resulting in the entire tree becoming riddled with beetles burrowing under the bark and eating the phloem, ultimately destroying the tree. Because of the commercial damage that is wreaked by *Ips* infestations, as well as by other species of beetles that also attack various species of trees, an enormous amount of research has been done on their aggregation pheromones in the hopes of being able to control outbreaks of these infestations.

Some experts on chemical communication limit the use of the word *pheromone* to cases in which all receivers respond in the same way and in a stereotyped manner to the chemical signal. For example, all honey bees in a hive respond in the same manner to the bee alarm pheromone, by increasing movement, searching for the threat, and immediately attacking it. This is why a person who is stung by a bee is likely to get multiple stings—the first bee that attacks releases an alarm pheromone that alerts all the other bees, and they then descend on the unfortunate person and continue to attack by stinging (however, the alarm pheromone is quite complex, with different components eliciting different responses—see Wager & Breed, 2000). According to this more restrictive definition, odors that communicate individual identity in mammals are not pheromones because different receivers can respond differently to the same odor based on their previous experience with the sender (friendly versus hostile, or familiar versus unfamiliar). For the sake of simplicity, in this chapter I define a pheromone as an intraspecific chemical signal, regardless of the type of response that it elicits.

Animals that use chemical communication may mark the substrate, objects in their environment (e.g., trees, rocks, twigs), or their conspecifics or even spread odors on themselves by self-grooming. As an example, during social encounters, male neotropical singing mice (*Scotynomys teguina*) rub their front paws on their abdominal scent gland and then ***allogroom*** females, spreading the male secretions all over the females' fur (Fernández Vargas et al., 2011). The function of this marking of females is not well understood. Interestingly, as their name implies, singing mice also produce acoustic signals in both audible and ultrasonic ranges.

Birds have always been considered to be a rather conspicuous exception to the wide use of chemical communication because they are believed to have poor olfactory abilities and are not know to use chemical signals. However, recent work in Ellen Ketterson's lab at Indiana University may change this perception. Danielle Whittaker, a postdoctoral student of Ketterson's, has demonstrated that the preen oil of dark-eyed juncos (*Junco hyemalis*) varies with population, individual identity, and sex (Whittaker et al., 2010). Moreover, both female and male juncos can distinguish the sex of an individual from its preen oil and prefer the preen oil odors of males (Whittaker et al., 2011). This ability to discriminate odors in this bird species may have important effects on their parental behavior (including potentially on reproductive success) because when preen oil was applied to the nests of females, they spent less time incubating their eggs if the odors were from unfamiliar individuals but not when they were their own odors (Whittaker et al., 2009). These are very exciting findings that may help to change our assumptions about the ability of some birds to communicate by means of chemical signals.

Tactile and Vibrational Communication

Tactile communication is very familiar to humans because we use it very frequently. When you hug or put your arm around a friend's shoulder to comfort him, you are using tactile communication—your gesture communicates your sympathy and concern without a need for words. In nonhuman animals, tactile communication can take many forms. Primates (monkeys and apes) use many tactile signals that are similar to those used by humans. Additionally, allogrooming in primates appears to serve dual functions—it removes parasites from the skin and fur of the individual being groomed, and it also strengthens and reinforces existing social bonds between the two individuals involved in this interaction (Nakamichi & Shizawa, 2003; Majolo et al., 2005; Dunbar, 2010). In most primate species, subordinate individuals groom more dominant ones, and the behavior also serves to acknowledge that the groomer recognizes the higher social status of the partner (Seyfarth, 1980; Schino, 2001). Other taxa also may use touch, including grooming, licking, or stroking, as part of displays that function in different contexts. Although there do not appear to be any rigorous scientific studies, it is a common belief that canids, including domestic dogs, use licking to reinforce friendly interactions and social bonds; however, at present this is only a popular conjecture. On the other hand, some insects, such as fruit flies (*Drosophila*), use licking as part of their courtship displays (Bastock & Manning, 1955; Spieth, 1974). Some insects, crustaceans, and spiders use tapping and stroking behaviors during courtship and mating. One interesting example is found in the brown recluse

(*Loxosceles reclusa*), a venomous spider found in parts of the southern Midwest of the United States. Courting males and females approach one another very slowly and then initiate a prolonged period of stroking motions; the male in particular extends his long front legs over the female and gently touches and strokes her prior to mating (Tang-Martínez, personal observation).

In some cases it is difficult to distinguish clearly between tactile and vibrational signals. For example, in web-building spiders, males entering a female's web may be mistaken for prey and get eaten by the female (Robinson, 1982; Elgar, 1991). Complicating matters even more, females in these species tend to be huge in comparison to the males, and it is very easy for a female to overpower a male. Consequently, courtship in these species is a very risky business, and males must be very careful in communicating their intentions to the female. Typically, males will stand at the edge of the web and begin slowly plucking the web, using certain species-specific patterns that indicate courtship (Robinson, 1982; Maklakov et al., 2003). Depending on the female's response, the male may very slowly approach the female, continuing to pluck as he does so. If all goes well, he will end up as a successful suitor rather than becoming a tasty snack for the female! Traditionally, the plucking signal of the male was considered a form of tactile communication. However, it can just as easily be interpreted as a vibrational signal because the plucking creates vibrations that are propagated through the web (Maklakov et al., 2003).

Multiple Modalities

I have discussed each modality of communication separately. However, in real life, an animal may often use more than one modality simultaneously. I have already mentioned bowerbirds using both acoustic and visual signals in their courtship display. Grouse, among many other species of birds, also combine visual and acoustic signals during courting behaviors—they strut or stomp while conspicuously displaying bright colors on their heads, body, or tails, while simultaneously making "booming" or drumming calls that can be heard over surprisingly long distances (Dantzker et al., 1999). In some mammals, scent communication is accompanied by conspicuous visual signals. For example, South American bush dog (*Speothos venaticus*) females perform a handstand as they urine-mark objects in their environment (Kleiman, 1972). Likewise, ring-tailed lemurs (*Lemur catta*) undulate their tails conspicuously at the same time that they release anal-gland scents; this seems to constitute a visual signal, as well as perhaps helping to waft the odor towards the intended receiver. Invertebrates, such as some jumping spiders, also combine modalities, sometimes using tactile and chemical cues simultaneously or both visual and vibrational cues (e.g., Girard et al., 2011).

CONTEXTS OF COMMUNICATION

From the information provided above, you probably already have a pretty good grasp of the types of situations in which animals use communication. Communication is a *sine qua non* for sociality—that is, all animals that are social by necessity must have some form of communication. However, even solitary (not living in groups) animals must communicate for the purpose of mating and reproductive behavior. (The only possible exception to this generalization might be species that reproduce by cloning, are hermaphroditic, and are not likely to encounter other individuals of their own species.)

In general, the contexts in which communication is used include, but are not limited to, sexual behavior, agonistic behavior, parental care, group coordination, and food and predators. I discuss these uses below.

Sexual Behavior

Signals used for reproduction include sex attractants (such as sex pheromones), **sexual excitants** (sometimes called aphrodisiacs), and displays that help the male and female to coordinate their behaviors so that mating can occur. In many species of crickets, for example, males produce two distinctly different songs during courtship (Zuk et al., 2008). The first song, called the calling song, attracts the female and is very different from one species to another (e.g., Gray & Cade, 2000; Fitzpatrick & Gray, 2001); once the female approaches, the male switches to a new song, called the courtship song, which seems to excite the female, making it more likely that she will accept him as a mate (but many courtship attempts, whether in crickets or other taxa, are unsuccessful because females may refuse to mate even after assiduous courtship on the part of males). Keep in mind also that, in some species, it is the females that produce signals to attract males. For example, it is the female in most moth species (e.g., silkworm moth [*Bombyx mori*], gypsy moth [*Portheria dispar*]) that produces the sex pheromone that attracts males, sometimes from distances of 2 km or more (Jacobson & Beroza, 1963; Wilson & Bossert, 1963). Likewise, in dogs it is the female that produces a powerful sex pheromone to attract males (Beach & Gilmore, 1949; Dunbar, 1977). As we saw in the case of the orb-web spiders, courtship behavior can be tricky, and, in most species, both the male and female may risk injury or attack if they proceed too quickly, before their potential mate is ready. Thus, courtship often involves complex sequences of signaling behaviors by both the male and the female. Sometimes these complex sequences are referred to as courtship dances, and these have been described in spiders, birds, fishes, and even in cephalopods, such as squids (e.g., Sauer et al., 1997; Von Hippel, 2000;

Trainer et al., 2002; Girard et al., 2011). In species that stay together for an extended period after mating, courtship signals and activities may also help to create and reinforce a ***pair bond***—the two individuals will stay together and interact in an amicable and nonaggressive manner and will cooperate in activities that help to increase their reproductive success (e.g., in parental care—see below).

Agonistic Behavior

In animal behavior we use the term ***agonistic*** to refer to interactions that involve both aggression (offensive signals and behaviors) and submission (defensive signals and behavior). The use of this specialized term makes sense if we consider what is really happening when two individuals are involved in what most nonspecialists would call an aggressive encounter. Imagine that you are physically fighting with another person—most likely your feelings (as well as the feelings of your opponent) will fluctuate between aggression and fear because you cannot be sure that opponent will not be a better fighter and beat you up. Your behavior is likely to go from advancing and threatening your opponent one instant to backing off and trying to protect yourself the next. Thus, you and the opponent are continuously going back and forth between aggression and submission. Thus, aggression only addresses one aspect of what happens in these types of interactions, and *agonism* becomes a more accurate word. Agonism can occur in many different contexts. Individuals, whether male or female, may compete for mates, food, territories, resources, or shelters, among other reasons. Sometimes competition can involve actual physical fights, but in many cases animals use only ***ritualized displays*** (see below) and signal until one individual backs down and the other gains access to the resource.

In almost all species, signals and displays that indicate threat or aggression have similar characteristics. They make the animal look bigger, and weapons are exposed. A direct stare also is considered a signal of aggression in many species. Think of a dog that is behaving very aggressively—it will stand tall with head held high, ears will be perked straight up, hackles on the back will be raised, tail will be up, teeth will be bared in a snarl, and the dog will likely stare directly at its opponent. By contrast, a dog showing submission will cower (body pressed close to the floor), ears will be pressed back against the head, tail will be hidden between the legs, the teeth will be hidden behind a grimace, and the head (and eyes) will be turned directly away from the opponent. In an extreme case, the submissive dog might even fall down to the ground completely and roll over on its back. Both of these submissive displays make the dog look as small and as nonthreatening as possible. The aggressive and submissive displays are virtually mirror images of each other. We call such displays that are exact opposites of one another ***antithetical*** displays because the posture in one case is the antithesis (exact opposite) of the other (Darwin, 1872). Images for antithetical aggressive and submissive displays in dogs and discussion by Darwin (including also other images of dog and cat postures) can be found at http://www.brocku.ca/MeadProject/Darwin/Darwin_1872_02.html.

I used the example of dogs because it is probably familiar to many of you. However, the same types of behaviors are seen in many other species across most of the animal kingdom. Ungulates, such as deer or antelope, may stand very tall, stare at their opponents, and swing their antlers or horns (weapons) threateningly in the air. Fishes expand their gill covers, making them look much bigger from the front; toads may puff up their bodies, making themselves appear alarmingly larger than they really are. Invertebrates also make themselves look larger and may display their weapons. For example, mantis shrimp (marine stomatopod crustaceans) rear up and perform "meral spreads" by opening and displaying their brightly colored raptorial appendages (powerful weapons that can smash down on an opponent, causing serious injury or death) in a very conspicuous threat display (Caldwell & Dingle, 1975). And, of course, the submissive displays tend to be opposites of the threat displays. See also section "Typical- and Variable-Intensity Displays," below, for more information on agonistic signals and displays.

Parental Care

Not all species of animals show parental care, but among those that do, communication looms big. Signals relevant to parental care include the begging or squeaking calls of baby birds and mammals, which generally garner the attention of the parents, who may then feed or protect their young. Larvae of some social insects (e.g., social bees and wasps) use special signals to beg food from the adults (Kaptein et al., 2005). Recognition between parents and their offspring can be critical to the survival of the young, and it is not surprising that many species in many different taxa have evolved signals that aid in this recognition. Young who become lost or isolated from their parent(s) typically produce signals that attract the parent and lead to the retrieval of the offspring (Smotherman et al., 1974). In birds, calls are often used for parent-offspring identification (reviewed in Halpin, 1991). In many other taxa, including some fishes, salamanders, and mammals, chemical signals play a similar role (reviewed in Halpin, 1991). Parents may also warn their offspring of danger. A case in point is a group of cichlid fishes (family Cichlidae) called mouthbrooders. In most of these species, the female carries the fertilized eggs in her mouth until the young hatch. (During this time, swallowing, coughing, and yawning are mostly suppressed so that the female does not accidentally swallow or disgorge the developing young.) After the fry (baby fish) hatch they continue to associate with the mother and return to her mouth to sleep or when they are in danger. In fact, the mother and offspring communicate with various signals to ensure that the fry return the mother's mouth when a threat is perceived (Baerends & Baerends-Von Roon, 1950; Balshine-Earn & Earn, 1998).

Group Coordination

Among social species, signals frequently serve to keep the group together or to coordinate behaviors. For example, in many species of primates, as members of the troop spread out to forage (whether in savannahs or in trees), individuals will emit **contact calls** to keep in touch with the group and allow others to determine where all group members are (e.g., Sugiura, 1998). Special signals that may be visual, acoustic, or chemical are also used by some species to initiate particular behaviors, such as group movement from one place to another or a coordinated attack on prey (e.g., Arnold & Zuberbühler, 2006). Among the truly social insects (termites, ants, social bees and wasps) chemical communication controls much of the group's behavior, ranging from parental care to alarm behavior to (in some cases) coordinated attacks on other groups. Among the army ants (e.g., genus *Eciton*), daily raids away from the nest or nomadic movement from one nest to another are exquisitely coordinated, primarily by chemical cues (Deneubourg et al., 1989). Thousands of ants literally march in files over the landscape (ground, branches, etc.), attacking and eating everything in their path, while simultaneously maintaining extraordinary group cohesion and order. I was lucky to witness such a raid on a visit to Costa Rica in 2012; this was an amazing experience because (at the risk of being anthropomorphic) the ants gave the impression of being unstoppable and moved as if they had one mind and one common goal.

Food and Predators

Many species use signals to communicate the existence or location of food and to warn conspecifics against predators (but note that there is some debate as to the function of so-called alarm signals in some species; see the section "Interspecific Communication" for a few alternative possibilities).

Signals related to the presence of food most frequently occur in species that are social or live in aggregations. We have already discussed the aggregation pheromone of bark beetles, which alerts other beetles that a suitable tree (food) has been located. Treehoppers generate vibrational signals that propagate along plant stems and announce that a good source of food has been encountered (Cocroft, 2005). Tent caterpillars lay chemical trails that lead other nestmates to food and then back to the nest (Fitzgerald & Webster, 1993). And when ants invade your home, think pheromones! Foraging ants leave scent trails that guide other ants to food—for example, on your kitchen counter (Morgan, 2008). As a result, the best way to get rid of the ants is to destroy their pheromone trails (unfortunately, this is not always easy). Birds and mammals will often use special vocalizations to alert group members to the presence of food (e.g., Caine et al., 1995; Elgar, 1986; Elowson et al., 1991).

Perhaps the most amazing example of communication about food is the ***waggle dance*** of the honey bee (Von Frisch, 1967). Worker bees (all females) belong to different castes, each one specialized for different tasks. Some workers are "scouts" whose job is to leave the hive, find food, and then come back to the hive and signal to the foragers where the food is. Foragers then follow the "instructions" of the scouts, find the food, and bring it back to the hive. After initially finding food, the scouts perform a waggle dance on a vertical surface inside the hive, and the foragers follow the dance by maintaining contact with the scouts as they dance (the inside of the hive is dark, so tactile cues are used). The waggle dance resembles a flattened figure eight with the scout bee going around one circle and then the other, waggling its abdomen as it moves through the straight portion (called the straight run) between the two circles. As the scout dances, the angle of the straight run in relation to straight up on the vertical indicates the exact direction of the food in relation to the position of the sun outside. Note that the straight run does not directly point to the food but rather indicates where the food is in relation to the sun; if the sun is not visible outside (as in a completely overcast sky) the foragers will not be able to use the waggle dance to determine direction. In addition to indicating the direction of the food, scouts also indicate the distance to the food by how many waggles they perform during the straight run (the more waggles, the longer it takes to do the straight run and the farther away the food is) and the quality or richness of the source by how long they continue performing the waggle dance (see Volume 1, Chapter 12). Interestingly, if the food is very close, scouts do not perform the waggle dance; instead, they perform a "round dance" that consists of moving around and around in a simple circle. Foragers leave the hive and fly in concentric circles in the vicinity of the hive until they locate the food. Images of the honey bee waggle dance and round dance can be found in these links.

http://media.wiley.com/mrw_images/els/articles/a0002940/image_n/nfg002.gifhttp://images.tutorvista.com/content/biotic-community/honey-bee-worker-waggle-dance.jpeg http://1.bp.blogspot.com/_pOW4pcoh2Z0/Sun4XSUM4PI/AAAAAAAA FE/0roWi1owCSc/s320/honeybees.jpg

Interestingly, the importance of the waggle dance has recently been challenged (Grüter & Farina, 2008, 2009). These authors claim that forager bees that follow the dances often disregard the direction and location information encoded by the dances. They suggest that the information on the location of food sources (i.e., spatial or navigational information) is used primarily as a back-up if other sources of information are not available or are inconsistent. In response, Axel Brockmann and Moushumi Sen Sarma (2009) argue that just because honey bees may use a variety of information (e.g., food odors brought back by scouts or their own previous experience), it does not mean that the waggle dance is not important or that the information it conveys is not relevant. Specifically, they suggest that the location information encoded in the waggle dance may be particularly important when food is patchily distributed and difficult to find, as is often the case in the tropics and subtropics where eight of the nine species of honey bees are found and where the waggle dance likely evolved.

Interestingly, this is not the first time that the conventional understanding of the waggle dance has been challenged. As far back as the 1960s, several scientists questioned whether the waggle dance really serves as a "language" that communicates location and/or distance to the food sources; instead they suggested that bees were using odors (rather than the information in the waggle dance) to find food (e.g., Johnson, 1967; Wenner, 1967). Von Frisch (1967) countered this criticism by pointing to a number of his own experiments that seemed to show very clearly that bees do, in fact, use the dances to locate food. Subsequently, James Gould (1975) conducted a series of elegant and exquisitely controlled experiments to test the assertions of the critics. He concluded that under different conditions (e.g., stress level of the bees and the abundance of flower sources the bees are using), honey bees could use *either* the information in the waggle dance or odors. However, he also found excellent evidence that bees are able to gain information from the dance and that, even when odors are used, the waggle dance is necessary to recruit the foragers. He also suggested that Wenner (1967) and Von Frisch (1967) obtained different result and reached different conclusions because of differences in methodology that resulted in their concentrating on two different stages of the recruitment process using the dance. Recently, Tania Munz (2005) has summarized the history of this early challenge to the waggle dance.

The recurring debates over the waggle dance of honey bees is a good example of how science works. Karl Von Frisch's discovery of the waggle dance more than 60 years ago won him the 1973 Nobel Prize in Physiology or Medicine. Moreover, the waggle dance has remained one of the most celebrated examples of complex communication in animals and appears in every textbook on animal behavior. Yet even the most widely accepted ideas in science can be challenged, tested, and revised—and this is an integral and necessary part of the practice of science. My own opinion of the current debate is that Brockmann and Sen Sarma (2009) are correct in their criticism of Grüter and Farina (2009), but it also is likely that honey bees use a variety of cues when foraging and are not limited only to the information encoded in the waggle dance. This should not be surprising because most animals do show flexibility in their behaviors—and, of course, this is the same conclusion that Gould reached in 1975.

The alarm pheromones of honey bees already have been discussed. Many other species produce alarm signals in the presence of potential predators. Best known are the alarm calls of birds and of mammals. These calls may elicit alertness (e.g., meerkats stand on their hind legs and look around for the predator [Manser, 1999]), evasive behaviors (e.g., ground squirrels immediately run and hide in their burrows [Leger & Owings, 1978]), or attack behaviors (e.g., birds may gather and gang up on the predator, a behavior know as *mobbing* [Grieser, 2009]). Fishes in some species, when attacked by a predator, release an *alarm substance* that causes other fishes nearby to quickly swim away from the area of the injured fish (Chivers & Smith, 1994).

These are certainly not the only contexts in which animals use signals and displays for communication. However, they are likely the most common.

INFORMATION CONTENT

Peter Marler (1961), one of the pioneers in the study of animal communication, proposed that most signals carry information about five different aspects of the sender: (1) species identity, (2) individual identity, (3) sexual identity, (4) physiological condition/motivation, and (5) information about the environment.

Species Identity

Most communication is intraspecific, and most species produce signals that differ from those of even other closely related species (e.g., Roelofs & Comeau, 1969; Muroyama & Thierry, 1998; Fitzpatrick & Gray, 2001; Guillete et al., 2010). Thus, almost by definition, most signals will carry information about the species of the sender. Depending on the situation, receivers may respond to this information in different ways. For example, courting males and females may use these signals to determine whether their prospective mate is a member of their own species (e.g., Blair, 1964; Roelofs & Comeau, 1969; Gray & Cade, 2000). If it is not, then typically the courtship is aborted and the animals continue searching for an appropriate conspecific partner. Signals that help prevent hybridization mistakes by identifying the species of the sender are known as **behavioral** (or **ethological**) **isolating mechanisms**.

Individual Identity

Many signals (but not all) also contain information about the individual identity of the sender (reviewed in Halpin, 1986; Vaché et al., 2001). Individual identity can be used in many different contexts. In many mammals, the individual odors of nest mates are learned during early life, and later used to discriminate between siblings and non-siblings (Halpin, 1991; Tang-Martínez, 2001). Likewise, parents may learn the individual odors of their offspring and then use the information to restrict parental care to their own young. Both of these cases are considered examples of **kin recognition**. Among social species it is often important to discriminate between neighbors and strangers, between friend and foe, and between individuals of higher and lower social rank. Ground squirrels have been found to respond very differently to the odors of their neighbors as compared to those of conspecifics from other social groups (Harris & Murie, 1982). Using playback experiments, it has been demonstrated that birds can distinguish between the songs of their territorial neighbors and unfamiliar intruders (e.g., Wunderlee, 1978; Lovell & Lein, 2004). Many ungulates (e.g., sheep [*Ovis aries*] and cattle [*Bos taurus*]) seem to use a variety of different cues, including visual, acoustic, and chemical, to determine the individual identity of conspecifics (e.g., Alexander & Shillito-Walser, 1978; Barfield et al., 1994; Ligout & Porter, 2004; Ligout et al., 2004). Even among invertebrates, mantis shrimp respond differently when presented with the odor of an individual they have previously defeated, as compared to that from an individual that has defeated them (Caldwell, 1985). In the first instance they approach the odor and perform meral displays; in the second, they immediately flee and hide. All of these examples (and there are many others) imply that animal can extract information on individual identity from the signals produced by conspecifics. Although somewhat different in nature, signals that convey information on individual identity may also allow for recognizing group members (e.g., Keil et al., 2012). These types of signals are of particular importance in certain social species. For example, social insect, such as some bees, are known to distinguish the odors of hive members from those of intruders (Breed, 1998). In this case the cue appears to hive-specific (a hive odor) and not an individual odor per se.

Sexual Identity

Some signals contain information on the sex of the sender. In some cases, the signal is produced by only one sex, which automatically provides information on the sender's sexual identity. For example, although in many tropical bird species both males and females sing, in most North American songbirds only males sing (Marler & Slabbekoorn, 2004; Slater & Mann, 2004).

In other species, specific signals indicate sex of the sender. Morphological structures or colors may vary greatly between males and females in some species (e.g., birds and insects), thus providing a clear visual signal that indicates sex (e.g., Badyaev & Hill, 2003; Emlen et al., 2005; Ribak et al., 2009). Neotropical singing mice are able to distinguish the sex of senders by their odors (Fernández-Vargas et al., 2008). Many other examples exist involving the various modalities of communication.

Physiological Condition/Motivation

The physiological condition of an individual will affect its motivation and, consequently, its behavior. When we speak of physiological condition, most frequently we are referring to the animal's hormonal profile and neural processes at any point in time. As examples, among vertebrates, high levels of testosterone in males are correlated with sexual and aggressive motivations and behaviors; in females an increase in certain hormones and a concomitant decrease in others bring about estrus and lead to the production of particular signals and displays that indicate the female is in reproductive condition (e.g., Vandenbergh, 1969; Dunbar, 1977; Fernández-Vargas et al., 2008). The songs and visual displays performed by the males of many bird species during courtship are signals that the male is physiologically ready and motivated to mate (Fusani, 2008). The same signals, conveying the same information (reproductive condition and motivation), may also signify threats when directed at other males. Likewise, females use signals that indicate their willingness to mate: female birds perform a conspicuous ***copulation solicitation display*** in which the female crouches and elevates the tail, exposing the cloaca (Harju, 1971; Yasukawa et al., 1987), and female rodents perform ***lordosis*** (a posture in which the female bends the body, raises its rump in the air, wiggles the ears, moves the tail to one side exposing the vagina, and produces squeak-like vocalizations [Beach, 1976]). In both instances these displays signal sexual readiness and are invitations for the male to mount. Sex pheromones produced by female mammals and insects also are examples of signals that indicate readiness to mate.

Threat and submissive displays, respectively, carry the message that the sender is motivated to attack or is more likely to behave submissively. During any agonistic encounter, the motivation of the two participants may vary rapidly, going from being very aggressive to being frightened and submissive. These motivational changes can be signaled almost from second to second by the displays of the opponents (see the section "Typical- and Variable-Intensity Displays").

Other motivational states, such as parental care or hunger (e.g., in offspring that depend on parents for food) also fall into this category (Redondo & Arias de Reyna, 1988). Of course, in general, every species will have its own signals and displays to communicate various physiological and motivational states. The previous discussions under the contexts of communication include many other relevant examples.

Information about the Environment

This category primarily includes signals about food and predators. The alarm signals of insects, fishes, birds, and mammals, some of which were discussed previously, are excellent examples. Both birds and mammals may also encode information about the type of predator detected, specifically if the predator is an aerial predator or a terrestrial predator (e.g., ground squirrels; Sherman, 1977, 1985). The responses of receivers differ depending on the type of predator. Some mammals, particularly ungulates, perform visual alarm displays (Bildstein, 1983). These include ***tail-flagging*** in deer and ***spronking*** or ***stotting*** in various antelope species (Caro, 1986). When deer detect a potential predator, they run while holding the tail straight up (with the underside showing) and moving it quickly from side to side; typically the contrasting coloration of the underside of the tail makes this tail-flagging particularly conspicuous. During spronking or stotting, antelope that have detected danger run, then stop and leap straight up into the air, come back down, run, and repeat the behavior. Deer and antelope that observe these behaviors scan the surroundings, presumably for predators, and then also take evasive actions (Caro, 1995).

The waggle dance of honey bees, the scent trails of ants, and the aggregation pheromones of bark beetles illustrate the complexity and variety of some signals relating to presence and location of food. Other food-relevant signals, for example in social vertebrates, may be much simpler and consist of calls and other vocalizations.

Information about Status

In addition to the messages identified by Marler (1961). signals also may convey information on status or quality of the sender. According to this view, male signals used during courtship convey information about the quality of the displaying male (e.g., his health, survival potential, strength, or stamina). and females use this information in selecting which male to mate with (Hill, 1991; Kayser & Hill, 2000). Likewise, signals used during agonistic encounters may not simply say, "I am feeling aggressive"; rather, they may communicate the status or fighting ability (i.e., the quality) of the sender (Pryke et al., 2001; Wyman et al., 2008). Note that in both cases, such signals would do more than simply communicate physiological condition or motivation.

Metacommunication

Another interesting category includes signals that serve to modify the meaning of other signals. The best-known example is the "play bow" of dogs (http://4 hpetpals.osu.edu/images/dogPlayful.jpg). When dogs signal that they want to play they assume a particular posture: head down, forelegs stretched out in front, rump in the air, tail wagging (Bekoff, 1995). The message is not only that the dog wants to play but also that all behavior that follows will be done in play and is not to be interpreted as real aggression. Thus, behaviors such as growling, barking, lunging, or jumping on the play partner are modified into play behaviors rather than maintaining their original threat function. Signals or displays that modify the meaning of other signals or displays are considered an example of ***metacommunication***—that is, communication about communication (Bekoff, 1995).

RITUALIZATION AND ORIGINS OF DISPLAYS

One question that early animal behaviorists struggled with was the evolutionary origin of displays. Many displays are very exaggerated and stereotyped, and one cannot help but wonder where they came from and what affected their evolution. Because we cannot go back in time and actually see what displays were like in the past and how they changed over time, we can only infer where different displays may have originated. Nonetheless, there is general agreement that many displays originally arose from behaviors that were not used for communication but rather had a different function in a different context (Daanje, 1950; Tinbergen, 1952; Morris, 1956; Lorenz, 1972).

Researchers suggest that there are several possible noncommunicative sources for many of the displays that we are most familiar with: ***intention movements***, ***autonomic behaviors***, ***displacement activities***, ***ambivalent behaviors***, and ***redirected behaviors***.

Intention Movements

Most animals, just before they perform a behavior, will engage in movements and activities that are predictive of what the animal is about to do.

A typical example is flight-intention movements in birds. Before a bird flies off, it will bend its legs, extend it head and body forward, and open its wings. This behavior has nothing to do with communication; the animal simply must assume this posture in preparation for flight. Many courtship displays in birds involve the male bending the legs, leaning forward, with head outstretched and wings open. However, in these cases, the male simply holds this posture in a very exaggerated manner while positioning himself in front of the female. This display is believed to be derived from the intention movement for flight (Daanje, 1950). However, the behavior has now become a stereotyped signal used in communication.

Autonomic Behaviors

When animals are cold or scared, they may fluff up their fur or feathers either to keep warm or as an automatic response to something frightening (this is where the expression "I was so scared that my hair was standing on end" comes from). Such behaviors are automatic physiological responses that are not under the direct control of the animal; they have nothing to do with communication. Now consider a dog that is signaling that it is highly motivated to behave aggressively—chances are that its hackles (the fur on its shoulders and upper back) will be raised. This "piloerection" is believed to be derived evolutionarily from the piloerection that happens as an autonomic response (Morris, 1956). Some scent marking behaviors (e.g., with urine or feces) are likewise believed to be derived from normal urination or defecation (Morris, 1956). Similarly, erection of the penis, sometimes made more conspicuous by bright coloration, is used by some primates as a threat display (Wickler, 1967). It has been proposed this signaling behavior is derived from automatic erection of the penis during copulation.

Displacement Activities

During conflict situations, animals will sometimes suddenly stop what they are doing and briefly perform another, apparently totally unrelated, behavior. For example, when two birds are fighting, one of them might suddenly stop and start preening itself for a few seconds before returning to the fight. These behaviors are called ***displacement activities*** and are thought to function to relieve high levels of tension that are likely to occur during conflict situations (Zeigler, 1964). In some birds (e.g., some ducks) courting males include a display that consists of very conspicuously preening their wing feathers; other related species simply point to the edge of the wing in an exaggerated and very stereotyped display (Lorenz, 1972). The area that is preened or pointed to is likely to be very colorful, which makes the display even more striking. This is considered an example of a courtship display that comes from a displacement activity.

Ambivalent Behaviors

In conflict situations, when animals are undecided and their motivation changes quickly from one instant to the next, they may perform ambivalent behaviors. For example, a bird attempting to defend her eggs from an approaching predator may charge the predator but then immediately run back towards the nest; an individual in an agonistic encounter with a conspecific may alternate between approaching its opponent and running away from it. In all such cases it appears that the animals cannot decide what to do, so it alternates between two mutually contradictory behaviors. A number of displays seem to have evolved from ambivalent behaviors. Niko Tinbergen (1951) described the courtship and breeding behavior of the stickleback (*Gasterosteun aculeatus*). Sticklebacks are small fish in which the male defends a territory, builds a nest, and courts passing females. If a female is interested, she will enter the territory, lay eggs in the male's nest, and leave; the male stays behind and takes care of the eggs. The courtship display of the male stickleback is called the "zig-zag dance" because the male will repeatedly alternate between approaching the female and then swimming away from her, doing a zig-zag towards and away from the female (Tinbergen, 1951). In gulls, there is a "facing away" or "head flagging" display (Tinbergen, 1959). Again during courtship, the male and the female rapidly alternate between looking at each other and then quickly looking away, moving the head repeatedly back and forth in an exaggerated and stereotyped manner. These displays are considered to be modified ambivalent behaviors that have evolved into displays.

Redirected Behaviors

Again, in conflict situations, one or both participants may suddenly attack objects in the environment (e.g., rocks or tree trunks) or even subordinate bystanders. In such cases, it is assumed that the animal is directing aggression at a "safe" target that is not capable of retaliating (Moynihan, 1955; Tinbergen, 1959). This in turn may relieve some of the aggression in the animal. During both agonistic and courtship displays, some species use behaviors reminiscent of redirected aggression. The previously discussed example of chimps slapping the ground and vigorously shaking branches (Muller & Mitani, 2005) might be another example of redirected aggression, although the exact motivation of such displays is not really known.

Ritualization occurs when behaviors that originally were used in a noncommunicative context evolve into behaviors (displays) used specifically for communication. When this happens, we say that the behavior has been ***emancipated*** from its original function and motivation (Huxley, 1923; Tinbergen, 1952; Morris, 1957; Blest, 1961). For example, a displacement activity, originally used to relieve tension during conflicts, evolves into a sexually motivated display that functions to promote mating. During the process of ritualization, displays tend to become increasingly exaggerated, stereotyped, and conspicuous. In some cases anatomical changes, for example exaggerated structures or colors, also evolve, making the display even more striking (Tinbergen, 1952; Morris, 1957).

TYPICAL- AND VARIABLE-INTENSITY DISPLAYS

Displays have three components: form, frequency, and duration. The form of the display is its "appearance"—in the case of visual displays, it is what the display actually looks like. The frequency refers to how many times the display is performed in a given unit of time, and the duration is how long the display lasts when it is performed.

Animal behaviorists distinguish between typical-intensity and variable ("graded")-intensity displays (Leyhausen, 1956; Morris, 1957). In a typical-intensity display, the form of the display does not change with subtle changes in the level of motivation of the sender. Images showing several typical-intensity courtship displays (in three different species of birds) can be found in the following links.

http://d3bdn46jsbygsl.cloudfront.net/144_snowy_egret.jpg
http://2.bp.blogspot.com/_bOKmjbY7wEo/S5gTdchiR4I/AAAAAAAACC0/iJqC_i4OfdI/s320/Greater+Sage+Grouse.jpg
http://standingoutinmyfield.files.wordpress.com/2012/06/blue-bird-of-paradise.jpg?w=529

For example, a courtship display consisting of the legs-bent, head-forward, open-wings posture (described above) is displayed in exactly the same, very stereotyped manner regardless of whether the sender is very highly motivated or only moderately motivated to mate. And if, during the display, the sender suddenly becomes more highly motivated, the appearance of the display will not change. Levels of motivation can, nonetheless, be communicated by the frequency or duration of the display. Consider an analogy of a doorbell. The sound of the doorbell never changes, regardless of whether the bell-ringer is very anxious to get you to come to the door or is very patient. The sound of the doorbell is the form. For example, a laid-back bell-ringer will ring the doorbell once for a relatively short duration and then wait. But a bell-ringer who is very anxious to get your attention can ring the doorbell repeatedly in a short period (increase in frequency) or can press the doorbell button and keep it pressed so that the doorbell rings for an extended period (increase in duration). This is exactly what happens in a typical-intensity display: the form is constant regardless of the exact motivational level of the sender, but level of motivation can still be communicated by the frequency or duration of the display. Although there are no set rules, many courtship displays tend to be of typical intensity, with the form communicating only a motivation to mate (examples in Morris, 1957; Hinde, 1970).

In contrast, in variable- or graded-intensity displays, the sender's level of motivation is indicated by continuous changes in the form of the display (e.g., Leyhausen, 1956). Images of variable- or graded-intensity displays can be found in the following links.

http://rint.rechten.rug.nl/rth/dennen/agr_fear1.jpg shows the continuum from extreme fear to extreme aggression in the facial expression in dogs.
http://rint.rechten.rug.nl/rth/dennen/agr_fear2a.jpg shows the continuum from extreme fear to extreme aggression in the body postures of cats.

For example, consider antithetical displays in a dog involved in an agonistic interaction. When the dog is highly aggressive it stands tall, ears up and tail up; when it is highly submissive, it cringes with the body lowered, the ears laid back flat against the head, and the tail tucked between its legs. But as the dog's motivation fluctuates between high levels of aggression and high levels of fear (submission), these postures will form a continuum with, for example, the tail going back and forth between being held high to an intermediate position, to down near the rump, to completely tucked between the legs. Therefore, even subtle changes in the sender's motivation can be communicated by changes in the form of the display. Changes in levels of motivation can still also be indicated by frequency and duration. Going back to the doorbell analogy, this would be similar to having a bank of doorbell buttons that vary gradually in the tone (form) depending on how anxious the ringer happens to be. In this case, the bell-ringer can select which button to use depending on her level of motivation and can go back and forth among the buttons as her motivation changes. Again, she can also indicate her level of motivation by varying the frequency or duration of ringing, regardless of which button she happens to be pushing. Many agonistic displays are of variable intensity with the form of the display changing rapidly as the motivation of the sender changes (examples in Leyhausen, 1956; Hinde, 1970).

In general it can be argued that typical-intensity displays convey clear and unambiguous information about the overall motivation of the sender (e.g., "I want to mate"). On the other hand, their form does not convey information about second-to-second changes in the motivation of the sender (but remember that such changes can still be conveyed by duration and frequency of the display). Variable-intensity displays are very rich in the information they communicate, indicating even very subtle changes in the motivation of the sender (e.g., "I am really aggressive—oops, I am a bit more scared now—and now I am *really* terrified and about to run away"). Some models suggest that agonistic displays are variable precisely because aggressive motivation varies so much during agonistic encounters (discussions in Wilson, 1975; Enquist, 1985; Bradbury & Vehrencamp, 1998). However, the down side of such a continuous and detailed flow of information is that the form of the display can change so quickly that the meaning may be more likely to be misunderstood.

IS COMMUNICATION HONEST?

So far, all the information I have provided is considered the traditional or classical perspective of animal communication. Richard Dawkins and John Krebs (1978) argued that this perspective misses the point because, in their view, what we have discussed as communication is really all about manipulation and deception—animals will "lie" (give out misinformation) to attain their goals. For example, they suggest that when an animal is engaged in an agonistic interaction it will do whatever is necessary to defeat the opponent. A subordinate individual has no interest in accurately communicating to the opponent that it is frightened and submissive; instead it should "bluff" the opponent by performing threat displays that signal high levels of aggression—the animal lies to obtain its goal of outcompeting the opponent. Similarly, according to Dawkins and Krebs, a male courting a female should not "tell the truth" about his social status or quality—instead he should pretend to be of very high status and quality even if, in reality, he is of inferior quality. The goal of the male is to mate with the female, and he will lie to manipulate the female into accepting him as a mate.

These ideas provoked a strong reaction. Robert Hinde (1981), a highly respected British animal behaviorist, suggested that Dawkins and Krebs had misrepresented the traditional view of communication and erected a 'straw man.' Both Hinde (1981) and W. John Smith, another student of communication (Smith, 1977, 1986) argued that Dawkins and Krebs's perspective did not make sense from an evolutionary perspective. For example, lying and deception could only evolve if there was little or no fitness cost to these behaviors. These ideas are illustrated by an example. Imagine an agonistic interaction, and suppose that the subordinate individual lies and pretends to be highly aggressive. If the truly dominant individual calls his bluff and attacks, the costs to the sender may be extreme—he may be badly injured or even killed. Thus the cost of lying should prevent its evolution. A number of other researchers, writing both prior to and following Dawkins and Krebs's 1978 paper, argued that the cost of signals makes them necessarily honest (Zahavi 1975, 1977; Enquist ,1985; Grafen, 1990). Others pointed out that if all communication truly is about lying and misinformation, then everyone "knows" that this is the case and this strategy would not be effective—lies would routinely be ignored (Wallace, 1973). Additionally, if some individuals are successful at lying, then populations should evolve better mechanisms of detecting lies such that lying could not become established in populations (Wallace 1973; Smith, 1986). These and other arguments culminated with an emphasis on the evolution of "honest signals."

Among the scientists mentioned above, Amotz Zahavi, an Israeli scientist, was arguably the most important because he proposed the **handicap principle** (Zahavi, 1975, 1977), which essentially contends that signals are, by necessity, honest. Consider the case of a courting male—assume a peacock displays his very long and colorful tail as a signal of his superior quality, as compared to males of inferior quality. Zahavi argued that the cost of producing the long, colorful tail is very high—the male had to use energy to produce the tail, he is more susceptible to predation because his tail makes him more conspicuous to predators, and he is less able to maneuver and to escape predators or survive storms. In other words, having such a long and colorful tail is a handicap, and only males of the most superior quality can afford to have such a handicap. Inferior males simply do not have what it takes to produce the tail and still survive, and therefore it is impossible for them to lie about their quality. Therefore, argued Zahavi, the display is an honest indicator of male quality. Although initially Zahavi's ideas were criticized, in more recent times they have gained wide support (e.g., Pomiankowski, 1987; Grafen, 1990).

Is there any evidence that animals do sometimes lie? This is a very difficult question that can lead to philosophical, semantic, and anthropomorphic arguments. For example, *how do we define lying? How do we know if an animal intends to lie? Is there a difference between lying in humans and in nonhuman animals?* Specifically, from a biological and evolutionary perspective, an animal does not have to *intentionally* provide misinformation. Instead we can reconceptualize the concepts of honesty and dishonesty of signals by thinking in terms of **reliable** versus **unreliable** information. Thus an animal may provide unreliable information without necessarily intending to do so. (See also the next section, "Do Signals Really Convey Information?") In reality, we cannot know if an animal "intends" to lie or deceive (because we have no way of determining what goes on in the mind of the animal, this question is not within the realm of science). Thus, although many researchers continue to use the words *lying* and *deception*, to avoid anthropomorphism and unwarranted assumptions, it is best to use the terms *reliable* versus *unreliable* information.

The traditional view of animal communication had always posited that signals usually provide only probabilistic and conditional information about the motivation of the sender (Smith 1977, 1986; Hinde, 1981). Both of these authors emphasize that communication is probabilistic (in the sense that signals indicate only that its probable—not certain—that a particular behavior or event will follow) and conditional (signals indicate that senders' behaviors depends on how the receiver responds). Thus, in the example of the two animals involved in an agonistic encounter, if animal A performs a highly aggressive threat display, it is not indicating to animal B that "no matter what you do, I am definitely going to attack you and beat you up." Instead, the message is "if you do X (condition), there is a high probability that I will attack." If B does X, animal A may attack 70 percent of the time, but it may retreat the other 30 percent; and if B does Y, animal A is much more likely not to attack but may still do so (Hinde, 1981; Smith, 1986). Using this probabilistic and conditional perspective, *does it make any sense to say that animal A lied to or deceived animal B because 30 percent of the time it does not attack even if B does X?* Hinde (1981) and Smith (1986) do not think so.

What is the empirical evidence? Most studies have yielded ambiguous results that are open to interpretation (e.g., Rohwer, 1977; Rohwer & Rohwer, 1978). However, a few studies do suggest that, at least in some cases, animals may misinform. In great tits (*Parus major*, a small European bird similar to chickadees), dominant individuals have priority of access to feeding sites and chase away subordinates. Anders Møller (1988) observed that when dominant individuals were monopolizing the food, some subordinates would give alarm calls, causing the dominants to fly away. The subordinates would then immediately land at the feeding platform and eat until the dominants returned. Although it is impossible to know what the subordinates were "thinking," these results can certainly be interpreted as a fairly clear case of "crying wolf" to gain access to a food resource. But note that this strategy should work only as long as it is not used too frequently. If subordinates continuously cry wolf, the dominants should quickly learn to ignore them because they are not giving honest or reliable signals that a predator has been seen. Interestingly, this behavior of giving false alarm calls to gain access to food or other resources seems to occur in a number of avian and primate species.

Some studies that have directly examined misinformation in situations where it might be expected to occur have instead found that signals tend to be honest. Hermit crabs (*Pagurus bernhardus*) use the shells of snails as a home and shelter. Because availability of shells is limited, hermit crabs in poor-quality shells (e.g.. shells that are too small for them) often try to evict other crabs from better-quality shells. When a challenger approaches a resident crab with the better-quality shell, the resident may perform a threat display to discourage the challenger from approaching closer. But some residents do not perform the threat display. Mark Laidre (2009) conducted an experiment in which model crabs (dead crabs in a challenge posture and placed inside a shell) challenged residents by approaching them. If threatened by the resident, the model would either flee or continue to approach until it was very close to the resident (of course the behavior of the challenger was remotely controlled by the experimenter). In almost all cases, the resident's response was consistent with its behavior on first approach (i.e., the response indicated by the resident). That is, if a resident threatened and the challenger continued to approach, the resident would almost always attack. If the resident did not threaten when first approached, it would almost always back down and run away if the challenger came closer. If the threatened residents had been "lying" about their intention and willingness to attack, then they should have backed down as soon as the challenger continued to approach. Moreover, crabs that subsequently ran away should have also threatened to deceive the challengers into thinking that they would be attacked. Therefore, this suggests that the threat display in this species is an honest signal. Unfortunately, few other studies on honesty versus dishonesty have been so carefully designed and controlled.

DO SIGNALS REALLY CONVEY INFORMATION?

In recent years a new controversy has arisen about whether responses to signals are due to information contained in the signals or are simply byproducts of sensory physiology or other biological processes. For example, in the túngara frog, males vocalize to attract females. It is well known that females in most species in the genus *Physalemus* prefer males that produce elaborations of the normal male vocalization (these can be added chucks, whines, or repetitions of parts of the normal call) (Ryan, 1990; Ryan & Rand, 1993). What is most interesting is that the males of some of these species do not normally produce such elaborations; yet, when the elaborations are added to the normal species calls, the females prefer these artificially modified calls to the normal calls. In some species, males have, in fact, evolved more complex calls that include, for example, multiple whines or chucks. Because these elaborations do not appear to correlate with any obvious male qualities, and females respond to them for apparently arbitrary reasons (a preexisting **sensory bias**), it would seem to make little sense to say that these vocalizations are signals that convey any meaningful information about the sender (e.g., Ryan & Rand, 1993).

In another species, the water mite (*Neumania papillator*), courting males attract females by producing vibrations in the water that seem to mimic the vibrations produced by prey (Proctor, 1991), suggesting that they are exploiting the females' predatory response. This, and some other examples from various species, have been considered examples of **sensory traps** (Christy, 1995) by which males attract females (for mating) by producing signals that mimic other stimuli occurring in different contexts to which females are adapted to respond (e.g., responses to food or to predators). As in the case of the *Physalemus* frogs, such signals do not seem to provide accurate information to the females.

Based on such examples, Michael Ryan and collaborators (Rendall et al., 2009; Owren et al., 2010) subsequently offered a critique to communication in general, claiming that there is little clear-cut evidence that supports the idea that most signals convey information (as asserted by the more traditional views of communication that I presented above). Instead they posit that the issue is not whether senders signal information *per se* but rather that the signals influence or manage the behavior of the receiver in predictable ways (see also Owings & Morton, 1997). In their view, the critical function of signaling is for the sender to influence the behavior of the receiver and not the transmission of encoded information. There have been several responses to these ideas. Robert Seyfarth and Dorothy Cheney (animal behaviorists who have done extensive work on alarm vocalizations in monkeys), as well as several collaborators (Seyfarth et al., 2010) offer a different perspective. They point out that information is best defined as the "reduction of uncertainty in the receiver"— that is, by perceiving the signal the receiver obtains more accurate knowledge about its environment. In other words, if the signal has a reliable relationship to a subsequent event or situation (e.g., alarm call equals "a predator is nearby"), then it is accurate to say the signal encodes information and that the receiver responds based on that information. It is irrelevant whether the sender gave the alarm call with the "intention" of warning others because even a signal that is produced as a result of arousal of the caller (e.g.. as a result of fear when it detects a predator) can still contain information that receivers can use. Specifically, they point out that in many species of birds and mammals, callers give different alarm calls when they detect different types of predators, such as terrestrial versus aerial predators (see discussion above and examples in Seyfarth et al., 2010). Moreover, the receivers respond differently to each type of call, and their avoidance responses are consistent with the type of predator. Thus, Seyfarth and colleagues (2010) assert that alarm calls do in fact contain information about the presence of a predator, the type of predator, and (in some cases) the level of threat posed by the predator. Thus, calls inform listeners about the state of their world. These authors conclude that, in general, animal signals encode real information and that the informational approach to communication has been, and continues to be, valuable in advancing our understanding.

Two other papers (Carazo & Font, 2010; Font & Carazo, 2010) reaffirm that signals contain information but also offer a novel definition, contending that information must be defined in terms of its adaptive significance to the receiver. Thus, Pau Carazo and Enrique Font emphasize the evolution of signaling and the selection pressures (costs and benefits) affecting both senders and receivers; signaling systems evolve precisely because receivers are able to make beneficial decisions based on the information they have obtained. They also argue that signals always have meaningful information but that this does not mean that they contain messages deliberately encoded by senders.

More recently, a theoretical paper (Ruxton & Schaeffer, 2011) attempts to resolve some of these controversies by offering a series of new definitions that the authors hope will clarify some of the ambiguities. Specifically, they suggest that sender ("informer" in their terminology) and receiver ("perceiver" in their terminology) engage in a trade; the informer influences the perceiver and, as a result, the perceiver undergoes a change in what it knows about its world—that is, it gains information as the result of the communication exchange. The authors then go on to offer new definitions for "influence" and "information" that they hope will resolve the debate.

In my opinion, it seems clear that in some cases (e.g., exploitation of preexisting sensory biases, as in the case of the *Physalemus* frogs; sensory traps) it is difficult to argue that any meaningful information is being provided by the sender. On the other hand, there are many cases where it is equally difficult to argue that signals do not encode information and that no information is exchanged. I agree that more careful and explicit definitions are necessary in order to avoid confusion. However, perhaps it is also important to realize that, in some ways, there is nothing special about animals being able to obtain information from conspecific cues and signals. In reality, information is all around us at all times, whether or not it is encoded in conspecific signals. For example, if we get up in the morning and the sky is heavily overcast and we hear thunder, we have information that the probability of rain is high (a reduction of uncertainty about the kind of weather that we can expect); certainly, no one would suggest that there is any intentionality involved. Likewise, a dog can urinate with no intention of encoding information or communicating anything to another dog. But if a second dog perceives the odor and can determine the sex and reproductive condition of the urinator, then it has gleaned meaningful information from the odor. Likewise, an escaping prisoner has no intention of leaving behind any information, but a bloodhound can obtain enough information from the odor left behind and follow the prisoner's trail. Although this discussion goes beyond animal signals and communication, I hope that it reinforces the point that intentionality is not necessary for information transfer and that signals and cues produced by conspecifics are only one way in which animals can obtain relevant information from their environment. Perhaps placing animal signaling systems in a broader perspective that emphasizes the acquisition and processing of information would avoid some of the conceptual and semantic debates that have arisen. For additional discussions and perspectives on the issues of signal reliability and deception, see also an excellent book by William Searcy and Steve Nowicki (2005),

ENVIRONMENT AND THE EVOLUTION OF COMMUNICATION

Traditionally, animal behaviorists have always been aware of the importance of the environment to the evolution of signals and the sensory modalities involved (Marler, 1961; Smith, 1977). For example, if we consider a fish species that evolved and lives in very muddy water, we would not expect it to have evolved visual signals because it is too difficult to see such signals in turbid water. Instead, these fish would be predicted to use acoustic, chemical, or electrical signals. The same is true of terrestrial species that inhabit areas of very heavy undergrowth where vision is impeded. On the other hand, aquatic species that live in bodies of water with rapid and unpredictable currents may be at a disadvantage if they were to use chemical signals because these would quickly be diluted and carried away by currents. Heavy vegetation may, under some circumstances, absorb and change acoustic signals, thereby attenuating them; in such cases one would expect species to use other modalities or to modify their acoustic signals to increase their effectiveness. Background noise, such as from heavy winds or even human traffic, can also interfere with acoustic signals; Henrik Brumm and Hans Slabbekoorn (2005) have reviewed the many ways that various taxa change their vocalizations to counteract the effects of noisy environments. Likewise, lizard species (genus *Anolis*) that live in windy areas with lots of rapidly moving vegetation have been found to speed up their territorial displays, which makes their displays more conspicuous (Ord et al., 2007).

John Endler (1992) proposed the concept of **sensory drive**, a complex model that includes some of the ideas discussed above about effects of the environment on signal evolution. However, Endler defines the environment as including both the physical environment and the presence of other animals, whether conspecifics or not. Specifically, Endler argues that the evolution of signals, the sensory systems (including both sensory organs and brain areas needed for processing signals) of receivers, and signaling behavior have coevolved. That is, changes in one set of traits have affected the evolution of the other sets of traits. Additionally, these traits also will be affected by the physical characteristics of the environment and by ecological interspecies processes such as predation.

Among many examples, in the túngara frog mentioned previously, the sensory organs of females appear to respond most strongly to certain frequencies of sound; males have evolved calls in those frequencies, and females therefore respond very strongly to these calls (Ryan & Rand, 1990). Phylogenetic analyses suggest that the sensory sensitivity of the females existed first and that it helped to shape the evolution of vocalizations in males (Ryan & Rand, 1993). Also in this system, bat-eating frogs use the vocalizations of males to locate them (Ryan et al., 1982). According to the sensory drive model, over time we should see modifications in male calls to counteract the ability of bats to find them. The final result should be a compromise between being able to effectively attract females and being able to avoid predation. Moreover, both females and bats can be expected to respond evolutionarily to any changes in male calls, setting up the potential for an "evolutionary arms race" between male frogs and their bat predators. The physical environment can also affect not only the structure of signals themselves but also signaling behavior. For example, there are certain bird species in which males perform courtship displays in specialized courtship arenas within tropical forests (e.g., Anciães & Prum, 2008). In these species, the males will choose to display only in microhabitats that have certain light conditions and at times of the day when their displays will be most visible with regards to the light conditions. Increased use of such specialized microhabitats for signaling may, in turn, affect the evolution of sensory systems to make communication more effective.

In summary, sensory drive, which is currently an important area of study, emphasizes that signals, signaling behaviors, and sensory systems do not evolve in a vacuum. Rather they evolve jointly, and their evolution is further shaped by the physical environment. The end result is signals, sensory systems, and signaling behaviors that are highly effective in a given environment.

INTERSPECIFIC COMMUNICATION

The discussion up to this point has concentrated on intraspecific communication. The obvious question is whether communication can also occur between members of different species. The answer is yes, although most communication does appear to be intraspecific. Nonetheless, interspecific communication has been demonstrated in several different contexts.

One context involves alarm calls. Several studies have shown that birds of different species respond to the calls of heterospecifics. For example, Magrath and colleagues (2009) recorded normal alarm calls from three Australian species and then conducted a playback experiments to determine the responses of the three species to heterospecific calls. Two species, the superb fairy wren (*Malurus cyaneus*) and the white-browed scrubwren (*Sericornis frontalis*), often flock together and have common predators. Each species recognized the "hawk alarm calls" of the other species and immediately sought cover. The situation with the third species, a honeyeater (*Phylidonyris novaehollandiae*), was more complicated. Honeyeaters only respond to scrubwren alarm calls, while fairy wrens and scrubwrens both respond to honeyeater calls. Primates also are known to respond to heterospecific alarm calls. In one experiment, Klaus Zuberbühler (2000) demonstrated that Diana and Campbell's monkeys (*Cercopithecus campbelli*) reacted with alarm to one another's calls. Even more interesting is the case of a bird, the yellow casqued hornbill (*Ceratogymna elata*), which responds to the "eagle alarm calls" of the Diana monkey but not to "leopard alarm calls"; both crowned eagles (*Stephanoaetus coronatus*) and leopards (*Panthera pardus*) prey on Diana monkeys, but only the eagle preys on hornbills (Rainey et al., 2004). Thus, the hornbill not only recognizes the alarm calls of a mammal but also discriminates appropriately between calls that indicate two different categories of predator.

Another area in which interspecific communication may be important is prey-predator interactions. In fact, some researchers have suggested that alarm calls in birds and mammals, and behaviors such as stotting and spronking in ungulates, actually communicate information to the predator. Alarm calling or spronking may be ***pursuit-deterrence*** signals, although this notion remains relatively controversial (Caro, 1995). The idea is that the sender is "saying" to the predator one of two things: (1) "I have seen you and am taking evasive action—you can't surprise me, so you are wasting your time pursuing me" or (2) "I am of such high quality that I can afford the risk of letting you know that I am here; I am of such good quality that I can escape, so you are wasting your time pursuing me."

Although there has been a great deal of discussion about pursuit-deterrence signals and behaviors, few studies have demonstrated that this type of communication actually occurs. One exception is a series of experiments conducted by Manuel Leal (1999) on a Puerto Rican lizard (*Anolis cristatellus*) and its snake predator (*Alsophis portaricencis*). Leal used carefully controlled experiments to demonstrate that the lizard displays when approached by the snake and that the display frequency of the lizard is positively correlated with its stamina in a running apparatus. This result suggests that lizards that display at higher frequencies are also more likely to be able to outrun the snake. Moreover, other observations and experiments indicate that when *Alsophis* observe a lizard displaying, they *may* be less likely to attack (Leal & Rodriguez-Robles, 1995). There are, however, some problems with this latter study because of small sample sizes and the inability to adequately control the hunger level and motivation of the snakes (Manuel S. Leal, personal communication). Nonetheless, this is one of the most convincing studies of possible prey-predator communication. Additionally, it provides evidence that the frequency of the lizard display may be an honest signal, accurately indicating running ability and the likelihood of escaping from a snake that attacks.

Another interesting example of what appears to be a pursuit-deterrence signal is described by Matthew Barbour and Rulon Clark (2012), in a ground squirrel–rattlesnake system. Under natural field conditions, California ground squirrels (*Otospermophilus beechei*) tail-flag (conspicuously wave their tail from side to side) when they detect a snake. Free-ranging rattlesnakes (*Crotalus oreganus oreganus*), which normally prey on these squirrels, are less likely to strike and more likely to leave the area when an adult squirrel approaches them and tail-flags. The authors conclude that tail-flagging possibly communicates that the squirrel is vigilant; this might indicate to the snake that the squirrel will not be an easy target; the snake may then go to another area where it might have better luck catching a less vigilant squirrel. Although many questions remain, it is also possible that tail-flagging is simply a "nervous" gesture by squirrels that have seen a snake; however, if both snakes and squirrels are able to extract relevant information, the behavior could have two effects: deterring snake attacks and alerting other squirrels to the presence of danger. Not all responses given by prey in response to predators always have such clear-cut beneficial results. For example, Jan Randall and Marjorie Matocq (1997) found that footdrums produced by banner-tailed kangaroo rats (*Dipodomys spectabilis*) frequently deter predator gopher snakes (*Pituophis melanoleucus affinis*) from stalking and attacking the footdrumming individual; higher amounts of footdrumming resulted in decreased rates of stalking. The conclusion is again that footdrumming acts as a deterrent by informing the snake that its intended prey is alert and vigilant. However, the exact results depended on the hunger level of the snakes; nonhungry snakes were more likely to avoid the footdrumming, while hungry snakes investigated and came closer. Moreover, because kangaroo rats also use footdrumming in territorial displays, it appears that snakes that are very hungry may actually eavesdrop on the territorial display and use the footdrums to locate and capture prey. Thus, this is a case in which there are not only important benefits but also potential costs even if the signal in some contexts serves to deter predators.

Finally, mutualistic interactions often require interspecific communication. Here I discuss only two examples among many. Honeyguides (e.g., *Indicator indicator*) are African birds that are known to guide honey badgers (*Mellivora capensis*) and humans to bee colonies (Dean & MacDonald, 1981; Isack & Reyer, 1989). The honeyguide, by making certain calls, communicates that it has found a beehive and recruits honey badgers or humans to follow it. Once at the hive, the badger or human tears into the hive and takes the honey; the honeyguide, which by itself is not capable of breaking open a beehive, also benefits because it obtains food in the form of wax and exposed larvae.

One of the most amazing mutualisms involves the cleaner fish (e.g., *Labroides dimidiatus*). These fishes, which are very small and live in coral reefs, set up "cleaning stations" attended by many different species of larger fish, including many large and voracious predators. "Customer fish" line up and wait their turn to be cleaned. *Labroides* starts by picking off dead cells and parasites from the body of the "customer" but eventually enters the mouth to remove debris and decaying food from between the teeth, entrance to the gills, and throat of the customer (Feder, 1966; Côté, 2000). During this time, the large predator fish rests in a trance-like state, with its mouth wide open while the cleaner goes in and out. Clearly, such a potentially dangerous interaction (especially for the cleaner) requires unequivocal communication between cleaner and customer. As might be expected, the cleaners and customers have evolved certain signals to ensure that the interaction is carried out in a mutually beneficial manner (Feder, 1966; Côté, 2000). Both cleaner and customer benefit from this mutualistic interaction because the cleaner gets food (debris, parasites) and the customer is rid of noxious pathogens and other irritants.

CONCLUSIONS

In this chapter I have attempted an introduction to a very vast field of study in animal behavior. In addition to the many areas that I have covered there are many others that I simply could not include because of space limitations. Similarly, my examples are not exhaustive, and many other, equally fascinating examples exist for virtually every topic that I discuss. My goal was to give you a general idea of what we mean by animal communication, introduce you to classical concepts and perspectives, and familiarize you with some of the more modern developments and debates. If nothing else, I hope that after reading this chapter you have become aware of the complexity of this field and of the need for much more research in this area.

ACKNOWLEDGMENTS

I am grateful to Ken Yasukawa for inviting me to write this chapter and for his careful reading of the original manuscript and many helpful suggestions for improvement and on references. I also thank the members of the University of Missouri–St. Louis's Animal Behavior Discussion Group, including most specially Aimee Dunlap and Stan Braude, for illuminating discussions on the issue of signals, information, and intentionality. Gabriel Francescoli generously e-mailed me a package of relevant references that were invaluable. I also thank Steve Glickman, Roy Caldwell, and the late George Barlow for encouraging and nurturing my interest in communication ever since my graduate school days. As always, I am grateful to Arlene Zarembka for her support and patience during the time that I was completely engrossed in writing this chapter, and most especially for her assistance in cross-checking references and proofing parts of the manuscript.

REFERENCES AND SUGGESTED READING

Alexander, G. & E. Shillito-Walser. (1977). Importance of visual cues from various body regions in maternal recognition of the young in Merino sheep (*Ovis aries*). *Applied Animal Ethology*, 3, 137–143.

Anciães, M. & R. O. Prum. (2008). Manakin display and visiting behaviour: A comparative test of sensory drive. *Animal Behaviour*, 75, 783–790.

Arnold, K. & K. Zuberbühler. (2006). Language evolution: Semantic combinations in primate calls. *Nature*, 441, 303.

Badyaev, A. V. & G. E. Hill. (2003). Avian sexual dichromatism in relation to phylogeny and ecology. *Annual Review of Ecology and Systematics*, 34, 27–49.

Baerends, G. P. & J. M. Baerends-Von Roon. (1950). An introduction to the study of the ethology of cichlid fishes. *Behaviour Supplement*, 1, 1–142.

Bakke, A. (1978). Aggregation pheromone components in the bark beetle *Ips acuminatus*. *Oikos*, 31, 184–188.

Balshine-Earn, S. & D. J. D. Earn. (1998). On the evolutionary pathway of parental care in mouth-brooding cichlid fish. *Proceedings of the Royal Society of London, B*, 265, 2217–2222.

Barbour, M. A. & R. W. Clark. (2012). Ground squirrel tail-flag displays alter both predatory strike and ambush site selection behaviours of rattlesnakes. *Proceedings of the Royal Society of London, B*, 279, 3827–3833.

Barfield, C. H., Z. Tang-Martínez, & J. M. Trainer. (1994). Domestic calves (*Bos taurus*) recognize their own mothers by auditory cues. *Behaviour*, 97, 257–264.

Bastock, M. & A. Manning. (1955). The courtship of *Drosophila melanogaster*. *Behaviour*, 8, 85–110.

Baylis, J. R. (1974). The behavior and ecology of *Herotilapia multispinosa*. *Zeitschrift für Tierpsychologie*, 34, 115–146.

Beach, F. A. (1976). Sexual attractivity, proceptivity and receptivity in female mammals. *Hormones and Behavior*, 7, 105–138.
Beach, F. A. & R. A. Gilmore. (1949). Responses of male dogs to urine from females in heat. *Journal of Mammalogy*, 30, 391–392.
Bekoff, M. (1995). Play signals as punctuation: The structure of social play in canids. *Behaviour*, 132, 419–429.
Bildstein, K. L. (1983). Why white-tailed deer flag their tails. *American Naturalist*, 121, 709–715.
Blair, W. F. (1964). Isolating mechanisms and interspecies interactions in anuran amphibians. *Quarterly Review of Biology*, 39, 334–344.
Blest, A. D. (1961). The concept of ritualization. In W. H. Thorpe & O.L. Zangwill (eds.), *Current Problems in Animal Behaviour* (pp. 102–124). Cambridge, UK: Cambridge University Press.
Boch, R. & D. A. Shearer. (1971). Chemical releasers of alarm behavior in the honeybee, *Apis mellifera*. *Journal of Insect Physiology*, 17, 2277–2285.
Borgia, G. (1986). Sexual selection in bowerbirds. *Scientific American*, 254, 92–101.
Bradbury, J. W. & S. L. Vehrencamp. (1998). *Principles of Animal Communication*. Sunderland, MA: Sinauer Associates.
Breed, M. D. (1998). Recognition pheromones of the honeybee. *BioScience*, 48, 463–470.
Brockmann, A. & M. Sen Sarma. (2009). Honeybee dance language: Is it overrated? *Trends in Ecology and Evolution*, 24, 583.
Brumm, H. & H. Slabbekoorn. (2005). Acoustic communication in noise. *Advances in the Study of Behavior*, 35, 151–209.
Byers, J. A. (1983). Sex-specific response to aggregation pheromone: Regulation of colonization density in the bark beetle *Ips paraconfusus*. *Journal of Chemical Ecology*, 9, 129–142.
Caine, N. G., R. L. Addington, & T. L. Windfelder. (1995). Factors affecting the rates of food calls given by red-bellied tamarins. *Animal Behaviour*, 50, 53–60.
Caldwell, R. L. (1985). A test of individual recognition in the stomatopod *Gonadactylus festae*. *Animal Behaviour*, 33, 101–106.
Caldwell, R. L. & H. Dingle. (1975). Ecology and evolution of agonistic behavior in stomatopods. *Naturwissenschaften*, 62, 214–222.
Carazo, P. & E. Font. (2010). Putting information back into biological communication. *Journal of Evolutionary Biology*, 23, 661–669.
Caro, T. M. (1986). The functions of stotting in Thompson's gazelles: Some tests of the predictions. *Animal Behaviour*, 54, 1147–1154.
Caro, T. M. (1995). Pursuit deterrence revisited. *Trends in Ecology and Evolution*, 10, 500–503.
Chivers, D. P. & R. J. F. Smith. (1994). Intra- and interspecific avoidance of areas marked with skin extract from brook sticklebacks (*Culea inconstans*) in a natural habitat. *Journal of Chemical Ecology*, 20, 1517–1523.
Christy, J. H. (1995). Mimicry, mate choice, and the sensory trap hypothesis. *American Naturalist*, 146, 171–181.

Cocroft, R. B. (2005). Vibrational communication facilitates cooperative foraging in a phloem-feeding insect. *Proceedings of the Royal Society of London, B*, 272, 1023–1029.

Cocroft, R. B. & R. L. Rodriguez. (2005). The behavioral ecology of insect vibrational communication. *BioScience*, 55, 323–334.

Côté, I. M. (2000). Evolution and ecology of cleaning symbioses in the sea. *Oceanographic Marine Biology Annual Review*, 38, 311–355.

Daanje, A. (1950). On locomotory movements in birds and the intention movements derived from them. *Behaviour*, 3, 48–98.

Dantzker, M. S., G. B. Deane, & J. W. Bradbury. (1999). Directional acoustic radiation in the strut display of male sage grouse *Centrocercus urophasianus*. *Journal of Experimental Biology*, 202, 2893–2909.

Darwin, C. (1872). *The Expression of the Emotions in Man and the Animals*. London: John Murray.

Dawkins, R. & J. R. Krebs. (1978). Animal signals: Information or manipulation. In J. R. Krebs & N. B. Davies (eds.), *Behavioural Ecology* (pp. 282–309). Oxford, UK: Blackwell Scientific Publications.

Dean, W. R. J. & I. A. W. MacDonald. (1981). A review of African birds feeding in association with mammals. *Ostrich*, 52, 135–155.

Deneubourg, J. L., S. Goss, N. Franks, & J. M. Pasteels. (1989). The blind leading the blind: Modeling chemically mediated army ant raid patterns. *Journal of Insect Behavior*, 2, 719–725.

Dunbar, I. F. (1977). Olfactory preferences in dogs: The response of male and female to conspecific odours. *Behavioral Biology*, 20, 471–481.

Dunbar, R. M. (2010). The social role of touch in humans and primates: Behavioural functions and neurobiological mechanisms. *Neuroscience and Biobehavioral Reviews*, 34, 260–268.

Ehret, G. (2005). Infant rodent ultrasounds—a gate to the understanding of sound communication. *Behavioral Genetics*, 35, 19–29.

Elgar, M. A. (1986). House sparrows establish foraging flocks by giving chirrup calls if the resources are divisible. *Animal Behaviour*, 34, 169–174.

Elgar, M. A. (1991). Sexual cannibalism, size dimorphism, and courtship behavior of orb-weaving spiders (Araneidae). *Evolution*, 45, 444–448.

Elowson, A. M., P. L. Tannenbaum, & C. T. Snowdon. (1991). Food-associated calls correlate with food preferences in cotton-top tamarins. *Animal Behaviour*, 42, 931–937.

Emlen, D. J., J. Hunt, & L. W. Simmons. (2005). Evolution of sexual dimorphism and male dimorphism in the expression of beetle horns: phylogenetic evidence for modularity, evolutionary lability, and constraint. *American Naturalist*, 166, S42–S68.

Endler, J. A. (1992). Signals, signal conditions and the direction of evolution. *American Naturalist*, 139, S125–S153.

Enquist, M. (1985). Communication during aggressive interactions with particular reference to variation in choice of behavior. *Animal Behaviour*, 33, 1152–1161.

Feder, H. M. (1966). Cleaning symbiosis in the marine environment. *Symbyosis*, 1, 327–380.

Fernández-Vargas, M., Z. Tang-Martínez, & S. M. Phelps. (2008). Olfactory responses of neotropical short-tailed singing mice (*Scotinomys teguina*) to odors of the mid-ventral sebaceous gland: Discrimination of conspecifics, gender, and female reproductive condition. *Journal of Chemical Ecology*, 34, 429–437.

Fernández-Vargas, M., Z. Tang-Martínez, & S. M. Phelps. (2011). Singing, allogrooming, and allomarking behavior during intra- and inter-sexual encounters in the neotropical short-tailed singing mouse (*Scotinomys teguina*). *Behaviour*, 148, 945–965.

Fitzgerald, T. D. & F. X. Webster. (1993). Identification and behavioral assays of the trail pheromone of the forest tent caterpillar, *Malacosoma disstria Hübner* (Lepidoptera: Lasiocampidae). *Canadian Journal of Zoology*, 71, 1511–1515.

Fitzpatrick, M. J. & D. A. Gray. (2001). Divergence between the courtship songs of the field crickets *Gryllus texensis* and *Gryllus rubens* (Orthoptera: Gryllidae). *Ethology*, 107, 1075–1085.

Font, E. & P. Carazo. (2010). Animals in translation: Why there is meaning (but probably no message) in animal communication. *Animal Behaviour*, 79, e1–e6.

Fusani, L. (2008). Testosterone control of male courtship in birds. *Hormones and Behavior*, 54, 227–233.

Gibson, J. S. & G. W. Uetz. (2008). Seismic communication and mate choice in wolf spiders: Components of male seismic signals and mating success. *Animal Behaviour*, 75, 1253–1262.

Girard, M. B., M. M. Kasumovic, & D. O. Elias. (2011). Multimodal courtship in the peacock spider *Maratus volans* (O.P.-Cambridge 1874). *PLoS ONE*, e25390.

Gould, J. L. (1975). Honey bee recruitment: The dance language controversy. *Science*, 189, 685–693.

Grafen, A. (1990). Biological signals as handicaps. *Journal of Theoretical Biology*, 144, 517–546.

Gray, D. & W. H. Cade. (2000). Sexual selection and speciation in field crickets. *Proceedings of the National Academy of Sciences, USA*, 97, 14449–14454.

Grieser, M. (2009). Mobbing calls signal predator category in a kin group-living bird species. *Proceedings of the Royal Society of London, B*, 276, 2887–2892.

Grüter, C. & W. M. Farina. (2008). The honeybee waggle dance: Can we follow the steps? *Trends in Ecology and Evolution*, 24, 242–247.

Grüter, C. & W. M. Farina. (2009). Informational conflicts created by the waggle dance. *Proceedings of the Royal Society of London, B*, 275, 1321–1327.

Guillette L. M., L. L. Bloomfield, E. R. Batty, M. W. R. Dawson, & C. B. Sturdy. (2010). Black-capped (*Poecile atricapillus*) and mountain chickadee (*Poecile gambeli*) contact calls contain species, sex, and individual identity features. *Journal of the Acoustical Society of America*, 127, 1116–1123.

Hagedorn, M. (1986). The ecology, courtship and mating of gymnotiform electric fish. In T. H. Bullock & W. Heiligenberg (eds.), *Electroreception* (pp. 497–525). New York: Wiley.

Halpin, Z. T. (1986). Individual odors in mammals: Origins and functions. *Advances in the Study of Behavior*, 16, 40–70.

Halpin, Z. T. (1991). Kin recognition cues of vertebrates. In P. G. Hepper (ed.), *Kin Recognition* (pp. 220–258). Cambridge, UK: Cambridge University Press.

Harju, H. J. (1971). Spruce grouse copulation. *Condor*, 73, 380–381.

Harris, M. A. & J. O. Murie. (1982). Responses to oral gland scents from different males in Columbian ground squirrels. *Animal Behaviour*, 30, 140–148.

Heth, G., E. Frankenberg, A. Raz, & E. Nevo. (1987). Vibrational communication in subterranean mole rats (*Spalax ehrenbergi*). *Behavioral Ecology and Sociobiology*, 21, 31–33.

Hill, G. E. (1991). Plumage coloration is a sexually selected indicator of male quality. *Nature*, 350, 337–339.

Hinde, R. A. (1970). *Animal Behavior: A Synthesis of Ethology and Comparative Psychology*. New York: McGraw Hill.

Hinde, R. A. (1981). Animal signals: Ethological and game theory approaches are not incompatible. *Animal Behaviour*, 29, 535–542.

Hunsaker, D. (1962). Ethological isolating mechanisms in the *Sceloporus torquatus* group of lizards. *Evolution*, 16, 62–74.

Huxley, J. (1923). Courtship activities in the red-throated diver *Colymbus stellatus pontopp*; together with a discussion on the evolution of courtship in birds. *Journal of the Linnean Society*, 2, 491–562.

Isack, H. A. & H. U. Reyer. (1989). Honeyguides and honey gatherers: Interspecific communication in a symbiotic relationship. *Science*, 243, 1343–1346.

Jacobson, M. & M. Beroza. (1963). Chemical insect attractant. *Science*, 140, 1367–1373.

Jenssen, T. A. (1970). Female responses to filmed displays of *Anolis nebulosus* (Sauria: Iguanidae). *Animal Behaviour*, 18, 640–647.

Johnson, D. L. (1967). Honey bees: Do they use the direction information contained in their dance maneuver? *Science*, 155, 844–847.

Kaptein, N., J. Billen, & B. Gobin. (2005) Larval begging for food enhances reproductive options in the ponerine ant *Gnamptogenys striatula*. *Animal Behaviour*, 69, 293–299.

Kayser, A. J. & G. E. Hill. (2000). Structurally based plumage coloration is an honest signal of quality in male blue grosbeaks. *Behavioral Ecology*, 11, 202–209.

Keil, N. M., S. Imfeld-Muller, J. Aschwanden, & B. Wechsler. (2012). Are head cues necessary for goats (*Capra hircus*) in recognizing group members? *Animal Cognition*, 15, 913–921.

Kleiman, D. (1972). Social behavior in the maned wolf and bush dog: A study in contrast. *Journal of Mammalogy*, 53, 791–806.

Laidre, M. E. (2009). How often do animals lie about their intentions? An experimental test. *American Naturalist*, 173, 337–346.

Leal, M. (1999). Honest signaling during prey-predator interactions in the lizard *Anolis cristatellus*. *Animal Behaviour*, 58, 521–526.

Leal, M. & J. A. Rodriguez-Robles. (1995). Antipredator responses of *Anolis cristatellus* (Sauria: Polychrotidae). *Copeia*, 1995, 155–161.

Leger, D. W & D. H. Owings. (1978). Responses to alarm calls by California ground squirrels: Effects of call structure and maternal status. *Behavioral Ecology and Sociobiology*, 3, 177–186.

Leyhausen, P. (1956). Das verhalten der katzen (Felidae). *Handbook of Zoology (Berlin)*, 8, 1–34.

Ligout S., M. Keller & R. H. Porter. (2004). The role of olfactory cues in the discrimination of agemates by lambs. *Animal Behaviour*, 68,785–792.

Ligout, S. & R. H. Porter. (2004). The role of visual cues in lambs' discrimination between individual agemates. *Behaviour*, 141, 617–632.

Lorenz, K. (1972). Comparative studies on the behavior of Anatinae. In P. H. Klopfer & J. P. Hailman (eds.), *Function and Evolution of Behavior: An Historical Sample from the Pen of Ethologists* (pp. 231–258). Reading, MA: Addison-Wesley.

Lovell, S. F. & M. R. Lein. (2004). Neighbor-stranger discrimination by song in a suboscine bird, the alder flycatcher, *Empidonax alnorum*. *Behavioral Ecology*. 15, 799–804.

Magrath, R. D., B. J. Pitcher, & J. L Gardner. (2009). An avian eavesdropping network: Alarm signal reliability and heterospecific response. *Behavioral Ecology*, 20, 745–752.

Majolo, B., R. Ventura, & N. Koyama. (2005). Postconflict behavior among male Japanese macaques. *International Journal of Primatology*, 26, 321–336.

Maklakov, A. A., T. Bilde, & Y. Lubin. (2003). Vibratory courtship in a web-building spider: Signaling quality or stimulating the female? *Animal Behaviour*, 66, 623–630.

Manser, M. B. (1999). Response of foraging group members to sentinel calls in suricates, *Suricata suricatta*. *Proceedings of the Royal Society of London, B*, 266, 1013–1019.

Marler, P. (1961). The logical analysis of animal communication. *Journal of Theoretical Biology*, 1, 295–317.

Marler, P. & H. Slabbekoorn. (2004). *Nature's Music: the Science of Birdsong*. San Diego, CA: Elsevier Academic Press.

Møller, A. P. (1988). False alarm calls as a means of resource usurpation in the great tit *Parus major*. *Ethology*, 79, 25–30.

Morgan, E. D. (2008). Trail pheromones of ants. *Physiological Entomology*, 34, 1–17.

Morris, D. (1956). The feather posture of birds and the problem of the origin of social signals. *Behaviour*, 9, 75–113.

Morris, D. (1957). "Typical intensity" and its relation to the problem of ritualization. *Behaviour*, 11, 1–12.

Moynihan, M. (1955). Some aspects of the reproductive behavior of the black-headed gull (*Larus rudibundus rudibundus* L.) and related species. *Behaviour Supplement*, 4, 1–201.

Muller, M. N. & J. C. Mitani. (2005). Conflict and cooperation in wild chimpanzees. *Advances in the Study of Behavior*, 35, 275–331.

Munz, T. (2005). The bee battles: Karl von Frisch, Adrian Wenner and the honey bee dance language controversy. *Journal of the History of Biology*, 38, 535–570.

Muroyama, Y. & B. Thierry. (1998). Species differences in male loud calls and their perception in Sulawesi macaques. *Primates*, 39, 115–126.

Nakamichi, M. & Y. Shizawa. (2003). Distribution of grooming among adult females in a large, free-ranging group of Japanese macaques. *International Journal of Primatology*, 24, 607–625.

Narins, P. M., E. R. Lewis, J. J. U. M. Jarvis, & J. O'Rain. (1997). The uses of seismic signals by fossorial southern African mammals: A neuroethological gold mine. *Brain Research Bulletin*, 44, 641–646.

Ord, T. J., R. A. Peters, B. Clucas, & J. A. Stamps. (2007). Lizards speed up visual displays in noisy motion habitats. *Proceedings of the Royal Society of London, B*, 274, 1057–1062.

Owings, D. H. & E. S. Morton. (1997). The role of information in communication: An assessment/ management approach. In D. H. Owings, M. D. Beecher, & N. S. Thompson (eds.), *Perspectives in Ethology*, Vol. 12: *Communication* (pp. 359–390). New York: Plenum Press.

Owren, M. J., D. Rendall, & M. J. Ryan. (2010). Redefining animal signaling: Influence versus information in communication. *Biology and Philosophy*, 25, 755–780.

Payne, K. (1998). *Silent Thunder: In the Presence of Elephants*. New York: Simon and Schuster.

Payne, K., W. R. Langbauer Jr., & E. M. Thomas. (1986). Infrasonic calls of the Asian elephant (*Elephas maximus*). *Behavioral Ecology and Sociobiology*, 18, 297–301.

Pomiankowski, A. (1987). Sexual selection: The handicap mechanism does work—sometimes. *Proceedings of the Royal Society of London, B*, 231, 123–145.

Proctor, H. C. (1991). Courtship in the water mite *Neumania papillator*: Males capitalize on female adaptations for predation. *Animal Behaviour*, 42, 589–598.

Pryke, S. R., M. J. Lawes, & S. Andersson. (2001). Agonisitic carotenoid signaling in male red-collared widowbirds: Aggression related to the color signal of both the territory owner and model intruder. *Animal Behaviour*, 62, 695–704.

Rado, R., J. Terkel, & Z. Wollberg. (1998). Seismic communication signals in the blind mole rat (*Spalax ehrenbergi*): Electrophysiological and behavioral evidence for their processing by the auditory system. *Journal of Comparative Physiology, A*, 183, 503–511.

Rainey, H. J., K. Zuberbühler, & P. J. B. Slater. (2004). Hornbills can distinguish between primate alarm calls. *Proceedings of the Royal Society of London, B*, 271, 755–759.

Randall, J. A. & E. R. Lewis. (1997). Seismic communication between the burrows of kangaroo rats, *Dipodomys spectabilis*. *Journal of Comparative Physiology, A*, 181, 525–531.

Randall, J. A. & M. D. Matocq. (1997). Why do kangaroo rats (*Dipodomys spectabilis*) footdrum at snakes? *Behavioral Ecology*, 8, 404–413.
Redondo, T. & L. Arias de Reyna. (1988). Locatability of begging calls in nestling altricial birds. *Animal Behaviour*, 36, 653–661.
Rendall, D., M. J. Owren, & M. J. Ryan. (2009). What do animal signals mean? *Animal Behaviour*, 78, 233–240.
Ribak, G., M. L. Pitts, G. S. Wilkinson, & J. G. Swallow. (2009). Wing shape, wing size and sexual dimorphism in eye span in stalk-eyed flies (Diopsidae). *Biological Journal of the Linnean Society*, 98, 860–871.
Robinson, M. H. (1982). Courtship and mating behaviors in spiders. *Annual Review of Entomology*, 27, 1–20.
Roelofs, W. L. & A. Comeau. (1969). Sex pheromone specificity: Taxonomic and evolutionary aspects in Lepidoptera. *Science*, 165, 398–400.
Rohwer, S. A. (1977). Status signaling in Harris sparrows: Some experiments in deception. *Behaviour*, 61, 107–129.
Rohwer, S. A. & F. C. Rohwer. (1978). Status signaling in Harris' sparrows: Experimental deceptions achieved. *Animal Behaviour*, 26, 1012–1022.
Ruxton, G. D. & H. M. Schaeffer. (2011). Resolving current disagreements and ambiguities in the terminology of animal communication. *Journal of Evolutionary Biology*, 24, 2574–2585.
Ryan, M. J. (1990). Signals, species, and sexual selection. *American Scientist*, 78, 46–52.
Ryan, M. J. & A. S. Rand. (1990). The sensory basis of sexual selection for complex calls in the túngara frog, *Physalaemus pustulosus* (sexual selection for sensory exploitation). *Evolution*, 44, 305–314.
Ryan, M. J. & A. S. Rand. (1993). Sexual selection and signal evolution: The ghosts of biases past. *Philosophical Transactions of the Royal Society, B*, 340, 187–195.
Ryan, M. J., M. D. Tuttle, & A. S. Rand. (1982). Sexual advertisement and bat predation in a neotropical frog. *American Naturalist*, 119, 136–139.
Sauer, W. H. H., M. J. Roberts, M. R. Lipinsky, M. J. Smale, R. T. Hanlon, D. M. Webber, et al. (1997). Choreography of the squids' "nuptial dance." *Biological Bulletin*, 192, 203–207.
Schino, G. (2001). Grooming, competition, and social rank among female primates: A meta-analysis. *Animal Behaviour*, 62, 265–271.
Searcy, W. A. & S. Nowicki. (2005). *The Evolution of Animal Communication: Reliability and Deception in Signaling Systems*. Princeton, NJ: Princeton University Press.
Seyfarth, R. M. (1980). The distribution of grooming and related behaviours among adult female vervet monkeys. *Animal Behaviour*, 28, 798–813.
Seyfarth, R. M, D. L. Cheney, T. Bergman, J. Fischer, K. Zuberbühler, & K. Hammerschmidt. (2010). The central importance of information in studies of animal communication. *Animal Behaviour*, 80, 3–8.
Sherman, P. W. (1977). Nepotism and the evolution of alarm calls. *Science*, 197, 1246–1253.

Sherman, P. W. (1985). Alarm calls of Belding's ground squirrels to aerial predators: Nepotism or self preservation? *Behavioral Ecology and Sociobiology*, 17, 313–323.

Sivalinghem, S., M. M. Kasumovic, A. C. Mason. M. C. B. Andrade, & D. O. Elias. (2010). Vibratory communication in the jumping spider *Phiddipus clarus*: Polyandry, male courtship signals, and mating success. *Behavioral Ecology*, 21, 1308–1314.

Slater, P. J. B. (1983). The study of communication. In T. R. Halliday & P. J. B. Slater (eds.), *Animal Behaviour*, Vol. 2: *Communication* (pp. 9–42). Oxford, UK: Blackwell Publishing.

Slater, P. J. B. & N. I. Mann. (2004). Why do the females of many bird species sing in the tropics? *Journal of Avian Biology*, 35, 289–294.

Smith, W. J. (1977). *The Behavior of Communicating: An Ethological Approach*. Cambridge, MA: Harvard University Press.

Smith, W. J. (1986). An "informational" perspective on manipulation. In R. W. Mitchell & N. S. Thompson (eds.), *Deception: Perspectives on Human and Nonhuman Deceit* (pp. 71–86). Albany: State University of New York Press.

Smotherman, W. P., R. W. Bell, J. Starzec, J. Elias, & T. A. Zackman. (1974). Maternal responses to infant vocalizations and olfactory cues in rats and mice. *Behavioral Biology*, 12, 55–66.

Spieth, H. T. (1974). Courtship behavior in *Drosophila*. *Annual Review of Entomology*, 19, 385–405.

Sugiura, H. (1998). Matching of acoustic features during the vocal exchange of coo calls by Japanese macaques. *Animal Behaviour*, 55, 673–687.

Tang-Martínez, Z. (2001). The mechanisms of kin discrimination and the evolution of kin recognition in vertebrates: A critical re-evaluation. *Behavioural Processes*, 53, 21–40.

Templeton, C. N. & E. Greene. (2007). Nuthatches eavesdrop on variations in heterospecific chickadee mobbing alarm calls. *Proceedings of the National Academy of Sciences, USA*, 104, 5479–5482.

Tinbergen, N. (1951). *The Study of Instinct*. London: Oxford Clarendon Press.

Tinbergen, N. (1952). "Derived" activities: Their causation, biological significance, origin and emancipation during evolution. *Quarterly Review of Biology*, 27, 1–32.

Tinbergen, N. (1959). Comparative studies of the behaviour of gulls (Laridae): A progress report. *Behaviour*, 15, 1–70.

Trainer, J. M., D. B. McDonald, & W. A. Learn. (2002). The development of coordinated singing in cooperatively displaying long-tailed manakins. *Behavioral Ecology*, 13, 65–69.

Vaché, M., J. Ferron, & P. Gouat. (2001). The ability of red squirrels (*Tamiasciurus hudsonicus*) to discriminate conspecific olfactory signatures. *Canadian Journal of Zoology*, 79, 1296–1300.

Vandenberg, J. G. (1969). Endocrine coordination in monkeys: Male sexual responses to the female. *Physiology and Behavior*, 4, 261–264.

Von Frisch, K. (1967). *The Dance Language and Orientation of Bees*. Cambridge, MA: Harvard University Press.
Von Hippel, F. A. (2000). Vigorously courting male sticklebacks are poor fathers. *Acta Ethologica*, 2, 83–89.
Wager, B. R. & M. D. Breed. (2000). Does honeybee sting alarm pheromone give orientation information to defensive bees? *Annals of the Entomological Society of America*, 93, 1329–1332.
Wallace, B. (1973). Misinformation, fitness, and selection. *American Naturalist*, 107, 1–7.
Wenner, A. M. (1967). Honey bees: Do they use the distance information contained in their dance maneuver? *Science*, 155, 847–849.
Whittaker, D. J., D. G. Reichard, A. L. Dapper, & E. D. Ketterson. (2009). Behavioral responses of nesting female dark-eyed juncos, *Junco hyemalis*, to hetero- and conspecific passerine preen oils. *Journal of Avian Biology*, 40, 579–583.
Whittaker, D. J., K. M. Richmond, A. K. Miller, R. Kiley, C. Bergeon Burns, J. Atwell, & E. D. Ketterson. (2011). Intraspecific preen oil odor preferences in dark-eyed juncos (*Junco hyemalis*). *Behavioral Ecology*, 22, 1256–1263.
Whittaker, D. J., H. A. Soini, J. W. Atwell, C. Hollars, M. V. Novotny, & E. D. Ketterson. (2010). Songbird chemosignals: Preen oil volatile compounds vary among individuals, sexes, and populations. *Behavioral Ecology*, 21, 608–614.
Wickler, W. (1967). Socio-sexual signals and their intraspecific imitation among primates. In D. Morris (ed.), *Primate Ethology* (pp. 69–147). London: Weidenfeld and Nicholson.
Wilson, E. O. (1975). *Sociobiology*. Cambridge, MA: Harvard University Press.
Wilson, E. O. & W. H. Bossert. (1963). Chemical communication among animals. *Recent Progress in Hormone Research*, 19, 673–716.
Wunderlee, J. M. (1978). Differential responses of territorial yellowthroats to the songs of neighbors and non-neighbors. *Auk*, 95, 389–395.
Wyatt, T. D. (2003). *Pheromones and Animal Behaviour: Communication by Smell and Taste*. Cambridge, UK: Cambridge University Press.
Wyman, M. T., M. S. Mooring, B. McCowan, M. C. T. Penedos, & L. A. Hart. (2008). Amplitude of bison bellows reflect male quality, physical condition and motivation. *Animal Behaviour*, 76, 1625–1639.
Yasukawa, K., R. L Knight, & S. K. Skagen. (1987). Is courtship intensity a signal of male parental care in red-winged blackbirds (*Agelaius phoeniceus*)? *Auk*, 104, 628–634.
Young, J. C., R. M. Silverstein, & M. C. Birch. (1973). Aggregation pheromone of the beetle *Ips confusus*: Isolation and identification. *Journal of Insect Physiology*, 11, 2273–2277.
Zahavi, A. (1975). Mate selection—selection for a handicap. *Journal of Theoretical Biology*, 53, 205–213.
Zahavi, A. (1977). The cost of honesty (further remarks on the handicap principle). *Journal of Theoretical Biology*, 67, 603–605.

Zeigler, H. P. (1964). Displacement activity and motivational theory: A case study in the history of ethology. *Psychological Bulletin*, 61, 362–376.

Zuberbühler, K. (2000). Interspecies semantic communication in two forest primates. *Proceedings of the Royal Society of London, B*, 267, 713–718.

Zuk, M., D. Rebar, & S. Primrose Scott. (2008). Courtship song is more variable than calling song in the field cricket *Teleogryllus oceanicus*. *Animal Behaviour*, 76, 1065–1071.

4

Mating Systems and the Measurement of Sexual Selection

Kenyon B. Mobley

INTRODUCTION

The process of *sexual selection* underlies the way that individuals choose mates and produce offspring in nature (Darwin, 1871). The opportunities and choices that an individual may make are governed by both physiological and phylogenetic constraints, the types of social interactions and potential-mate encounters an individual experiences, the method of parental care employed, and a whole host of ecological factors that may influence the breeding biology of individuals within a population. The sum total of these mating and reproductive opportunities and the methods used to estimate the strength and direction of sexual selection are the subject of this chapter: mating systems and the measurement of sexual selection.

Mating-system research is a rapidly growing field with the specific aim to understand the rich variety of reproductive behaviors, sex roles, and expression of secondary sexual traits that abound in nature. Historically, the term *mating system* refers to a behavioral strategy employed by a population of animals to obtain mates (Emlen & Oring, 1977), including various aspects of sexual selection such as mate choice and mate competition, parental care, and ecological factors that may influence the degree of multiple mating within a population. Today, interest in mating-system research has expanded considerably and now encompasses various alternative mating behaviors and strategies, genetic mating systems, and a theoretical framework for estimating the strength and direction of sexual selection based on quantitative selection theory. Still, the field of mating systems holds true to the original goal of understanding the patterns of mating and reproductive success of a population and how these patterns, in turn, are regulated by the environmental conditions that may ultimately govern the expression of such behaviors.

In this chapter I will provide a brief historical overview of animal mating systems and how our understanding of sexual selection has grown since Charles Darwin's (1871) treatise on the subject in *The Descent of Man and Selection in Relation to Sex*. Because the literature can be daunting and laden with different perspectives, a second goal of this chapter is to provide a user-friendly guide to help navigate the more commonly used methods to quantify sexual selection as well as the advantages and drawbacks to each. The objective here is to provide students and researchers a firm foundation from which to investigate questions concerning mating systems and the measurement of sexual selection in their own particular system of choice.

A BRIEF HISTORY OF MATING SYSTEMS

Among the first to recognize the importance of mating systems was Darwin, who, among his other important contributions to the field of sexual selection, introduced the founding principles of mating-system evolution. He noted,

> That some relation exists between polygamy and the development of secondary sexual characters, appears nearly certain; and this supports the view that a numerical preponderance of males would be eminently favourable to the action of sexual selection. (Darwin, 1871, pp. 254–255)

The first principle that can be deduced from this passage is that a causal relationship exists between the mating system and the presence of secondary sexual characters. Here, Darwin recognized that the degree to which secondary sexual characters are expressed depends strongly on the degree of multiple mating within a population. The second principle that Darwin identified is the relationship between the strength of sexual selection and the sex ratio. He understood that if the sex ratio was skewed toward one sex (males in this instance) then the other sex would become limiting, and therefore ***intrasexual competition*** would increase on the nonlimiting sex. Thus, a greater bias towards the nonlimiting sex should result in stronger sexual selection acting on the nonlimiting sex (Trivers, 1972; Emlen & Oring, 1977;

Clutton-Brock, 1991). Taken together, these ideas provided the raw materials for the theoretical framework for modern-day mating-system research.

A renewed interest in sexual selection occurred in the latter part of the twentieth century with the introduction of two theories, **_anisogamy_** and **_parental investment theory_**. The theory of anisogamy (Parker et al., 1972) posited that the differential investment into the production of gametes by males and females gave rise to the evolution of sexual selection and conflict between the sexes. The theory of anisogamy is predicated on the fact that males produce many small, motile sperm whereas females produce large, non-motile, and nutritious eggs. As a result of this differential investment, males are predicted to compete more intensely for fertilization opportunities, generating a male-biased operational sex ratio under normal conditions. Although both theories speak more to the underlying differences in the sexes rather than the actual mechanisms of sexual selection (Kokko & Jennions, 2008), they set the stage for further development of theories concerning the fundamental conflict between the sexes driving sexual selection and mating system evolution.

The Environmental Potential for Polygamy

The first cogent attempt to synthesize mating-system theory was advanced by Stephen Emlen and Lewis Oring (1977). In their treatise, they outlined a theoretical framework for discussing mating systems that is still in place today. The crux of their theoretical framework rests on the ability of certain individuals in a population to monopolize or limit access of competitors to potential mates. If a few individuals of one sex can monopolize several mates, then variance in the number of mates, or mating success, will increase. This variation is quantifiable, as we will see later in the chapter, and divides individuals of each sex into "winners" or "losers," or, in keeping with Emlen and Oring's original terminology, the "haves" and "have nots." Thus, a precondition for the evolution of various mating systems is the ability for some individuals to monopolize mates. Because the ability to monopolize mates is theoretically constrained by phylogenetic factors, such as reproductive physiology and parental care strategies, as well as ecological factors such as mate availability, they coined the concept of the **_environmental potential for polygamy_** to capture the essence of both the ecological conditions necessary for and the ability to capitalize on multiple mating.

To illustrate their point of the environmental potential for polygamy, Emlen and Oring (1977) devised a graphical model in which two critical components, namely the temporal availability of mates and the spatial distribution of resources, influence the outcome of the mating system for a particular population. For example, if the spatial distribution of resources such as nesting sites is uniform and all individuals are reproductively receptive at the same time, then the ability of a single individual to monopolize multiple mates is diminished. If, on the other hand, mating is asynchronous such that only a few individuals are receptive at a particular time and resources are spatially clumped, for instance around a particularly nice patch of resources, then the ability of a particular individual to monopolize multiple mates is greatly increased. The resulting mating systems predicted to evolve from these scenarios are **monogamy** and **polygamy** (either **polygyny** or **polyandry**, depending on the monopolizing sex), respectively. While these two scenarios lie at the two extremes of the environmental potential for polygamy continuum, one can envision a multitude of situations where both the temporal availability of mates and the spatial distribution of resources are variable and may differ between the sexes, thereby setting the stage for the evolution of different mating systems. An important point here is that because temporal availability of mates and spatial distributions of resources are likely to fluctuate in a particular habitat or population, this model also has the implicit assumption that mating systems can be flexible and are not necessarily fixed for a population.

The final major contribution by Emlen and Oring (1977) was the elaboration on the concept of the **operational sex ratio**, which quickly became a cornerstone of mating-system theory. Originally described in Emlen (1976, p. 309) as "the ratio of receptive females to potential mating males at one time," the operational sex ratio was presented as an instantaneous measure of both the intensity and direction of mate competition, and hence sexual selection. Emlen and Oring (1977) also point out that skewed operational sex ratios have the ability to manipulate which mating system would be most likely to evolve in various conditions, such as polygyny under male-biased conditions and polyandry under female-biased conditions.

In the time since Emlen and Oring's (1977) publication, two major developments led to increased interest in mating-system research and, ultimately, a new appreciation of the intricacies of sexual selection. The first such advancement is a theoretical framework for quantifying sexual selection that combines elements of mating-system theory with selection theory, thereby bridging the gap between sexual selection, mating-system organization, and quantitative selection theory (Wade, 1979; Arnold & Wade, 1984; Arnold & Duvall, 1994; Shuster & Wade, 2003; Jones, 2009). The second such advancement was the technological development of molecular markers that opened a floodgate of molecular inquiries into patterns of parentage or **genetic mating systems** in natural populations (Avise, 2004; Jones et al., 2005). The combination of these two major advancements has solidified the field of mating systems, propelling research into the twenty-first century.

A Theoretical Quantitative Framework

In order to discuss the first major advance in mating-system research in recent times, namely the synthesis of a theoretical quantitative framework to measure population-level sexual selection, we need to go back in time to Angus J. Bateman's classic study on the mating behavior of *Drosophila melanogaster*, rediscovered by Trivers (1972). Based on several simple yet elegant experiments using visible phenotypic markers of both sexes, Bateman (1948) demonstrated that male *Drosophila* exhibited higher variance in mating success, or the number of mates per male, than did females. Similarly, he demonstrated that males exhibited higher variance in fertility, or the number of offspring per male, than did females. Finally, Bateman postulated that the strength of sexual selection on males was driven by a higher correlation between mating success and reproductive success, famously stating, "Variance in number of mates is, therefore, the only important cause of the difference in the variance in fertility" (Bateman, 1948, p. 364). These core principles, namely the variance in mating success, the variance in reproductive success, and the relationship between mating and reproductive success, were dubbed **Bateman's principles** (Wade, 1979; Arnold & Wade, 1984; Arnold, 1994; Arnold & Duvall, 1994), and a new era of quantitative appraisals of mating-system organization was born.

After the initial introduction of Bateman's principles as a method to quantify sexual selection, a flurry of critical reviews rapidly followed, shedding doubt on whether one can distinguish sexual selection from random mating patterns using quantitative approaches (Sutherland, 1985, 1987; Koening & Albano, 1986; Hubbell & Johnson, 1987) or whether quantification of sexual selection is even interesting (Grafen, 1987). To add to this confusion, various other methods to measure mating and reproductive inequality were also proposed to quantify mating systems (Kokko et al., 1999), although at the time few of these measures were actually tested and compared empirically (Fairbairn & Wilby, 2001).

This building controversy was met with the timely publication of "Mating Systems and Strategies" by Stephen Shuster and Michael Wade (2003), which attempted to synthesize the burgeoning field of quantitative mating systems. The authors' approach combined a theoretical framework for classifying mating systems based on the temporal availability of mates and the spatial distribution of resources (*sensu* Emlen & Oring, 1977) with a new metric for quantifying the direction and intensity of sexual selection, I_{mates}, or the sex difference in the opportunity for selection. Further refinements to the theory have been added by Adam Jones (2009), who proposed an upper limit to the standardized selection differential in units of phenotypic standard deviations (s'_{max}) and clarified some earlier misconceptions on the actual maximum strength of selection.

Today, quantitative methods to measure sexual selection and mating systems remain a controversial topic and are subject to vigorous debate (e.g., Klug, Heuschele, et al., 2010; Krakauer et al., 2011). Although some progress has been made to synthesize the various methods, attempts to reach a consensus on the most appropriate methods appear to be unattainable in the near future. Theoretical advancements to mating-system theory have thus far outpaced rigorous empirical testing, although recent work is beginning to bridge the knowledge gap by investigating different quantitative mating-system measurements under various ecological scenarios (Jones et al., 2005; Mills et al., 2007; Mobley & Jones, 2007, 2009; Barreto & Avise, 2010; Fitze & Le Galliard, 2011). Further empirical testing and refinement of theory is clearly warranted and will be necessary to settle this dispute in the future.

Molecular Markers and Genetic Mating Systems

The discovery and development of molecular ***DNA fingerprinting*** techniques, such as ***microsatellite DNA*** and ***amplified fragment length polymorphism*** (***AFLP***) markers, have allowed for the investigation of parentage and kinship within natural populations in unprecedented detail (Avise, 2004). This discovery then fueled a molecular revolution in behavioral ecology. For example, the discovery of ***extra-pair paternity*** in socially monogamous birds illustrated the disparity between the ***social mating system*** (i.e., social interactions and parental care strategies) and the realized genetic mating system in many species (Dunn et al., 2001; Hasselquist & Sherman, 2001; Griffith et al., 2002). Overturning this paradigm is especially relevant to mating systems because the seemingly widespread occurrence of gaudy ***secondary sexual traits*** in birds was a conundrum to Darwin and subsequent researchers. To his credit, Darwin proposed a verbal model of how sexual selection can operate in monogamous species based on the temporal variability in breeding timing of individuals of variable quality (Darwin, 1871). This theory is often called the Darwin-Fisher theory of sexual selection, based on the mathematical proof of the theory by R. A. Fisher (Fisher, 1915; Kirkpatrick et al., 1990; Jones & Ratterman, 2009). Although the Darwin-Fisher theory of sexual selection may help explain the existence of sexual selection in species that are genetically monogamous, the application of molecular markers now provides ample support for the idea that the degree of extra-pair paternity explains the existence of elaborate traits in a variety of socially monogamous bird species. This in turn reaffirms Darwin's original notion that a link between polygamy and secondary sexual characters exists (Darwin, 1871; Møller & Birkhead,

1994; Petrie & Kempenaers, 1998; Griffith et al., 2002; Westneat & Stewart, 2003).

Today, it is generally recognized that a complete understanding of sexual selection in the wild requires some knowledge of the genetic mating system. Their widespread availability and low cost, along with advances in parentage analysis (Jones & Ardren, 2003; Jones et al., 2010), have made molecular markers easy to apply to many different animal mating systems. Coupled with new quantitative approaches for the measurement of sexual selection, we now have a framework for testing specific hypotheses concerning the evolution of mating systems, the direction of sexual selection, and selection on individual traits associated with mating and reproductive success.

This brief synthesis highlights the important methodological and theoretical advances in the field of mating systems to date. However, it should be noted that this treatment glosses over significant contributions of many others that have increased our overall understanding of mating systems. For example, other noteworthy milestones include Fisher's contributions to sex-ratio theory (Fisher, 1930), James F. Crow's formulation of the opportunity for selection (Crow, 1958, 1962), and the mathematical formulation of spatial distribution analysis by mean crowding by Monte Lloyd (1967), just to name a few.

TERMINOLOGY

Despite over 150 years of research in mating systems, the field suffers from a lack of generally accepted terminology, and attempts to disambiguate various definitions have met with limited success (Emlen & Oring, 1977; Reynolds, 1996). For example, if you initiate a literature search for "mating systems," you are likely to find an equal number of references concerning animal mating systems and plant mating systems. A quick perusal would suggest zoologists and botanists have divergent views of what comprises a mating system, complete with separate lexicons (Reynolds, 1996; Sakai & Westneat, 2001). Although the emphases of plant and animal mating systems may appear to be different, in actuality both are concerned with patterns of parentage within populations. Animal mating systems are primarily concerned with patterns of sexual selection within populations (Emlen & Oring, 1977; Reynolds, 1996; Jones et al., 2004), while plant mating systems focus on rates and consequences of selfing (Sakai & Westneat, 2001; Shuster, 2009). Thus, animal and plant mating systems can be unified with the concept of the ***parental table***, which summarizes the genetic parentage of all progeny in a breeding population by ascribing the number of progeny produced by each potential pairing regardless of whether the offspring are produced by selfing or outcrossing

(Arnold & Duvall, 1994). Another method for unifying plant and animal mating systems is via ***spatial and temporal mean crowding*** (Shuster, 2009), a mathematical interpretation of the original spatial and temporal distributions of resources as discussed in Emlen and Oring (1977).

For the purposes of this chapter, I will use the general definition of a mating system as the behavioral strategy of a population employed in obtaining mates and reproductive opportunities, following Emlen and Oring (1977, Box 4.1). At the most basic level, animal mating systems can be divided into monogamous (having a single mate) or polygamous (having more than one mate) (Box 4.1). Because *polygamy* is not a particularly useful term with respect to the direction of sexual selection, it can be broken down into further subcategories to reflect whether males mate multiply (polygyny, stronger sexual selection

---

Box 4.1
Definitions of commonly used terms to describe mating systems

Mating system—The general behavioral strategy of a population employed in obtaining mates and reproductive opportunities (e.g., monogamy, polygamy). The mating system may also be referred to as the mating pattern or breeding system.

Social mating system—The social behavioral strategy of a population, including mate monopolization strategies, the presence and characteristics of pair bonds, and the patterns of parental care provided by each sex (e.g., social monogamy, resource defense polygamy, promiscuity).

Genetic mating system—Actual patterns of parentage within a population, including all precopulatory and postcopulatory sexual selection (e.g., genetic monogamy, genetic polyandry).

Breeding bout—The minimum unit of time required to produce viable offspring.

Monogamy—Each sex mates a maximum of once during a breeding bout.

Polygamy—A sex or both sexes mate more than once during a breeding bout.

Polygyny—Males have a variable number of mates during a breeding bout; females have a maximum of one mate.

Polyandry—Females have a variable number of mates during a breeding bout; males have a maximum of one mate.

Polygynandry—The condition where both males and females have a variable number of mates during a breeding bout. Sometimes also referred to polyandrogyny when males are more variable than females and polygynandry when females are more variable than males in the number of mates.

in males), females mate multiply (polyandry, stronger sexual selection in females), or both sexes mate multiply (***polygynandry***, variable strength of sexual selection in both sexes, Box 4.1). Mating systems can also be classified based on the nature and extent of their social relationships (social mating systems) or genetic relationships (genetic mating systems; Box 4.1). The social mating system includes various behavioral aspects of mating-system organization such as mate monopolization and parental care strategies, while the genetic mating system encapsulates the actual patterns of parentage within a population.

To properly discuss and compare mating systems, it is important to consider the amount of time required for a male and female to produce viable offspring, here defined as a ***breeding bout*** (Box 4.1). A breeding bout is roughly equivalent to a selection episode (Lande & Arnold, 1983), or the period of interest in which selection can be measured. For some species with parental care, a breeding bout may include the time required to raise offspring and additional time required until an adult is physiologically ready to mate again. This inequality in the amount of time, as well as the physiological ability of one sex to produce more offspring than the other, sets the stage for the evolution of different strategies for obtaining sexual partners between the sexes. It should be noted that a breeding bout in many species is variable, and the amount of time that is required to produce offspring is dependent upon a variety of physiological parameters such as condition or age of the individual as well as various ecological parameters such as temperature, food availability, or population density (Kvarnemo & Ahnesjö, 1996). Additional issues arise when individuals care for multiple broods of offspring and when the time for a particular brood to develop from a parental pairing is difficult to measure. In these instances, exact parentage data would be one way to distinguish the amount of time that a successful pairing took to produce viable offspring.

Establishing a common currency of reproductive time is important to help identify and clarify mating-system terminology between studies. In particular, considerable confusion arises when mating systems are considered under various time scales. For example, a species where males and females mate once per breeding bout but may change partners between breeding bouts would be considered monogamous per breeding bout but polygamous during a mating season or a lifetime. As the behavioral strategy employed to obtain mates during a single breeding bout would be monogamous, this would be the most appropriate term to describe the species in this instance. While a breeding bout may be a difficult concept to apply in some species, it is still important to conceptualize a minimum unit of reproduction to clarify the relative number of breeding events within a reproductive lifetime.

The lexicon of social mating systems has been reviewed extensively, and precise definitions, as well as the social and ecological contexts under which social mating systems have evolved, exist (Emlen & Oring, 1977; Thornhill & Alcock, 1983; Reynolds, 1996; Shuster & Wade, 2003). Genetic mating systems, on the other hand, have received considerably less attention in the literature. To illustrate how to apply genetic-mating-system terminology, let us take the breeding biology of the family Syngnathidae (pipefish, seahorses, and seadragons) as an example. Syngnathids have a specialized reproductive mode where females transfer eggs to the male during copulation and the male provides all care to the developing offspring. Within this family, several genetic mating systems have evolved in different species, including monogamy, polyandry, and polygynandry (Jones & Avise, 2001; Wilson et al., 2003; Coleman & Jones, 2011; Mobley et al., 2011). In the seahorses (*Hippocampus* spp.), males and females develop extensive pair bonds and male seahorses do not accept eggs from additional females. Thus, during a single pregnancy, a male is truly genetically monogamous as judged by several parentage studies (Jones et al., 1998; Wilson & Martin-Smith, 2007). In all but a few rare occasions, females do not breed a second time during the pregnancy, as it takes nearly the same amount of time for a female to mature a new clutch of eggs as a male takes to brood and release offspring. Therefore females can also be considered genetically monogamous (Foster & Vincent, 2004; Mobley et al., 2011). In pipefishes of the genus *Syngnathus*, females generally produce more eggs than a single male can brood during a single male pregnancy (Berglund et al., 1986, 1989; Scobell et al., 2009), and hence the potential for polygamous mating systems, both polyandry and polygynandry, is expected to evolve. Indeed, several species of *Syngnathus* exhibit polygynandry where both males and females mate multiply during a single pregnancy, as well as polyandry where females mate multiply but males mate with just one female during a pregnancy (reviewed in Jones & Avise, 2001; Mobley et al., 2011).

Today, it is known that both males and females in the majority of taxonomic groups can mate multiply such that strict genetic monogamy, polygyny, and polyandry are rare, and therefore knowledge of different levels of qualitative and quantitative measurements may be more useful to compare between sexes, populations, and species.

Measuring Sexual Selection

A quick perusal of the mating-system literature shows that there are many methods developed and advocated over the years. Measurements of sexual selection fall into two main categories: (1) qualitative measurements that predict the direction and intensity of mate competition and hence sexual selection and (2) quantitative measurements that capture the direction and intensity of sexual selection using variance-based, quantitative theoretical approaches (Table 4.1). In general, if you are interested in understanding how demography and social interactions influence the direction and intensity of mate competition, parental investment, and the organization of social mating systems, then qualitative measurements of the mating system such as the operational sex ratio and the potential reproductive rate are typically employed. If, on the other hand, you would like to investigate how a particular trait is related to mating and reproductive success, or whether or not sexual selection is likely to occur in your own particular population or species, then quantitative measurements of sexual selection based on variance in mating and reproductive success would be more appropriate.

Qualitative Measurements of Sexual Selection

There have been several qualitative measurements of the direction and intensity of sexual selection developed over the years, including the operational sex ratio and the potential reproductive rate. Additional qualitative measures of sexual selection such as ***mean spatial crowding*** (m^*) and ***mean temporal crowding*** (t^*) focus on how ecological factors such as mate-encounter rates, the distribution of individuals in breeding territories, and breeding synchrony influence mating-system organization (Shuster & Wade, 2003; Shuster, 2009). Because these measures generally predict the outcome of sexual selection by estimating the outcome of mating competition, these measures may be considered qualitative measurements of sexual selection to differentiate them from measurements that are related to quantitative selection theory.

The Operational Sex Ratio

The operational sex ratio predicts the direction and intensity of sexual selection based on mating competition as judged by the number of adults ready to mate in a population at a given time (Emlen & Oring, 1977). A bias in the operational sex ratio should reflect the strength of this competition and therefore should translate into higher interspecific competition for access to the limiting sex (Emlen & Oring, 1977; Kvarnemo & Ahnesjö, 1996; Weir et al., 2011). The operational sex ratio can also be used to predict which sex will compete for access to mates, thereby determining the sex roles for each sex (Gwynne, 1991). Thus, whether a population exhibits ***conventional sex roles***, that is, the condition when males compete more intensely for access to females, or ***sex-role reversal***, or the condition that females compete more

Table 4.1 Qualitative and quantitative methods commonly used to measure sexual selection and mating system variation.

Method	Description	Notes	Citation
Qualitative methods			
Operational sex ratio (OSR)	The ratio of males to females ready to mate in a population at a given time	Predicts the direction and intensity of mate competition based on a population sample of adults.	Emlen, 1976; Emlen & Oring, 1977
Potential reproductive rate (PRR)	The maximum number of independent offspring that parents can produce per unit of time	Predicts the direction of sexual selection in species with biparental care.	Clutton-Brock & Vincent, 1991; Clutton-Brock & Parker, 1992
Mean spatial crowding (m^*)	The number of available mates or competitors that an average individual experiences per unit of space	Predicts the direction and intensity of mate competition based on the spatial clustering of individuals.	Lloyd, 1967; Shuster & Wade, 2003
Mean temporal crowding (t^*)	The number of available mates or competitors that an average individual experiences per unit of time	Predicts the direction and intensity of mate competition based on the temporal receptivity of individuals.	Shuster & Wade, 2003
Quantitative methods			
Opportunity for selection (I)	Variance in reproductive success divided by the mean reproductive success squared	Standardized variance in reproductive success. Must be > 0 for selection to operate.	Crow, 1958; Wade, 1979; Arnold & Wade, 1984

Table 4.1 (Continued)

Opportunity for sexual selection (I_S)	Variance in mating success divided by the mean mating success squared	Standardized variance in mating success. Must be > 0 for selection to operate.	Wade & Arnold, 1980
Bateman gradient (β_{SS})	The slope of the weighted least-squares regression of reproductive success on mating success	Linear relationship between mating and reproductive success. Must be > 0 for selection to operate.	Arnold & Duvall, 1994
Sex difference in sexual selection (I_{mates})	Difference of the opportunity for selection of males (I_{males}) and the opportunity for selection of females ($I_{females}$) at a sex ratio of unity	In purely polygynous mating systems, the sex difference in the opportunity for selection I represents the reproductive success attributable to mating success.	Shuster & Wade, 2003
Maximum standardized selection differential (s'_{max})	The product of the standardized Bateman gradient (β'_{SS}) and the square root of the opportunity of sexual selection (I_S)	The maximum potential for trait evolution attributable to precopulatory sexual selection per generation in units of phenotypic standard deviations.	Jones, 2009
Morisita's index (I_δ)	Observed variance in mating success corrected by the expected variance in mating success when all mate acquisition probabilities are equal.	Measure of variance based on a Poisson distribution.	Morisita, 1962; Kokko et al., 1999

(continued)

Table 4.1 (Continued)

Method	Description	Notes	Citation
Index of resource monopolization (Q)	Ratio of observed variance in mating success to the maximum possible variance in mating success corrected by the expected variance when all mate acquisition probabilities are equal.	Measure of variance based on a Poisson distribution.	Ruzzante et al., 1996; Kokko et al., 1999
Sexual selection on traits			
Standardized selection differential (s')	Covariance between trait values and relative reproductive success.	The strength of selection on a trait relating to reproductive success in the univariate case.	Lande, 1979; Lande & Arnold 1983
Standardized mating differential (m')	Covariance between trait values and relative mating success.	The strength of sexual selection on a trait relating to mating success in the univariate case.	Jones 2009
Standardized selection gradient (β')	The partial regression of relative reproductive success or relative mating success on a trait.	For multiple trait analysis, the overall contribution of a particular trait to the sexual selection process while holding all other traits constant.	Lande, 1979; Lande & Arnold, 1983; Arnold & Duvall, 1994

intensely for access to males, is largely dependent upon the operational sex ratio (Vincent et al., 1992; Kvarnemo & Ahnesjö, 1996).

The operational sex ratio is an instantaneous ratio of ready-to-mate males to females in a given population (Emlen & Oring, 1977; Kvarnemo & Ahnesjö, 1996). For many reasons, not all adults can mate at any particular point in time. For example, individuals may be excluded from mating due to processes associated with inter- or intrasexual competition, or perhaps they may choose to delay reproduction until they acquire the proper prerequisites to mating such as a territory or sufficient food resources (Kvarnemo, 1996; Ahnesjö et al., 2001). Moreover, adults may need sufficient time after mating to become prepared to reproduce again. Therefore calculations of the operational sex ratio should include all individuals that are able to mate in a population but not those that are physiologically (but not environmentally) unprepared to mate at a given time.

Mathematically, the operational sex ratio can be expressed either as a simple ratio of males to females (m/f) (Emlen & Oring, 1977) or as a relative ratio of males to total number of adults in a population [$m/(m+f)$] (Kvarnemo & Ahnesjö, 1996). While both expressions are often used interchangeably throughout the literature, it should be noted that they are not equivalent and have different mathematical properties. The simple ratio resembles the relative ratio near equality (i.e., equal numbers of males and females) but has asymptotic ends at highly skewed operational sex ratios (Figure 4.1). Thus, the simple ratio should be scaled logarithmically to help normalize the data (Figure 4.1). However, to aid comparisons between studies and for any linear analysis of the operational sex ratio, the relative expression is preferable because of its linear properties (Figure 4.1). Significant deviations from an operational sex ratio of equality are generally tested with either a χ^2 goodness-of-fit or a similar test (e.g., Fisher's exact test), which compares the numbers of observed individuals with the number of expected individuals under a sex ratio of equality (or a skewed adult sex ratio in some instances). Because this ratio is calculated on the actual numbers of individuals, these tests are independent of the two methods to calculate the operational sex ratio (Figure 4.1).

Several variants of the operational sex ratio theme can found in the literature. First, the operational sex ratio can be expressed as the ratio of the amount of "time in" invested by each sex (Clutton-Brock & Parker, 1992). Here, *time in* is equal to the amount of time when an adult is sexually active or capable of mating, and *time out* is the amount of time resting or preparing for mating. Thus, the direction and intensity of mating competition can arise when sexes invest differentially in such factors as parental care or spend more time

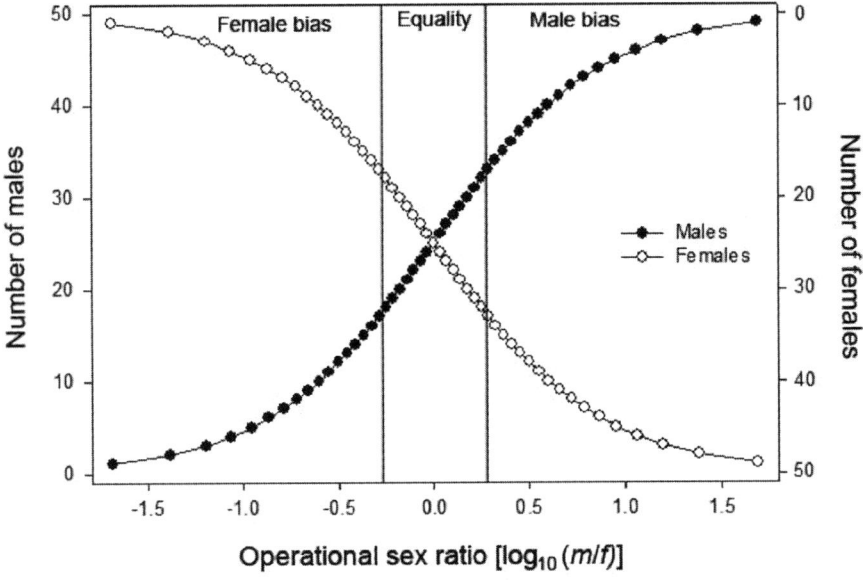

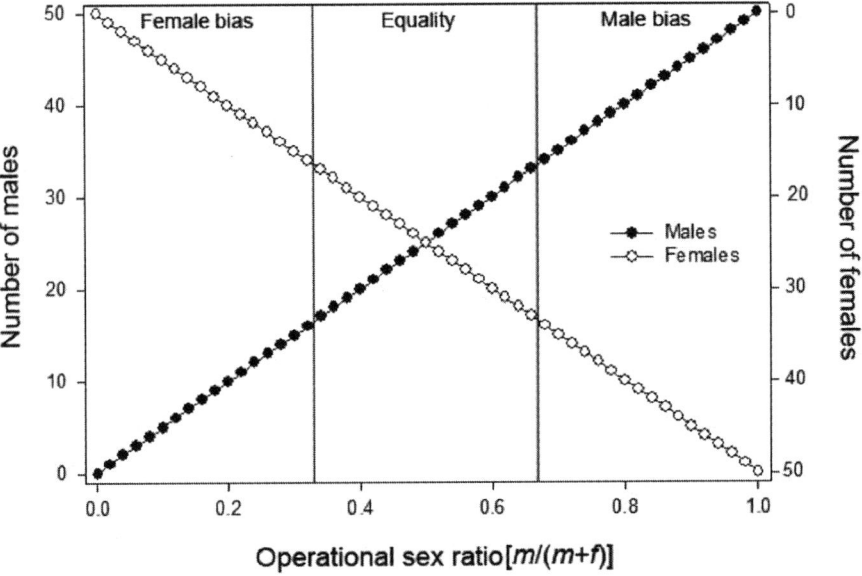

Figure 4.1. Properties of the operational sex ratio of a population of 50 sexually mature adults calculated using a simple ratio [$\log_{10}(m/f)$] or as a relative ratio [$m/(m/f)$]. As the operational sex ratio becomes more skewed, the simple ratio is asymptotic, whereas the relative ratio remains linear. The zone of equality is calculated as an operational sex ratio that does not deviate significantly from a sex ratio of unity (χ^2 two-tailed P < 0.05). Note that the zone of inequality corresponds to the same ratio of m/f in both calculations.

available for mating than the opposite sex. The operational sex ratio can also be expressed as a competitor-to-resource ratio (CRR), or the number of potential competitors divided by the number of resource units, when resource units are members of the opposite sex (Grant et al., 2001). The advantage of this formulation is that it allows for the direct comparison of behavior of males and females on a similar scale (e.g., Clark & Grant, 2010). Additionally, the ratio of males and females qualified to mate (Q) has been advocated in situations where individuals need to acquire a specific resource such as a nest site prior to mating (Ahnesjö et al., 2001). The authors argue that the quantity Q provides a more accurate measure of operational sex ratio by excluding individuals that are physically able to mate but are not qualified to mate (Ahnesjö et al., 2001). However, one should bear in mind that this measure offers an incomplete picture of sexual selection as individuals that do not hold a resource are excluded via competitive interactions and, as such, are actually part of mate competition (Shuster & Wade, 2003).

Empirical studies support that the operational sex ratio correctly predicts the strength and direction of sexual selection in many populations and species (reviewed in Thornhill & Alcock, 1983; Andersson, 1994; Kvarnemo & Ahnesjö, 1996; Mitani et al., 1996; Weir et al., 2011). A recent meta-analysis of 27 cross-taxa empirical studies comparing the operational sex ratio with mate competition demonstrated that the operational sex ratio correctly predicts the direction of aggression, courtship, and mate-guarding behaviors but not necessarily postcopulatory mate competition such as sperm competition (Weir et al., 2011).

Despite the overwhelming empirical support for this theory, the operational sex ratio has been scrutinized for several reasons. Chief among the criticisms is that the mathematical formulation of the operational sex ratio may not correctly predict the intensity of mate competition and is decoupled from mating systems (Kokko & Jennions, 2008; Shuster, 2009). For example, two hypothetical breeding populations of five ready-to-mate males and females will have an identical operational sex ratio of unity whether a single male monopolizes all five females (polygyny) or all five males have a single exclusive mate (monogamy). Although this example is extreme, it clearly illustrates that the operational sex ratio does not correctly predict the strength of sexual selection among populations with disparate mating systems (Shuster, 2009).

Additional misunderstandings may arise when considering how the operational sex ratio is quantified in various species. For instance, depending on the species of interest it may be difficult to tell whether an individual is in fact ready to mate without invasive techniques, adding imprecision to the measurement, particularly in small populations. Moreover, because the operational sex ratio is an instantaneous measure, calculation at different time intervals in mobile organisms may lead to faulty conclusions. For example, take a small patch of suitable breeding habitat where females maintain a territory around nests while males, having no predefined breeding territory, may enter or leave the breeding territory at will. Thus, on any given day, the operational sex ratio is completely dependent upon the presence or absence of males, which may or may not reflect actual mating competition. Further, new males that enter the territory at the same time that males leave the territory will give similar estimates of operational sex ratio despite the potential for males to have different qualities such that female competition for good mates may change the mate-competition dynamics. Lastly, one should bear in mind that mating competition may not a have a simple relationship with operational sex ratio. For example, at highly biased operational sex ratios competition for mates may be so fierce that it is an uneconomical strategy for members of the competing sex. In these cases, operational sex ratio forms a dome-shaped curve with competitive aggression (Grant et al., 2001; Weir et al., 2011). In spite of these shortcomings, the operational sex ratio is a useful metric to formulate hypotheses concerning mate competition and continues to be a highly lauded theory in behavioral ecology.

The Potential Reproductive Rate

Tim Clutton-Brock and colleagues (1991, 1992) introduced the concept of ***potential reproductive rate*** as a method of identifying which sex limits reproduction based on parental investment theory. The potential reproductive rate is equivalent to the maximum number of independent offspring that parents can produce per unit of time (Clutton-Brock & Vincent, 1991). This method was originally developed to calculate the direction of sexual selection in species with biparental care where characterization of the relative parental investment of each sex is not immediately obvious. In principle, the sex that produces fewer offspring per unit of time is the limiting sex, and therefore sexual selection should act more intensely on the nonlimiting sex. In practice, measuring the potential reproductive rate requires detailed information concerning the duration of care, the size of clutches and broods, the time between clutches, and the maximum rate of reproduction of males and females within a population (Clutton-Brock & Vincent, 1991; Clutton-Brock & Parker, 1992).

To date, the potential reproductive rate has garnered theoretical and empirical support from various animal mating systems (Clutton-Brock & Vincent, 1991; Clutton-Brock & Parker, 1992), although there are some notable exceptions (e.g., Sogabe & Yanagisawa, 2007; Maurer et al., 2011). However, the potential reproductive rate has been criticized on several fronts. For instance, the original calculation of the potential reproductive rate is dependent upon the rate of reproduction of the most fecund female and most virile male in a population, and therefore information gathered on all individuals within a population is not utilized (Arnold & Duvall, 1994; Ahnesjö et al., 2001). An additional consideration is that the potential reproductive rate can depend upon many ecological variables (Kvarnemo & Ahnesjö, 1996) such that only through careful experimental manipulations that control for various environmental factors can one measure the potential reproductive rate accurately (Ahnesjö et al., 2001; Scobell et al., 2009). This, in turn, requires a focal individual to be mated with an excess of mates in the absence of mate competition—a situation unlikely to occur under natural conditions. Thus, if understanding sexual selection in nature is our goal, then the potential reproductive rate is not a particularly useful metric.

Spatial and Temporal Mean Crowding

The intensity of sexual selection and mate competition are predicted to fluctuate if the spatial distribution and temporal receptivity of mates varies in nature (Emlen & Oring, 1977; Shuster & Wade, 2003). This relationship has been interpreted mathematically by applying a measure of density dependence termed *mean crowding* to both the spatial and temporal distribution of potential mates or same-sex competitors (Lloyd, 1967; Wade, 1995; Shuster & Wade, 2003). Mean spatial crowding (m^*) can be calculated as $m^* = \bar{m} + [(\sigma_m^2/\bar{m}) - 1]$, where $\bar{m}$ is the mean density of individuals per unit of area (patch size, quadrat size) and σ_m^2 is the variance of receptive mates per unit of area. Similarly, mean temporal crowding (t^*) is equivalent to $t^* = \bar{t} + [(\sigma_t^2/\bar{t}) - 1]$, where $\bar{t}$ is the mean number of receptive mates per unit of time (or time interval) and σ_t^2 is the variance of receptive mates per unit of time. Mean crowding can be thought of as a linear measurement of how aggregated a resource is, either spatially or temporally, and using the relationship of m^* to $\bar{m}$ to estimate patchiness helps to visualize this relationship. High levels of patchiness $m^* > \bar{m}$ reflect high aggregations of available mates in time or space, and stronger sexual selection should be experienced in patchy aggregations due to increased competition. Low levels of patchiness ($m^* < \bar{m}$) show a more even distribution of mates or competitors and hence weaker sexual selection.

Probably the most important consideration in applying mean crowding to a particular study system is the problem of scale (e.g., Levin, 1992). Lloyd (1967) commented in his original proposal for mean crowding that the scale would need to be determined empirically for each case but gave no real guidelines as to how this might be accomplished. For behavioral studies, the theoretical sampling scale for mean spatial crowding might best be approximated by territory size or perhaps mean perception or interaction distance (i.e., the mean distance at which an individual either notices or interacts with another individual, respectively). However, this may be difficult to investigate empirically in many species, so one might often need to make an educated guess as to what the scale might actually be for one's own particular study system point (Lloyd, 1967; Shuster & Wade, 2003). Shuster and Wade (2003) recommend a spatial scale that gives the greatest patchiness as a reasonable starting point. Mean temporal crowding, on the other hand, can best be defined in terms of the temporal availability of mates within a breeding season. If all individuals of one sex are available to breed at one particular time, then this would represent a maximum receptivity, and competition would be limited. Conversely, minimum receptivity represents a situation where all individuals of one sex become sexually active sequentially such that the ability of the opposite sex to monopolize mates becomes realistic in time. Thus, maximum and minimum breeding receptivity for a particular population set an upper and lower limit to mean temporal crowding, and scaling should reflect the mean receptivity (Shuster & Wade, 2003).

Mean spatial crowding has been applied to various mating systems, and studies show varying support for the link between mean spatial crowding and competitive interactions, depending on the species and social mating system (Wade, 1995; Kwiatkowski & Sullivan, 2002; Kelly, 2008; Pomfret & Knell, 2008; Casalini et al., 2010). For example, species with ***resource-defense social mating systems*** do not always show a relationship with mean spatial crowding and the intensity of mate competition (Kwiatkowski & Sullivan, 2002; Kelly, 2008; Casalini et al., 2010), and this discrepancy is likely due to different behavioral strategies employed under various competitive situations. Temporal crowding, on the other hand, has not been widely applied to empirical research, although this approach could be especially rewarding in understanding breeding synchrony in a variety of species. Mean spatial and temporal crowding are also part of a theoretical framework for classifying mating systems with the ***sex difference in sexual selection*** (I_{mates}; Shuster & Wade, 2003), so we return to these measures in that section.

Quantitative Measurements of Sexual Selection

Darwin recognized that sexual selection arises from differences in reproductive success caused by competition for access to mates (Darwin, 1871; Andersson, 1994). Quantitative measurements of sexual selection are based on the underlying principle that because all offspring are from one mother and one father (i.e., the Fisherian condition; Queller, 1997), if an individual mates and produces offspring with more than one mate, than the variance in mating success is not equal among members of that sex, given a sex ratio of equality. This nonrandom variance in Darwinian fitness generated through differential mating success is a mathematical formulation connected directly with Darwin's original definition of sexual selection.

Quantitative measurements of sexual selection, particularly if they are combined with precise genetic parentage information, offer an instantaneous snapshot of all precopulatory sexual selection in a population without any special reference or need to quantify behavioral mate competition or specific traits of interest. However, an underlying assumption in all variance-based methods is that both mating and reproductive success can be accurately measured for a reasonable number of individuals within a population. This may pose a real problem in some species for a variety of reasons. For example, it may be difficult to assess with accuracy parentage in populations where numbers of adults are too large to completely sample or when the number of adults and progeny are too few to provide enough statistical power (Jones & Ardren, 2003). Furthermore, variance-based approaches require careful estimation of the number of adults that are ready to mate but do not breed in a population. These individuals that are able but are excluded from breeding are sometimes referred to as the zero class because they have zero mating success (Shuster & Wade, 2003). While there are no hard-and-fast rules concerning who to include in the zero class, researchers need to make a special effort to clarify and justify who they include as these individuals can have a disproportionate effect on the observed variance in mating and reproductive success (Shuster & Wade, 2003; Klug, Lindström, et al., 2010; Fitze & Le Galliard, 2011). Finally, researchers should clarify which offspring they include when measuring reproductive success. Although the argument can be made for including only offspring that reproduce, for many species this requires too much time or expense to be a viable option. Therefore the number of developed embryos or independent offspring produced during a breeding bout would be a reasonable estimate of reproductive success.

Bateman's Principles

The renewed interest in Bateman's work in the 1970s provided a theoretical framework for sexual selection centered on both the variation in mating and reproductive success. Based in large part on the verbal arguments made by Bateman (1948), three specific points were articulated and are now referred to as Bateman's principles (Arnold, 1994). To reiterate, the three principles are (1) the sex experiencing the strongest sexual selection has a greater variance in reproductive success, (2) the sex experiencing the strongest sexual selection has a greater variance in mating success, and (3) the relationship between reproductive success and mating success increases with the intensity of sexual selection (Arnold & Duvall, 1994). Each of these three principles correspond to the mathematical formulation of the *opportunity for selection* (I), the *opportunity for sexual selection* (I_s) and the *sexual selection gradient* (or *Bateman's gradient*; β_{ss}), respectively. Together, Bateman's principles provide a means to predict the potential for sexual selection to occur in a sex within a generation without necessarily knowing which traits may be under selection.

Opportunity Measures

The first of Bateman's principles, the opportunity for selection, is equivalent to the strength of selection arising from the standard variance in fitness attributable to reproductive success for each sex and is sometimes referred to as *reproductive skew*. The opportunity for selection can be calculated as the variance in reproductive success divided by the mean reproductive success squared, or $I = \sigma^2 / \overline{X}_{rs}^2$, where σ^2 is the variance in number of offspring and $\overline{X}_{rs}$ is the mean number of offspring per individual in a population (Crow, 1958, 1962; Wade, 1979). An alternative and numerically identical approach to estimating I is to calculate the variance in relative reproductive success by dividing each individual's absolute reproductive success by the mean reproductive success for each sex (Jones, 2009).

The opportunity for selection is equivalent to the rate that absolute fitness increases in a population relative to the standing variance in absolute fitness, assuming all variance is due to additive genetic effects (Jones, 2009). Thus, I represents a theoretical maximum rate at which selection may act on a population in terms of offspring production (Wade, 1979). However, Jones (2009) demonstrated that the true theoretical maximum response to selection of any trait is actually $\sqrt{I}$, which has the added advantage of being in units of trait phenotypic standard deviations. The change in fitness from one generation to the next is likely to be less than predicted by I (or $\sqrt{I}$) because not all variance in fitness in a population is due to additive genetic variance (Jones, 2009). Therefore researchers often refer to I (and I_s) as a mathematical upper limit for selection operating in a single generation and hence an "opportunity" measure.

The opportunity for sexual selection (I_s) is similar in all respects to the calculation of I with the exception that it focuses on the standard variance in mating success for each sex in a population (Wade, 1979). Thus, I_s is calculated as the variance in mating success divided by the mean mating success squared, or

$I = \sigma^2/\overline{X}_{ms}^2$, where σ^2_{ms} is the variance in number of mates and $\overline{X}_{ms}$ is the mean number of mates per individual in a population.

The simplest and best method to compare I or I_s statistically between two groups (sexes, times, populations, etc.) is via an unequal variance test on relative reproductive or mating success, respectively. Several unequal variance tests exist, such as Bartlett's or Levene's, although Levene's test is preferable because it is more robust to the assumptions of normality and therefore offers a more conservative criterion (Sokal & Rohlf, 1995). Opportunity measures are the most commonly used measures to quantify sexual selection and enjoy theoretical support in various species and mating systems (reviewed in Searcy & Yasukawa, 1995; Shuster & Wade, 2003; Mills et al., 2007; Jones, 2009; Klug, Heuschele, et al., 2010). The advantages of opportunity measures include the facts that they are unitless measures that aid comparisons between different sexes, populations, and species and that the measures are directly linked with selection theory (Wade & Shuster, 2005; Jones, 2009). However, opportunity measures are harshly scrutinized, and criticisms include (1) the nonindependence of variance on population means, (2) the absence of a direct link to traits of interest, (3) introduction of variance through random processes, (4) the potential for opportunity measures to correctly identify sexual selection when the strength of sexual selection is weak, and (5) that they do not include information concerning mate quality (Downhower et al., 1987; Grafen, 1987; Ruzzante et al., 1996; Fairbairn & Wilby, 2001; Klug, Heuschele, et al., 2010). While criticism 1 has largely been dismissed (Nonacs, 2003; Jones, 2009) and criticism 2 is reconciled through selection differentials on traits of interest (Lande & Arnold, 1983; Jones, 2009; Klug, Heuschele, et al., 2010), criticisms 3, 4, and 5 present more difficult challenges. The grievance with criticism 3 (i.e., random processes introducing variance in opportunity measures) can be ameliorated by incorporating appropriate null models to determine whether variances are greater than expected by chance and by including power analyses (Klug, Heuschele, et al., 2010). Criticism 4, aimed at correctly identifying sexual selection when selection is weak, can be partially reconciled with a multiple-test approach, and authors now advocate using both opportunity measures, Bateman gradient approaches and the maximum standardized strength of selection, whenever possible (Jones, 2009; Krakauer et al., 2011). However, when selection is weak, it will always be difficult to differentiate between actual selection and stochastic processes, regardless of what method you use to estimate sexual selection. The remaining criticism applies to all qualitative and quantitative measures proposed thus far, and therefore more empirical and theoretical work to include issues concerning mate quality and other sources of variation is required to solve this issue.

The Bateman Gradient

Stevan Arnold and David Duvall (1994) proposed that the strength of sexual selection can be measured by the relationship between mating success and reproductive success as quantified by the sexual selection gradient, β_{ss}. The sexual selection gradient is equivalent to the slope of the weighted regression line comparing reproductive success to mating success (Arnold & Duvall, 1994). This specific relationship was renamed the Bateman gradient by Malte Andersson and Yoh Iwasa (1996) to differentiate it from selection gradients on phenotypic traits (Lande & Arnold, 1983).

Not only did the Bateman gradient represent a step forward by clarifying the relationship between mating success and Darwinian fitness expressed as offspring production, it provided a statistically rigorous and visually appealing method to demonstrate the strength and direction of sexual selection (Figure 4.2). The steeper the slope of the Bateman gradient experienced by a sex, the stronger sexual selection is likely to operate on that sex. Another way to interpret β_{ss} is that a sex that is limited in reproduction by extrinsic factors such as access to mating opportunities should have a steeper β_{ss} slope while a shallower slope suggests that a sex is limited by reproduction based on intrinsic factors such as the potential to produce gametes or provide parental care (Bateman, 1948; Arnold & Duvall, 1994). Because mating success necessarily needs to translate into increased fitness, this relationship must be positive in order for sexual selection to operate (Arnold & Duvall, 1994).

The Bateman gradient can be calculated on a particular sex using regression analysis and is usually compared between groups with covariance methods such as an analysis of covariance (e.g., Mobley & Jones, 2007, 2009; Barreto & Avise, 2010). A similar measure, the standardized Bateman gradient (β'_{ss}), equivalent to the slope of the weighted least regression line comparing relative reproductive success to relative mating success, is also currently used to compare different groups. This measurement is preferable to the uncorrected estimates of β_{ss} as it standardizes differences in the means of the two measures, allowing for more interpretable comparisons between groups of interest (Jones, 2009).

Bateman gradients have been applied in a variety of species, and there is general agreement with estimates of β_{ss} and the direction and intensity of sexual selection predicted by theory (e.g., Jones et al., 2005; Kvarnemo et al., 2007; Webster et al., 2007). Bateman gradients have also been employed to help explain patterns of sex-role reversal (Jones et al., 2000), anisogamy (Bjork & Pitnick, 2006), and postcopulatory sexual selection (Lorch, 2002) and even have been ascribed as a method to quantify sexual selection in hermaphrodites (Arnold, 1994; Anthes et al., 2010). Bateman gradients have also been

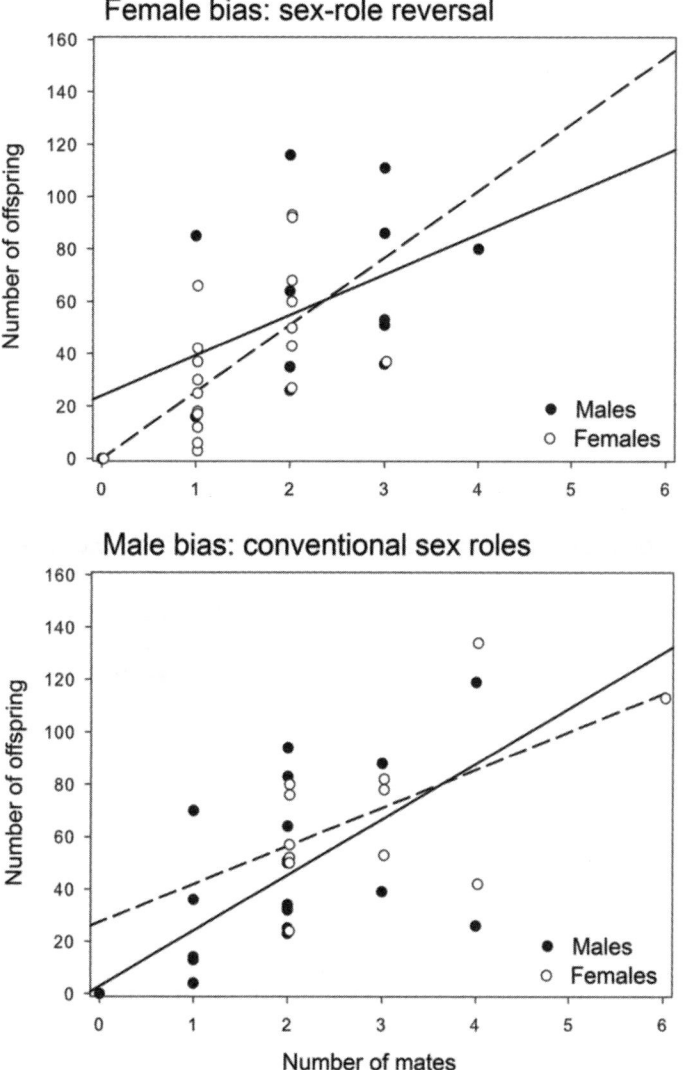

Figure 4.2. Bateman gradients for male and female broad-nosed pipefish, *Syngnathus typhle*, in two experimental breeding populations (data from Jones et al., 2005). Male Bateman gradients are represented by solid lines and closed circles, and female Bateman gradients are represented by dashed lines and open circles. In nature, the species is sex-role reversed and the operational sex ratio is female-biased. In the female-biased treatment, the strength of sexual selection is greater in females than in males as evidenced by the more positive slope of the female Bateman gradient. Under male-biased experimental conditions, the strength of sexual selection among males is greater than among females as evidenced by a more positive slope of the male Bateman gradient. All Bateman gradients are significantly greater than zero ($P < 0.001$), demonstrating that sexual selection is likely operating in both sexes. Note that Bateman gradients are not standardized.

criticized as too simplistic a view because they do not incorporate information such as variation in mate quality, the difficulty to procure matings, or the costs associated with developing traits that confer a mating advantage (Jennions & Kokko, 2010; Klug, Heuschele, et al., 2010; Fitze & Le Galliard, 2011). However, these criticisms have been appreciated by previous authors who propose Bateman gradients as one method to estimate whether or not sexual selection could be occurring without a priori knowledge of the targets of selection (Arnold & Duvall, 1994; Jones et al., 2004; Jones, 2009; Krakauer et al., 2011).

Sex Difference in Sexual Selection

Closely related to Bateman's principles, the sex difference in sexual selection (I_{mates}) has also been used as a measurement of sexual selection (Wade, 1979, 1995; Wade & Arnold, 1980; Shuster & Wade, 2003). The origin of this measure can be traced to Wade (1979), who considered the source of variation in reproductive success from males under a strictly polygynous mating system. It was determined that all variance in male mating success is equivalent to the sex difference in standardized reproductive success of males. Mathematically this relationship can be expressed as $I_{males} = (1/R_o)I_{females} + I_{mates}$ [or $I_{mates} = I_{males} - (1/R_o)I_{females}$] when R_o is equivalent to the sex ratio expressed as males/females. The derivation of I_{mates} can be found in both Wade (1979) and Shuster and Wade (2003).

In mating systems that deviate from strict polygyny, the variance in male mating success can no longer be described as a simple function of female reproductive success as females may split their reproductive output and invest differently in the offspring of several males. Thus in polyandrous and polygynandrous mating systems, a more generalized equation is recommended: $I_{males} = QI_{clutch} + I_s$, where Q is the reciprocal of mean male mating success, I_{clutch} is the standardized variance in female clutch size (i.e., the number of offspring produced with each male), and I_s is the opportunity for sexual selection on males (Jones, 2009). This slightly more complicated formula requires precise information on female clutch size, which can be estimated from parentage analysis. However, whenever female multiple mating is infrequent, this calculation approximates the more general equation for strict polygyny.

The sex difference in sexual selection has been applied extensively to a robust theoretical framework combining I_{mates} in conjunction with mean spatial and temporal crowding (m^*, t^*; see above) as a means to classify various existing mating systems (Shuster & Wade, 2003; Wade & Shuster, 2004; Shuster, 2009). By categorizing both the temporal and spatial variation in conjunction with the strength and direction of sexual selection, this framework can help explain the myriad of social and genetic mating systems in existence including sex-role reversal, alternative mating behaviors, and plant mating systems (Shuster & Wade, 2003; Shuster, 2009).

Maximum Standardized Selection Differential

Jones (2009) proposed a method that combines elements of Bateman's principles into a new metric for estimating the maximum strength of sexual selection experienced in units of phenotypic standard deviations. The ***maximum standardized selection differential*** is calculated as the product of the standardized Bateman gradient and the square root of the opportunity for sexual selection ($s'_{max} = \beta'_{ss}\sqrt{I_s}$). This measure is equivalent to the maximum rate of phenotypic evolution for a given trait in one generation and therefore is similar to other opportunity measures in that it makes no assumptions about the heritability of traits or fitness (Jones, 2009). This new metric has the advantage of focusing specifically on selection on fitness generated by differential mating success and therefore is more informative than either I or I_s alone.

So far, the maximum standardized sexual selection differential has only been applied in a handful of studies but adheres well to theoretical predictions. For instance, values of s'_{max} are greater in males than in females in species that have conventional sex roles (Serbezov et al., 2010) and display markedly similar values in species that appear to lack strong sexual selection between sexes (Barreto & Avise, 2010). Further application of this technique will demonstrate whether it is a useful metric to determine the direction and intensity of sexual selection.

The Morisita Index and the Index of Resource Monopolization

Several additional variance-based methods have been introduced as alternative measures of the potential for sexual selection, including the ***index of resource monopolization*** (Q; Ruzzante et al., 1996) and the ***Morisita index*** (I_δ; Morisita, 1962; Kokko et al., 1999). The index of resource monopolization is defined as $Q = [(\sigma^2 - \bar{x})/(n\bar{x}^2 - \bar{x})]$, and the Morisita index is $I_\delta = n\left[(\sum_{i=1}^{n} x_i^2 - n\bar{x})/((n\bar{x})^2 - n\bar{x})\right]$, where n is the number of individuals, x_i is the value of the resource (i.e., either mating or reproductive success) for the ith individual, $\bar{x}$ is the mean value across n individuals, and σ^2 is the variance. The argument for these methods is that these measures based on expected Poisson distributions are not as sensitive to population means and therefore are a better metric of sexual selection (Kokko et al., 1999; Fairbairn & Wilby, 2001). However, this criticism has been dismissed as fallacious (Nonacs, 2003). Empirical studies have demonstrated a disparity between the theoretical expectations of sexual selection theory and estimates of Q and I_δ (Jones et al., 2004; Croshaw, 2010), and both of these methods are heavily criticized for their redundancy (Jones et al., 2004; Mills et al., 2007). Perhaps the greatest drawback to applying I_δ and Q is a lack of connection to selection theory (Jones et al., 2004; Croshaw, 2010). As a result of the above concerns, neither of these methods have enjoyed widespread acceptance for quantifying sexual selection.

Selection on Sexually Selected Traits

Measuring sexual selection on traits of interest is a relatively straightforward task in principle. One can collect a sample of breeding adults and offspring from a population, measure traits that might be under sexual selection, and quantify the relationship between the trait and either mating success or reproductive success. However, one must bear in mind that an underlying assumption is that the correct target of selection has been identified. If you do not find sexual selection operating on a particular trait, then sexual selection may still exist, just not on that particular trait at the time of sampling. Likewise, if you do see sexual selection imparted on a trait, then the trait may be correlated with the actual target of sexual selection and not under sexual selection itself. Thus, the combination of trait analysis and other quantitative approaches such as I, I_s, β_{ss}, and s'_{max} is highly recommended (Jones et al., 2004; Shuster, 2009; Krakauer et al., 2011; but see Klug, Heuschele, et al., 2010, for a contrasting opinion).

The metric commonly used to measure selection on traits of interest is the standardized selection differential (s'), equivalent to the covariance between a trait and either mating or reproductive success (Lande & Arnold, 1983). For this measure, trait values are standardized by variance to have a mean of zero and a variance of one [$(z - \bar{z})/$(standard deviation of z), where z is an individual's trait value and $\bar{z}$ is the trait mean] so that traits can be compared in units of phenotypic standard deviations (Lande & Arnold, 1983). In this manner, selection differentials are an intuitive measure of selection as they predict how much the average phenotype evolves from one generation to the next, assuming that the genetic variance is completely additive (Lande, 1979; Lande & Arnold, 1983). To help differentiate between the selection differentials of particular traits to either mating success and reproductive success, the standardized mating differential (m') has been coined for the specific selection differential of traits on relative mating success (Jones, 2009).

Under the simplified scenario of one trait acting on mating and reproductive success, analyses are straightforward and linear regression models are generally applied (e.g., Jones et al., 2004, 2005). If traits do not satisfy the assumptions of normality or equal variances, transformations (e.g., ln, $\log_{10}$) should be applied before the trait has been standardized by variance. In the more complicated case where multiple traits may be under selection simultaneously, the standardized selection gradient (β') is calculated as the partial regression of a standardized trait value on either relative mating success or reproductive success (s', m') while holding the values of other traits constant (Lande & Arnold, 1983; Arnold & Duvall, 1994). Thus, the standardized selection gradient can be calculated using multiple regression techniques on traits or on principal-components scores (e.g., Jones et al., 2004; Head et al., 2008; Kelly, 2008; Hunt et al., 2009). However, because complicated genetic relationships can give rise to multidimensional or rugged fitness surfaces, other methods such as canonical analyses may be required to disentangle complex nonlinear relationships of multiple traits under selection (Phillips & Arnold, 1989; Blows & Brooks, 2003).

Surprisingly, methods to quantify sexual selection on traits enjoy considerable agreement despite having similar issues with random processes that may generate spurious correlations between traits and mating or reproductive success, as do other quantitative mating system measures (Klug, Heuschele, et al., 2010; Krakauer et al., 2011). However, because selection differentials and gradients are correlational analyses, it is sometimes difficult to disentangle true causative relationships between the trait and the measure of selection (Krakauer et al., 2011). Thus, convincing evidence for traits under selection is generally the result of manipulative experiments rather than field observations (Jones et al., 2004, 2005; Kelly, 2008; Hunt et al., 2009).

TWO EMPIRICAL EXAMPLES

To demonstrate how various measures of sexual selection behave under different selection regimes, I present two case studies where mating competition was manipulated by altering the operational sex ratio of breeding adults (Table 4.2). In both experiments, adults were placed in replicated artificial breeding populations at different sex ratios and allowed to mate freely over a period of time. For simplicity, all data from experimental treatments are pooled together, and therefore mating-system estimates represent the composite score across all individuals within the same sex-ratio treatment.

The first example set comes from a study conducted by Jones and colleagues (2004) on rough-skinned newts (*Taricha granulosa*). These salamanders are native to the West Coast of North America and migrate to breeding ponds during winter months, where mating and egg laying take place. This species is sexually dimorphic, and males are typically larger and possess a large

Table 4.2 Mating-system estimates from two experimental datasets. The first dataset represents artificial breeding populations of the rough-skinned newt, *Taricha granulosa*, at two different operational sex ratios [OSR = $m/(m+f)$]. This species is polygynandrous and displays conventional sex roles in natural populations. Selection and mating differentials are given for tail height. The second dataset represents artificial breeding populations of the broad-nosed pipefish, *Syngnathus typhle*, at three different adult sex ratios. This species is polygynandrous and sex-role reversed in natural populations. Selection and mating differentials are given for body length. Data are reproduced with permission from Jones et al. (2004) and Jones et al. (2005; with permission).

Conventional sex roles: *Taricha granulosa*

Treatment	OSR	Sex	N	$\bar{X}_{ms}$	σ_{ms}	I_s	$\sqrt{I_s}$	$\bar{X}\bar{X}_{rs}$	σ_{rs}	I	$\sqrt{I}$	β_{ss}	β'_{ss}	s'_{max}	I^*_{mates}	m'	s'
Male bias	0.75	Male	48	0.44	0.46	2.42	1.56	59.2	11899.2	3.39	1.84	141.5	1.04	1.62	3.28	0.14	0.36
		Female	16	1.31	0.50	0.29	0.54	177.7	10801.7	0.34	0.58	20.0	0.15	0.08		0.05	−0.14
Even	0.52	Male	48	1.35	0.87	0.48	0.69	153.2	20282.9	0.86	0.93	114.6	1.01	0.70	0.57	0.20	0.36
		Female	45	1.49	0.66	0.30	0.55	164.9	8634.0	0.32	0.57	22.4	0.20	0.11		0.03	−0.10

Reversed sex roles: *Syngnathus typhle*

Treatment	OSR	Sex	N	$\bar{X}_{ms}$	σ_{ms}	I_s	$\sqrt{I_s}$	$\bar{X}\bar{X}_{rs}$	σ_{rs}	I	$\sqrt{I}$	β_{ss}	β'_{ss}	s'_{max}	I^*_{mates}	m'	s'
Female bias	0.24	Male	13	2.23	1.19	0.24	0.49	58.4	1286.9	0.38	0.61	15.4	0.59	0.29	−6.57	0.05	0.37
		Female	42	0.69	0.71	1.48	1.22	18.1	702.3	2.15	1.47	27.9	1.07	1.30		0.04	0.23
Even	0.48	Male	20	1.70	0.85	0.30	0.54	41.0	1162.4	0.69	0.83	19.0	0.79	0.43	−0.03	0.14	0.51
		Female	22	1.36	0.81	0.44	0.66	34.0	763.3	0.66	0.81	17.5	0.70	0.46		0.25	0.38
Male bias	0.73	Male	32	1.31	1.45	0.84	0.92	30.8	1181.0	1.25	1.12	29.4	1.25	1.15	1.17	0.32	0.60
		Female	12	2.92	1.54	0.18	0.43	70.1	942.8	0.19	0.44	14.5	0.61	0.26		−0.03	0.22

*I_{mates} is calculated using the general formula ascribed in Shuster & Wade (2003).

tail crest, which appears to be under sexual selection (Janzen & Brodie, 1989; Jones et al., 2002). Males mate multiply, and females can lay multiple clutches of mixed paternity, and therefore the species is considered to be polygynandrous (Jones et al., 2002).

In this experiment, Jones and colleagues (2005) mated breeding populations of either a male-biased treatment of six males and three females or an even-sex-ratio treatment of six males and six females (Table 4.2). Under male-biased conditions, we would expect that sexual selection would be strongest among males either through male-male competition or female choice. Thus, mating-system estimates such as I_s, I, β'_{ss}, and s'_{max} as well as selection differentials (m', s') on male tail height should be higher on males in the male-biased treatment over the even-sex-ratio treatment.

Under male-biased sex ratios, females mated on average three times as much as males, leading to higher estimates of $\overline{X}_{ms}$ for females although σ_{ms} was similar for both sexes. Male $\overline{X}_{rs}$ was likewise reduced, and both sexes shared similar estimates of σ_{rs}. Under a sex ratio of equality, male and female $\overline{X}_{ms}$ and $\overline{X}_{rs}$ were similar, and σ_{ms} was also comparable, leading to similar values of I_s. However, because some males produced more offspring than others in the even-sex-ratio treatment, the variance in male reproductive success (σ_{rs}) was twice the σ_{rs} for females, leading to higher estimates of I, β_{ss}, β'_{ss}, and s'_{max} for males. As expected, in male-biased cases with conventional sex ratios, the strengths of sexual selection as measured by I_s, I, β_{ss}, β'_{ss}, and s'_{max} were all greater in males than in females and greater in females than in males in the even-sex-ratio treatment. It is interesting to note that female mating-system estimates were surprisingly similar between the two treatments, strongly suggesting that females are limited in their reproductive potential and that the addition of more mates does not substantially influence mating and reproductive success. Finally, the sex difference in the opportunity for selection (I_{mates}) shows that the strength of sexual selection is stronger in males under male-biased conditions.

For selection on the sexually selected trait of tail height, both m' and s' were similar among males in both operational sex ratios despite a nearly double potential for sexual selection on traits as judged by s'_{max} (Table 4.2). Females, on the other hand, had small estimates of both m' and s', indicating sexual selection to be weak or nonexistent in females. In summary, all mating-system estimates, including those specifically on a trait under selection, agree with the prediction that sexual selection in this species is stronger on males, although a more strongly biased operational sex ratio towards males does not necessarily increase sexual selection on male traits.

The second example comes from a study conducted by Jones and colleagues (2005) on the broad-nosed pipefish (*Syngnathus typhle*). Males and females mate multiply within a single pregnancy, and thus the particular population of interest is polygynandrous with respect to the genetic mating system (Jones et al., 1999). In the wild, this species generally encounters highly female-biased operational sex ratios and is sex-role reversed in relation to the direction of sexual selection (Berglund et al., 1986; Vincent et al., 1994). Therefore the a priori expectation is that sexual selection should act most strongly on females under conditions of an operational sex ratio of equality.

In this experiment, a change in operational sex ratio from male bias to female bias showed an increase in male $\bar{X}_{ms}$ and a concomitant decrease in female $\bar{X}_{ms}$ (Table 4.2). These changes in means were primarily driven by many males that did not mate in the male-biased treatment. An opposite relationship, namely a decrease in male $\bar{X}_{rs}$ and an increase in female $\bar{X}_{rs}$, was apparent with an increasing female bias. While the male σ_{ms} was variable between treatments, female σ_{ms} decreased with decreasing female bias, and the σ_{rs} of both males and females remained similar between treatments with some slight variations. As the operational sex ratio shifts from a female bias to a male bias, the resulting mating-system estimates I_s, I, β_{ss}, β'_{ss}, and s'_{max} all show a clear pattern of increasing in males while decreasing in females (Table 4.2, Figure 4.2). The sex difference in the opportunity for selection (I_{mates}) also shows a clear pattern of increasing from female bias to male bias, demonstrating that this measurement correctly predicts sex-role reversal as evidenced by negative values encountered at sex ratios of equality and female bias.

Interestingly, selection differentials on body size demonstrate conflicting patterns between the sexes. The highest m' and s' on male body size are found in the male-biased treatment, and this is in line with the expectation that larger males enjoy higher mating and reproductive success because they can carry more offspring in their brood pouch. Selection on female body size, on the other hand, appears to be highest in the even-sex-ratio treatment, suggesting stabilizing selection on female body size (Jones et al., 2005).

FUTURE CHALLENGES

Thus far we have explored the history and current theory concerning mating systems and the evolution of sexual selection. Although much progress has been made, still more work is required in order to gain a deeper appreciation for the intricacies of mating-system evolution. In particular, uncovering the causal link between social and genetic mating systems, investigating how mating systems vary both spatially and temporally, understanding the myriad of social and ecological influences on mating system organization, and differentiating between precopulatory and postcopulatory sexual selection are logical next steps in the field. Here, I briefly introduce these topics as potential lines of further research.

Social versus Genetic Mating Systems

The degree to which social mating systems influence genetic mating systems and vice versa is still an unresolved question in behavioral ecology. For some species of seahorses, the social mating system may appear promiscuous despite a purely monogamous genetic mating system (Wilson & Martin-Smith, 2007). On the other hand, in many species of socially monogamous birds, high levels of extra-pair copulations demonstrate that, despite the maintenance of stable pair bonds, a portion of the population cuckolds their primary mates and solicits copulations from potentially better mates (Hasselquist & Sherman, 2001; Griffith et al., 2002). As illustrated by these two examples, reconciling these two disparate types of mating systems seems to depend upon many variables such as the strength of social bonds, the degree of extra-pair mating, the presence of alternative mating behaviors, and other ecological factors.

One approach to disentangle the relationship between social and genetic mating systems can be through careful experimentation and manipulation of the social interactions of adults in a breeding population. For example, several manipulative experiments demonstrate a clear relationship between the operational sex ratio and various measures of the genetic mating system (e.g., Jones et al., 2000; Mills & Reynolds, 2003; Jones et al., 2004; Mills et al., 2007; Reichard, Ondračková, et al., 2008). These studies are the most informative in understanding the link between demographic variables and the actual strength of sexual selection. Comparative phylogenetic studies also have the potential to unearth how social and genetic mating systems may evolve in concert. For instance, studies such as those that compare sexual selection among socially cooperative versus solitary breeding birds suggest that sexual selection on cooperative breeders is reduced via social interactions. The nature of these cooperative social interactions, in turn, influences the evolution of mate competition and sexually selected traits within these species (Rubenstein & Lovette, 2009). While the strength of the relationship between social and genetic mating systems is quite likely to differ on a species-to-species basis, more studies along these lines are warranted and necessary to gain a greater understanding of how these systems are linked and feedback under different evolutionary scenarios.

Mating-System Variation

Sexual selection can manifest considerable variation between populations and at different times (Mobley & Jones, 2007; Cockburn et al., 2008; Gosden

& Svensson, 2008; Cornwallis & Uller, 2009; Mobley & Jones, 2009). Certainly part of this variation is due to both spatial and temporal environmental heterogeneity. Variation may also arise because behavioral traits and responses to selection may be plastic in space and time (Cornwallis & Uller, 2009). Populations that experience disparate intensities of sexual selection may theoretically drive patterns of speciation, strongly implying that mating-system variation is important to the speciation process (Payne & Krakauer, 1997; Kraaijeveld et al., 2007). Thus, understanding how and why mating systems vary is critical to decoding the process of sexual selection and the evolution of different mating systems (Emlen & Oring, 1977; Shuster & Wade, 2003; Cornwallis & Uller, 2009).

Thus far, only a few studies concerning variation in mating systems have been conducted, and therefore the underlying causes of mating-system variation are not well known. Often, mating-system studies focus on one exemplar population at one specific time, and therefore the degree to which the mating system varies is a black box in many cases. Part of the problem with conducting studies within and among populations is the large expense and work required to completely characterize the mating systems. However, studies along these lines would pay dividends for elucidating specific factors that influence mating-system organization by mapping mating-system parameters to specific details of the timing, habitat, and social interactions encountered over various spatial or temporal scales. For example, studies that investigate mating systems in multiple populations over broad geographic scales have identified a wide array of ecological factors that may shape local mating-system dynamics (e.g., Mobley & Jones, 2007, 2009).

Despite the relatively more studied underlying ecological causes for sexual selection within seasons or mating periods (see "Ecological Mating-System Influences"), there is surprisingly little known about what ecological variables may drive temporal patterns of sexual selection between mating seasons or years (Cornwallis & Uller, 2009). Recent studies suggest that variation between years can be greater than within years (Cockburn et al., 2008; Kasumovic et al., 2008), leading to the prediction that year-specific factors may influence differences in sexual selection. Year-specific factors may be the result of fluctuations in climate (Twiss et al., 2007; Cockburn et al., 2008) or due to demographic processes (Kokko & López-Sepulcre, 2007; Gosden & Svensson, 2008; Kasumovic et al., 2008). However, variation between years might not be attributable to any particular environmental variable measured, suggesting that in some cases sexual selection may fluctuate between years for unknown reasons (Lehtonen et al., 2009). Thus, comparing variation in mating and reproductive success within and between years should illuminate whether sexual selection varies due to specific ecological variables or results from stochastic processes such as environmental heterogeneity (Cornwallis & Uller, 2009).

Ecological Mating-System Influences

A wide range of ecological variables have the potential to influence mating and reproductive behaviors of individuals within a population, and understanding what these variables are is the first step in better understanding how species adapt and respond to fluctuating environmental conditions. Ecological factors can be divided into two main categories: (1) *abiotic* variables—extrinsic nonbiological properties of the environment and habitat that the population occupies—and (2) *biotic* variables, which include biological properties of the habitat extrinsic to the focal population as well as demographic properties intrinsic to the population itself.

Among abiotic factors that affect mating systems, temperature is the most studied and best understood, particularly because of its strong effect on community productivity, reproductive timing, and limiting reproductive rates, particularly in poikilothermic organisms (Ahnesjö, 1995, 2008; Kvarnemo, 1996; Fischer et al., 2003; Olsson et al., 2011). Other abiotic factors that may affect sexual selection include climate (Cockburn et al., 2008) and anthropogenic pollution (Lane et al., 2011), but such factors remain an understudied aspect of mating systems. In contrast, a wide range of biotic factors affecting the strength and direction of sexual selection have been identified, including (but not limited to) factors that affect food availability and hence fecundity and mate quality (Kvarnemo, 1997; Turner & McCarty, 1998); habitat structure and fragmentation (Turner & McCarty, 1998; Reichard, Ondračková, et al., 2008), parasite load (Fitze et al., 2004), resource competition (Martin & Martin, 2001), and predation (Kelly et al., 1999; Bronikowski et al., 2002).

Still another gap in our knowledge is how and to what degree demographic processes influence various aspects of sexual selection and mating-system evolution. By far the most studied demographic factor explored to date is sex ratio, but other factors such as population density, the length and synchrony of the breeding season, and effective breeding population size are likely be important to the sexual selection process (Griffith et al., 1999; Jirotkul, 1999; Spottiswoode & Møller, 2004; Kokko & Rankin, 2006; Mobley & Jones, 2007, 2009; Reichard, Smith, et al., 2008). Fluctuations in population density have the ability to affect mate encounter rates that influence processes such as mate competition, mate choice, and parental care decision making by providing individuals with more or better choices of potential mates (Kokko & Rankin, 2006). Variation in population density also can expose individuals to differing levels of competition for access to mates among the same sex (Kokko & Rankin, 2006). Moreover, population density may govern the expression and evolution of behaviors or traits of one sex, which in turn influence the opposite sex via social and ecogenetic feedback mechanisms (Kokko & López-Sepulcre, 2007; Alonzo, 2009). Thus the future challenge is to test the effect of demographic factors both within and between sexes to better understand how these are related to the sexual selection process.

Postcopulatory Sexual Selection

Notwithstanding his deep appreciation for precopulatory sexual selection including mate competition and mate choice, Darwin did not comment on the potential for *postcopulatory sexual selection*, or selection that takes place during or after mating, to occur (Eberhard, 2009). Postcopulatory sexual selection includes the phenomenon of gamete competition as well as the process of cryptic mate choice (Parker, 1970; Eberhard, 1996; Birkhead & Møller, 1998; Simmons, 2001). While these processes are usually referred to as *sperm competition* and *cryptic female choice*, I opt for the more general definitions to include the potential for opposite patterns to evolve in sex-role-reversed species (i.e., egg competition and cryptic male choice) (Partridge et al., 2009; Paczolt & Jones, 2010).

Postcopulatory processes have the potential to affect sexual selection in many species, but the degree to which postcopulatory sexual selection influences mating-system evolution is currently unknown. In many instances we cannot a priori distinguish between pre- and postcopulatory sexual selection, and therefore a particular challenge for future studies is to partition the relative influences of these two selective forces. One method to tackle this problem is to design an experiment (or series of experiments) that can readily distinguish between the two processes. These experiments could be complemented by parentage analysis utilizing molecular markers to assign unambiguous parentage to all offspring. In this manner, one can partition the variation experienced by the two forces, making the relative contribution of each more easily understood.

To date, there have been only a few attempts to quantify the total variance attributed to precopulatory and postcopulatory processes in the same species despite the potential for postcopulatory sexual selection to be strong (Hunt et al., 2009; Simmons & Beveridge, 2010). One example comes from a study on broad-nosed pipefish where the potential for postcopulatory sexual selection appears to be weak compared to all precopulatory sexual selection experienced (Partridge et al., 2009). Beetles, on the other hand, demonstrate a negative correlation between precopulatory weapons and postcopulatory testes size, indicating that a trade-off between the two selective forces is likely among these species (Simmons & Emlen, 2006). For the most complete picture on the relative effects of precopulatory versus postcopulatory sexual selection, more studies along these lines are warranted and are necessary to resolve this outstanding issue.

CONCLUSIONS

The classification of animals by various reproductive behaviors related to the sexual selection process has garnered much interest in the times since Darwin's formulation of the theory of sexual selection. As we have seen in this chapter, the field of animal mating systems is dynamic and prone to contentious debate, making it one of the most exciting fields in animal behavior. Although we have made much progress, a dizzying array of questions related to mating systems and sexual selection is still left outstanding. For example, *what accounts for the stunning diversity of social and genetic mating systems? Why do some lineages experience strong sexual selection and others do not? How might divergence in mating systems contribute to the speciation process? Why do various forms of parental care evolve? Why are some species sex-role reversed? What are the demographic and ecological factors that influence mating system organization?* While definitive answers to these questions seem out of reach at the moment, one thing is clear from our continued study of sexual selection—it is a powerful evolutionary process that explains a staggering variety of behavioral and morphological traits in animals. In the future, the study of mating systems will continue to grow, and further theoretical refinement in concert with carefully conceived behavioral experiments and field surveys should lead to the ultimate goal of understanding how sexual selection operates in nature.

ACKNOWLEDGMENTS

I thank Malte Andersson, Tonje Aronsen, Anders Berglund, Inês Braga-Gonçalves, Charlotta Kvarnemo, Gunilla Rosenqvist, Sebastian Wacker, and Bob Wong for insightful discussions. I am particularly indebted to Adam Jones and Steve Shuster for providing valuable advice and encouragement throughout the years.

REFERENCES AND SUGGESTED READING

Ahnesjö, I. (1995). Temperature affects male and female potential reproductive rates differently in the sex-role reversed pipefish, *Syngnathus typhle*. *Behavioral Ecology*, 6, 229–233.

Ahnesjö, I. (2008). Behavioural temperature preference in a brooding male pipefish *Syngnathus typhle*. *Journal of Fish Biology*, 73, 1039–1045.

Ahnesjö, I., C. Kvarnemo, & S. Merilaita. (2001). Using potential reproductive rates to predict mating competition among individuals qualified to mate. *Behavioral Ecology*, 12, 397–401.

Alonzo, S. H. (2009). Social and coevolutionary feedbacks between mating and parental investment. *Trends in Ecology and Evolution*, 25, 99–108.

Andersson, M. (1994). *Sexual Selection*. Princeton, NJ: Princeton University Press.

Andersson, M. & Y. Iwasa. (1996). Sexual selection. *Trends in Ecology and Evolution*, 11, A53–A58.

Anthes, N., D. Patrice, J. R. Auld, J. N. A. Hoffer, P. Jarne, J. M. Koene, et al. (2010). Bateman gradients in hermaphrodites: An extended approach to quantify sexual selection. *American Naturalist*, 176, 249–263.

Arnold, S. J. (1994). Bateman principles and the measurement of sexual selection in plants and animals. *American Naturalist*, 144, S126–S149.

Arnold, S. J. & D. Duvall. (1994). Animal mating systems: A synthesis based on selection theory. *American Naturalist*, 143, 317–348.

Arnold, S. J. & M. J. Wade. (1984). On the measurement of natural and sexual selection: Theory. *Evolution*, 38, 709–719.

Avise, J. C. (2004). *Molecular Markers, Natural History and Evolution*. Sunderland, MA: Sinauer Associates.

Barreto, F. S. & J. C. Avise. (2010). Quantitative measures of sexual selection reveal no evidence for sex-role reversal in a sea spider with prolonged paternal care. *Proceedings of the Royal Society of London, B*, 277, 2951–2956.

Bateman, A. J. (1948). Intra-sexual selection in *Drosophila*. *Heredity*, 2, 349–368.

Berglund, A., G. Rosenqvist, & I. Svensson. (1986). Reversed sex-roles and parental energy investment in zygotes of two pipefish (*Syngnathidae*) species. *Marine Ecology—Progress Series*, 29, 209–215.

Berglund, A., G. Rosenqvist, & I. Svensson. (1989). Reproductive success of females limited by males in two pipefish species. *American Naturalist*, 133, 506–516.

Birkhead, T. R. & A. P. Møller. (1998). *Sperm Competition and Sexual Selection*. San Diego, CA: Academic Press.

Bjork, A. & S. Pitnick. (2006). Intensity of sexual selection along the anisogamy-isogamy continuum. *Nature*, 7094, 742–745.

Blows, M. W. & R. Brooks. (2003). Measuring nonlinear selection. *American Naturalist*, 162, 815–820.

Bronikowski, A. M., M. E. Clark, F. H. Rodd, & D. N. Reznick. (2002). Population-dynamic consequences of predator-induced life history variation in the guppy (*Poecilia reticulata*). *Ecology*, 83, 2194–2204.

Casalini, M., M. Reichard, & C. Smith. (2010). The effect of crowding and density on male mating behaviour in the rose bitterling (*Rhodeus ocellatus*). *Behaviour*, 147, 1035–1050.

Clark, L. & J. W. A. Grant. (2010). Intrasexual competition and courtship in female and male Japanese medaka, *Oryzias latipes*: Effects of operational sex ratio and density. *Animal Behaviour*, 80, 707–712.

Clutton-Brock, T. H. (1991). *The Evolution of Parental Care*. Princeton, NJ: Princeton University Press.

Clutton-Brock, T. H. & G. A. Parker. (1992). Potential reproductive rates and the operation of sexual selection. *Quarterly Review of Biology*, 67, 437–456.

Clutton-Brock, T. H. & A. J. C. Vincent. (1991). Sexual selection and the potential reproductive rates of males and females. *Nature*, 351, 58–60.

Cockburn, A., H. L. Osmond, & M. C. Double. (2008). Swingin' in the rain: Condition dependence and sexual selection in a capricious world. *Proceedings of the Royal Society of London, B*, 275, 605–612.

Coleman, S. W. & A. G. Jones. (2011). Patterns of multiple paternity and maternity in fishes. *Biological Journal of the Linnean Society*, 103, 735–760.

Cornwallis, C. K. & T. Uller. (2009). Towards an evolutionary ecology of sexual traits. *Trends in Ecology and Evolution*, 25, 145–152.

Croshaw, D. A. (2010). Quantifying sexual selection: A comparison of competing indices with mating system data from a terrestrially breeding salamander. *Biological Journal of the Linnean Society*, 99, 73–83.

Crow, J. F. (1958). Some possibilities of measuring selection intensity in man. *Human Biology*, 30, 1–13.

Crow, J. F. (1962). Population genetics: Selection. In W. J. Burdette (ed.), *Methodology in Human Genetics* (pp. 53–75). San Francisco: Holden-Day.

Darwin, C. (1871). *The Descent of Man and Selection in Relation to Sex*. London: John Murray.

Downhower, J. F., L. S. Blumer, & L. B. Brown. (1987). Opportunity for selection: An appropriate measure for evaluating variation in the potential for selection? *Evolution*, 41, 1395–1400.

Dunn, P. O., L. A. Whittingham, & T. E. Pitcher. (2001). Mating systems, sperm competition, and the evolution of sexual dimorphism in birds. *Evolution*, 55, 161–175.

Eberhard, W. G. (1996). *Female Control: Sexual Selection by Cryptic Female Choice*. Princeton, NJ: Princeton University Press.

Eberhard, W. G. (2009). Postcopulatory sexual selection: Darwin's omission and its consequences. *Proceedings of the National Academy of Sciences, USA*, 106, 10025–10032.

Emlen, S. T. (1976). Lek organization and mating strategies in bullfrog. *Behavioral Ecology and Sociobiology*, 1, 283–313.

Emlen, S. T. & L. W. Oring. (1977). Ecology, sexual selection and the evolution of mating systems. *Science*, 197, 215–223.

Fairbairn, D. J. & A. E. Wilby. (2001). Inequality of opportunity: Measuring the potential for sexual selection. *Evolutionary Ecology Research*, 3, 667–686.
Fischer, K., P. M. Brakefield, & B. J. Zwaan. (2003). Plasticity in butterfly egg size: Why larger offspring at lower temperatures? *Ecology*, 84, 3138–3147.
Fisher, R. A. (1915). The evolution of sexual preference. *Eugenics Review*, 7, 184–192.
Fisher, R. A. (1930). *The Genetical Theory of Natural Selection*. Oxford, UK: Clarendon Press.
Fitze, P. S. & J. F. Le Galliard. (2011). Inconsistency between different measures of sexual selection. *American Naturalist*, 178, 256–268.
Fitze, P. S., B. Tschirren, & H. Richner. (2004). Life history and fitness consequences of ectoparasites. *Journal of Animal Ecology*, 73, 216–226.
Foster, S. J. & A. Vincent. (2004). Life history and ecology of seahorses: Implications for conservation and management. *Journal of Fish Biology*, 65, 1–61.
Gosden, T. P. & E. I. Svensson. (2008). Spatial and temporal dynamics in a sexual selection mosaic. *Evolution*, 62, 845–856.
Grafen, A. (1987). Measuring sexual selection: Why bother? In J. W. Bradbury & M. B. Andersson (eds.), *Sexual Selection: Testing the Alternatives* (pp. 221–233). Hoboken, NJ: John Wiley and Sons.
Grant, J. W. A., C. L. Gaboury, & H. L. Levitt. (2001). Competitor-to-resource ratio, a general formulation of operational sex ratio, as a predictor of competitive aggression in Japanese medaka (Pisces: Oryziidae). *Behavioral Ecology*, 11, 670–675.
Griffith, S. C., I. P. F. Owens, & K. A. Thuman. (2002). Extra pair paternity in birds: A review of interspecific variation and adaptive function. *Molecular Ecology*, 11, 2195–2212.
Griffith, S. C., I. R. K. Stewart, D. A. Dawson, I. P. F. Owens, & T. Burke. (1999). Contrasting levels of extra-pair paternity in mainland and island populations of the house sparrow (*Passer domesticus*): Is there an "island effect"? *Biological Journal of the Linnean Society*, 68, 303–316.
Gwynne, D. T. (1991). Sexual competition among females: What causes courtship-role reversal? *Trends in Ecology and Evolution*, 6, 118–121.
Hasselquist, D. & P. W. Sherman. (2001). Social mating systems and extrapair fertilizations in passerine birds. *Behavioral Ecology*, 12, 457–466.
Head, M. L., A. K. Lindholm, & R. Brooks. (2008). Operational sex ratio and density do not affect directional selection on male sexual ornaments and behavior. *Evolution*, 62, 135–144.
Hubbell, S. P. & L. K. Johnson. (1987). Environmental variance in lifetime mating success, mate choice, and sexual selection. *American Naturalist*, 130, 91–112.
Hunt, J., C. J. Breuker, J. A. Sadowki, & A. J. Moore. (2009). Male-male competition, female mate choice and their interaction: Determining total sexual selection. *Journal of Evolutionary Biology*, 22, 13–26.
Janzen, F. J. & E. D. I. Brodie. (1989). Tall tails and sexy males: Sexual behavior of rough-skinned newts (*Taricha granulosa*) in a natural breeding pond. *Copeia*, 1989, 1068–1071.

Jennions, M. D. & H. Kokko. (2010). Sexual selection. In D. F. Westneat & C. W. Fox (eds.), *Evolutionary Behavioral Ecology* (pp. 343–364). Oxford, UK: Oxford University Press.

Jirotkul, M. (1999). Population density influences male-male competition in guppies. *Animal Behaviour*, 58, 1169–1175.

Jones, A. G. (2009). On the opportunity for sexual selection, the Bateman gradient and the maximum intensity of sexual selection. *Evolution*, 63, 1673–1684.

Jones, A. G. & W. R. Ardren. (2003). Methods of parentage analysis in natural populations. *Molecular Ecology*, 12, 2511–2523.

Jones, A. G., J. R. Arguello, & S. J. Arnold. (2002). Validation of Bateman's principles: A genetic study of sexual selection and mating patterns in the rough-skinned newt. *Proceedings of the Royal Society of London, B*, 269, 2533–2539.

Jones, A. G., J. R. Arguello, & S. J. Arnold. (2004). Molecular parentage analysis in experimental newt populations: The response of mating system measures to variation in the operational sex ratio. *American Naturalist*, 164, 444–456.

Jones, A. G. & J. C. Avise. (2001). Mating systems and sexual selection in male-pregnant pipefishes and seahorses: Insights from microsatellite-based studies of maternity. *Journal of Heredity*, 92, 150–158.

Jones, A. G., C. Kvarnemo, G. I. Moore, L. W. Simmons, & J. C. Avise. (1998). Microsatellite evidence for monogamy and sex-biased recombination in the Western Australian seahorse *Hippocampus angustus*. *Molecular Ecology*, 7, 1497–1505.

Jones, A. G. & N. L. Ratterman. (2009). Mate choice and sexual selection: What have we learned since Darwin? *Proceedings of the National Academy of Sciences, USA*, 106, 10001–10008.

Jones, A. G., G. Rosenqvist, A. Berglund, S. J. Arnold, & J. C. Avise. (2000). The Bateman gradient and the cause of sexual selection in a sex-role-reversed pipefish. *Proceedings of the Royal Society of London, B*, 267, 677–680.

Jones, A. G., G. Rosenqvist, A. Berglund, & J. C. Avise. (1999). The genetic mating system of a sex-role-reversed pipefish (*Syngnathus typhle*): A molecular inquiry. *Behavioral Ecology and Sociobiology*, 46, 357–365.

Jones, A. G., G. Rosenqvist, A. Berglund, & J. C. Avise. (2005). The measurement of sexual selection using Bateman's principles: An experimental test in the sex-role-reversed pipefish *Syngnathus typhle*. *Integrative and Comparative Biology*, 45, 874–884.

Jones, A. G., C. M. Small, K. A. Paczolt, & N. Ratterman. (2010). A practical guide to methods of parentage analysis. *Molecular Ecology Resources*, 10, 6–30.

Kasumovic, M. M., M. J. Bruce, M. C. B. Andrade, & M. E. Herberstein. (2008). Spatial and temporal demographic variation drives within-season fluctuations in sexual selection. *Evolution*, 62, 2316–2325.

Kelly, C. D. (2008). Identifying a causal agent of sexual selection on weaponry in an insect. *Behavioral Ecology*, 172, 184–192.

Kelly, C. D., J. G. J. Godin, & J. M. Wright. (1999). Geographical variation in multiple paternity within natural populations of the guppy (*Poecilia reticulata*). *Proceedings of the Royal Society of London, B*, 266, 2403–2408.

Kirkpatrick, M., T. D. Price, & S. J. Arnold. (1990). The Darwin-Fisher theory of sexual selection in monogamous birds. *Evolution*, 44, 180–193.

Klug, H., J. Heuschele, M. D. Jennions, & H. Kokko. (2010). The mismeasurement of sexual selection. *Journal of Evolutionary Biology*, 23, 447–462.

Klug, H., K. Lindström, & H. Kokko. (2010). Who to include in measures of sexual selection is no trivial matter. *Ecology Letters*, 13, 1094–1102.

Koening, W. D. & S. S. Albano. (1986). On the measurement of sexual selection. *American Naturalist*, 72, 358-382.

Kokko, H. & M. D. Jennions. (2008). Parental investment, sexual selection and sex ratios. *Journal of Evolutionary Biology*, 21, 919–948.

Kokko, H. & A. López-Sepulcre. (2007). The ecogenetic link between demography and evolution: Can we bridge the gap between theory and data? *Ecology Letters*, 10, 773–782.

Kokko, H., A. Mackenzie, J. D. Reynolds, K. Lindström, & W. J. Sutherland. (1999). Measures of inequality are not equal. *American Naturalist*, 72, 358–381.

Kokko, H. & D. J. Rankin. (2006). Lonely hearts or sex in the city? Density-dependent effects in mating systems. *Philosophical Transactions of the Royal Society of London, B*, 361, 319–334.

Kraaijeveld, K., F. J. L. Kraaijeveld-Smit, & J. Komdeur. (2007). The evolution of mutual ornamentation. *Animal Behaviour*, 74, 657–677.

Krakauer, A. H., M. S. Webster, E. H. Duvall, A. G. Jones, & S. M. Shuster. (2011). The opportunity for sexual selection: Not mismeasured, just misunderstood. *Journal of Evolutionary Biology*, 24, 2064–2071.

Kvarnemo, C. (1996). Temperature affects operational sex ratio and intensity of male-male competition: Experimental study of sand gobies, *Pomatoschistus minutus*. *Behavioral Ecology*, 7, 208–212.

Kvarnemo, C. (1997). Food affects the potential reproductive rates of sand goby females but not of males. *Behavioral Ecology*, 8, 605–611.

Kvarnemo, C. & I. Ahnesjö. (1996). The dynamics of operational sex ratios and competition for mates. *Trends in Ecology and Evolution*, 11, 404–408.

Kvarnemo, C., G. I. Moore, & A. G. Jones. (2007). Sexually selected females in the monogamous Western Australian seahorse. *Proceedings of the Royal Society of London, B*, 274, 521–525.

Kwiatkowski, M. A. & B. K. Sullivan. (2002). Geographic variation in sexual selection among populations of an iguanid lizard *Sauromalus obesus* (=*ater*). *Evolution*, 56, 2039–2051.

Lande, R. (1979). Quantitative genetical analysis of multivariate evolution, applied to brain:body size allometery. *Evolution*, 33, 402–416.

Lande, R. & S. J. Arnold. (1983). The measurement of selection on correlated characters. *Evolution*, 37, 1210–1226.

Lane, J. E., M. N. K. Forrest, & C. K. R. Willis. (2011). Anthropogenic influences on natural animal mating systems. *Animal Behaviour*, 81, 909–917.

Lehtonen, T., B. B. M. Wong, & K. Lindström. (2009). Fluctuating mate preferences in a marine fish. *Biology Letters*, 6, 21–23.

Levin, S. A. (1992). The problem of pattern and scale in ecology: The Robert H. MacArthur Award Lecture. *Ecology*, 73, 1943–1967.

Lloyd, M. (1967). Mean crowding. *Journal of Animal Ecology*, 36, 1030.

Lorch, P. D. (2002). Understanding reversals in the relative strength of sexual selection on males and females: A role for sperm competition? *American Naturalist*, 159, 645–657.

Martin, P. R. & T. E. Martin. (2001). Ecological and fitness consequences of species coexistence: A removal experiment with wood warblers. *Ecology*, 82, 189–206.

Maurer, G., M. C. Double, O. Milenkaya, M. Süsser, & R. D. Magrath. (2011). Breaking the rules: Sex roles and genetic mating system of the pheasant coucal. *Oecologia*, 167, 413–425.

Mills, S. C., A. Grapputo, E. Koskela, & T. Mappes. (2007). Quantitative measure of sexual selection with respect to the operational sex ratio: a comparison of selection indices. *Proceedings of the Royal Society of London, B*, 274, 143–150.

Mills, S. C. & J. D. Reynolds. (2003). Operational sex ratio and alternative reproductive behaviours in the European bitterling, *Rhodeus sericeus*. *Behavioral Ecology and Sociobiology*, 54, 98–104.

Mitani, J. C., J. GrosLouis, & A. F. Richards. (1996). Sexual dimorphism, the operational sex ratio, and the intensity of male competition in polygynous primates. *American Naturalist*, 147, 966–980.

Mobley, K. B. & A. G. Jones. (2007). Geographical variation in the mating system of the dusky pipefish (*Syngnathus floridae*). *Molecular Ecology*, 16, 2596–2606.

Mobley, K. B. & A. G. Jones. (2009). Environmental, demographic and genetic mating system variation among five geographically distinct dusky pipefish (*Syngnathus floridae*) populations. *Molecular Ecology*, 18, 1476–1490.

Mobley, K. B., C. M. Small, & A. G. Jones. (2011). The genetics and genomics of Syngnathidae (pipefishes, seahorses and seadragons). *Journal of Fish Biology*, 78, 1624–1646.

Møller, A. P. & T. R. Birkhead. (1994). The evolution of plumage brightness in birds is related to extrapair paternity. *Evolution*, 48, 1089–1100.

Morisita, M. (1962). *I*-index, a measure of dispersion of individuals. *Research in Population Ecology*, 4, 1–7.

Nonacs, P. (2003). Measuring the reliability of skew indices: Is there one best index? *Animal Behaviour*, 65, 615–627.

Olsson, M., E. Wapstra, T. Schwartz, T. Madsen, B. Ujvari, & T. Uller. (2011). In hot pursuit: Fluctuating mating system and sexual selection in sand lizards. *Evolution*, 65, 574–583.

Paczolt, K. A. & A. G. Jones. (2010). Postcopulatory sexual selection and sexual conflict in the evolution of male pregnancy. *Nature*, 464, 401–404.

Parker, G. A. (1970). Sperm competition and its evolutionary consequences in insects. *Biological Reviews of the Cambridge Philosophical Society*, 45, 525–567.
Parker, G. A., R. R. Baker, & V. G. F. Smith. (1972). The origin and evolution of gamete dimorphism and the male-female phenomenon. *Journal of Theoretical Biology*, 36, 529–553.
Partridge, C., I. Ahnesjö, C. Kvarnemo, K. B. Mobley, A. Berglund, & A. G. Jones. (2009). The effect of perceived female parasite load on post-copulatory male choice in a sex-role-reversed pipefish. *Behavioral Ecology and Sociobiology*, 63, 345–354.
Payne, R. J. H. & D. C. Krakauer. (1997). Sexual selection, space, and speciation. *Evolution*, 51, 1–9.
Petrie, M. & B. Kempenaers. (1998). Extra-pair paternity in birds: Explaining variation between species and populations. *Trends in Ecology and Evolution*, 13, 52–58.
Phillips, P. C. & S. J. Arnold. (1989). Visualizing multivariate selection. *Evolution*, 43, 1209–1222.
Pomfret, J. C. & R. J. Knell. (2008). Crowding, sex ratio and horn evolution in a South African beetle community. *Proceedings of the Royal Society of London, B*, 275, 315–321.
Queller, D. C. (1997). Why do females care more than males? *Proceedings of the Royal Society of London, B*, 264, 1555–1557.
Reichard, M., M. Ondračková, A. Bryjová, & J. Bryja. (2008). Breeding resource distribution affects selection gradients on male phenotypic traits: Experimental study on lifetime reproductive success in the bitterling fish (*Rhodeus amarus*). *Evolution*, 63, 377–390.
Reichard, M., C. Smith, & J. Bryja. (2008). Seasonal change in the opportunity for sexual selection. *Molecular Ecology*, 17, 642–651.
Reynolds, J. D. (1996). Animal breeding systems. *Trends in Ecology and Evolution*, 11, A68–A72.
Rubenstein, D. R. & I. J. Lovette. (2009). Reproductive skew and selection on female ornamentation in social species. *Nature*, 462, 786–790.
Ruzzante, D. E., D. C. Hamilton, D. L. Kramer, & J. W. A. Grant. (1996). Scaling of the variance and the quantification of resource monopolization. *Behavioral Ecology*, 7, 199–207.
Sakai, A. K. & D. F. Westneat. (2001). Mating systems. In C. W. Fox, D. A. Roff, & D. J. Fairbairn (eds.), *Evolutionary Ecology: Concepts and Case Studies* (pp. 193–206). Oxford, UK: Oxford University Press.
Scobell, S. K., A. M. Fudickar, & R. Knapp. (2009). Potential reproductive rate of a sex-role reversed pipefish over several bouts of mating. *Animal Behaviour*, 78, 747–753.
Searcy, W. A. & K. Yasukawa. (1995). *Polygyny and Sexual Selection in Red-winged Blackbirds*. Princeton, NJ: Princeton University Press.
Serbezov, D., L. Bernatchez, E. M. Olsen, & L. A. Vøllestad. (2010). Mating patterns and determinants of individual reproductive success in brown trout (*Salmo*

trutta) revealed by parentage analysis of an entire steam living population. *Molecular Ecology*, 19, 3193–3205.

Shuster, S. M. (2009). Sexual selection and mating systems. *Proceedings of the National Academy of Sciences, USA*, 106, 10009–10016.

Shuster, S. M. & M. J. Wade. (2003). *Mating Systems and Strategies*. Princton, NJ: Princeton University Press.

Simmons, L. W. (2001). *Sperm Competition and Its Evolutionary Consequences in Insects*. Princeton, NJ: Princeton University Press.

Simmons, L. W. & M. Beveridge. (2010). The strength of postcopulatory sexual selection within natural populations of field crickets. *Behavioral Ecology*, 21, 1179–1185.

Simmons, L. W. & D. J. Emlen. (2006). Evolutionary trade-off between weapons and testes. *Proceedings of the National Academy of Sciences, USA*, 103, 16346–16351.

Sogabe, A. & Y. Yanagisawa. (2007). Sex-role reversal of a monogamous pipefish without higher potential reproductive rate in females. *Proceedings of the Royal Society of London, B*, 274, 2959–2963.

Sokal, R. R. & F. J. Rohlf. (1995). *Biometry*. New York: W. H. Freeman and Company.

Spottiswoode, C. & A. P. Møller. (2004). Extrapair paternity, migration and breeding synchrony in birds. *Behavioral Ecology and Sociobiology*, 15, 41–57.

Sutherland, W. J. (1985). Chance can produce a sex difference in variance in mating success and explain Bateman's data. *Animal Behaviour*, 33, 1349–1352.

Sutherland, W. J. (1987). Random and deterministic components of variance in mating success. In J. W. Bradbury & M. Andersson (eds.), *Sexual Selection: Testing the Alternatives* (pp. 209–219). Chichester: Wiley.

Thornhill, R. & J. Alcock. (1983). *The Evolution of Insect Mating Systems*. Cambridge, MA: Harvard University Press.

Trivers, R. L. (1972). Parental investment and sexual selection. In B. Campbell (ed.), *Sexual Selection and the Descent of Man 1871–1971* (pp. 136–179). London: Heinemann.

Turner, A. M. & J. P. McCarty. (1998). Resource availability, breeding site selection, and reproductive success of red-winged blackbirds. *Oecologia*, 113, 140–146.

Twiss, S. D., C. Thomas, V. Poland, J. A. Graves, & P. Pomeroy. (2007). The impact of climatic variation on the opportunity for sexual selection. *Biology Letters*, 3, 12–15.

Vincent, A., I. Ahnesjo, & A. Berglund. (1994). Operational sex ratios and behavioral sex differences in a pipefish population. *Behavioral Ecology and Sociobiology*, 34, 435–442.

Vincent, A., I. Ahnesjo, A. Berglund, & G. Rosenqvist. (1992). Pipefishes and seahorses: Are they all sex-role reversed? *Trends in Ecology and Evolution*, 7, 237–241.

Wade, M. J. (1979). Sexual selection and variance in reproductive success. *American Naturalist*, 114, 742–747.

Wade, M. J. (1995). The ecology of sexual selection: Mean crowding of females and resource-defense polygyny. *Evolutionary Ecology*, 9, 118–124.

Wade, M. J. & S. J. Arnold. (1980). The intensity of sexual selection in relation to male sexual behavior, female choice and sperm precedence. *Animal Behaviour*, 28, 446–461.

Wade, M. J. & S. M. Shuster. (2004). Sexual selection: Harem size and the variance in male reproductive success. *American Naturalist*, 164, E83–E89.

Wade, M. J. & S. M. Shuster. (2005). Don't throw Bateman out with the bathwater! *Integrative and Comparative Biology*, 45, 945–951.

Webster, M. S., K. A. Tarvin, E. M. Tuttle, & S. Pruett-Jones. (2007). Promiscuity drives sexual selection in a socially monogamous bird. *Evolution*, 61, 2205–2211.

Weir, L. K., J. W. A. Grant, & J. A. Hutchings. (2011). The influence of operational sex ratio on the intensity of competition for mates. *American Naturalist*, 177, 167–176.

Westneat, D. F. & I. R. K. Stewart. (2003). Extra-pair paternity in birds: causes, correlates, and conflict. *Annual Review of Ecology, Evolution and Systematics*, 34, 365–396.

Wilson, A. B., I. Ahnesjö, A. C. J. Vincent, & A. Meyer. (2003). The dynamics of male brooding, mating patterns, and sex roles in pipefishes and seahorses (family Syngnathidae). *Evolution*, 57, 1374–1386.

Wilson, A. B. & K. M. Martin-Smith. (2007). Genetic monogamy despite social promiscuity in the pot-bellied seahorse (*Hippocampus abdominalis*). *Molecular Ecology*, 16, 2345–2352.

5

The Evolution of Ornaments and Armaments

Geoffrey E. Hill

INTRODUCTION

When Darwin conceived his theory of *evolution* through *natural selection* (Darwin, 1859), he immediately recognized the power of the idea. Natural selection provided a unifying explanation for a tremendous range of natural phenomena. Unfortunately, and much to Darwin's consternation, the umbrella of evolution by natural selection could not quite accommodate the whole of biological diversity. In particular, it could not adequately explain two of the most fantastic features of animals: *ornaments* and *armaments* (Figure 5.1).

Within the realm of *behavioral ecology*, an ornament is a trait that enhances the appearance (in the broad sense that potentially involves any sensory perception) of an animal, sometimes to the detriment of survival. Elongated feathers, dewlaps, manes, fleshy wattles, bright coloration, songs, roars, conspicuous movements, and an array of additional sense-stimulating traits of animals appear to be ornaments (Figure 5.2). An armament, in contrast, is a trait that can serve as an aid in contests with conspecifics but that is not required for foraging or protection from predators. Armaments include traits like horns, antlers, spurs, and—most commonly—a big, strong body. To explain ornaments and armaments, Darwin had to conceive of a mechanism for evolution beyond natural selection. Whereas Darwin (1859) proposed that natural selection promotes the maintenance and spread of traits that enhance survival or fecundity, he presented *sexual selection* as a process that promotes the maintenance and spread of traits that enhance access to mates (Darwin, 1871). Darwin reasoned that, for a sexually reproducing organism, staying alive and having the capacity to produce offspring are for naught without access to a mate.

Figure 5.1. The large antlers of this bull elk (*Cervus canadensis*) certainly appear to be armaments. Research indicates, however, that females assess the size and symmetry of antlers during mate choice, so, like many traits, elk antlers may function as both armaments and ornaments.

Sexual selection is now accepted among evolutionary biologists as the primary force that shapes ornaments and armaments, but the specific processes by which these types of traits evolve remains hotly debated. In this chapter I will summarize current theory concerning how ornaments and armaments evolved and how they function in animal populations. Sexual selection can occur in plants, fungi, and single-celled organisms (Andersson, 1994; Skogsmyr & Lankinen, 2002) as well as in animals, but no one has proposed that organisms other than animals evolve ornaments or armaments. Hence, the focus of this chapter will be sexual selection in animals.

SEXUAL SELECTION AND ANIMAL BEHAVIOR

Ornaments and armaments tend to be morphological traits, so it may seem curious that the study of such traits is pursued primarily by students of animal behavior. However, it is behavior, and in particular sensory perception linked

Figure 5.2. The facial structures of this mandrill (*Mandrillus sphinx*) are the most colorful traits displayed by any primate and are among the most colorful traits among mammals. They likely function as both armaments and ornaments but have not been thoroughly studied.

to behavioral responsiveness, that connects the drive to find a mate with the evolution of traits such as red feathers and pheromones. Obviously, female choice of mates and physical contests between males play out as behavioral interactions, but more subtly, sensory perception and behavioral responsiveness also play roles in contests for mates when there is assessment prior to physical confrontation. Because humans are visual animals, we tend to think of sensory assessment in terms of vision, but animals use a wide range of signaling modes to present their ornaments and armaments, including pheromonal, chemical, electrical, acoustical, and tactile. An ornament can be a plume of volatile alkaloids released by a moth (Iyengar et al., 2001) just as much as it can be the visually stunning train of a peacock (Petrie et al., 1991).

FEMALE CHOICE AND MALE-MALE COMPETITION

Sexual selection occurs whenever there is competition among members of one sex for sexual access to members of the opposite sex. One important factor in determining the potential for sexual selection is investment in offspring, including everything from the deposition of nutrients in eggs to the nourishment of a growing embryo to the instruction an adult-sized offspring in hunting techniques (Ligon, 1999). The sex that invests the most in offspring becomes a resource that is competed for by the sex that invests less in offspring (Trivers, 1972). By definition, females have larger gametes and males have smaller gametes, so the initial investment in offspring by females is greater than males. In animals with ***parental care***, males can make up the disparity in initial investment by investing in offspring as much as or even more than females, but for most animal species, the pattern of greater female investment in offspring holds through reproduction (Ligon, 1999). The end result is that females are the choosy sex in most species of animals while males are the displaying and competing sex (Trivers, 1972). For the remainder of this chapter I will assume that ornaments are the result of female mate choice (technically termed ***intersexual selection***). Likewise, I will assume that armaments evolve via male-male competition (technically termed ***intrasexual selection***). In the unusual cases in which males provide more parental care than females, females often have more ornaments and armaments and males are the choosy sex (Ligon, 1999).

Other circumstances that promote sexual selection include greater availability of potential mates of one sex compared to the other. Such a disparity in partner availability can arise strictly from numerically greater representation of one sex (called a ***skewed sex ratio***; e.g., Hill et al., 1994). Mating systems that enable one sex to monopolize more than one member of the opposite sex can also skew the availability of mates and drive sexual selection even when an equal number of males and females are present (Andersson, 1994; Ligon, 1999). Such a skew in **operational sex ratio** occurs most commonly under a **polygynous** mating system in which individual males mate with more than one female (Ligon, 1999). **Extra-pair copulation** can also increase the availability of a few males to many females and hence alter the operational sex ratio of a population (Westneat & Stewart, 2003). Other circumstances that allow individuals of one sex to monopolize more than one individual of the opposite sex, such as a temporally staggered period of sexual receptivity, also lead to sexual selection (Stutchbury & Morton, 1995).

MATE CHOICE VERSUS COMPETITION

For the purposes of explanation, it is useful to think of female choice and male-male competition as distinct processes that lead to the evolution of distinct sets of secondary sexual traits: ornaments and armaments. For many animals, however, the processes of choosing mates and competing for access to mates act simultaneously on the same trait (Berglund et al., 1996). Consider, as an example, ring-necked pheasants (*Phasianus colchicus*). Males have dagger-like projections from their tarsi (lower legs) called spurs that are used to eviscerate rivals in fights. For millennia, humans have staged fights between pheasants and other fowl with spurs on their legs because the birds fight brutally, slashing each other with their spurs. It would seem safe to surmise that spurs are armaments that evolved through the advantages they bestowed in male-male competition (Davison, 1985). It turns out, however, that females assess the size of spurs when choosing a mate (von Schantz et al., 1989), and choice by females selects for larger spurs. So the spurs of pheasants—as well as the **secondary sexual traits** of many other species of animals—are simultaneously armaments and ornaments. To keep discussions tractable in this chapter, I will refer to particular traits as being either armaments or ornaments with the understanding that such a dichotomous characterization is commonly an oversimplification.

How sexual selection can give rise to armaments and ornaments is a focal topic of research within **evolutionary ecology** and behavioral ecology, but explaining ornaments has proven far more challenging than explaining armaments. Long before there was a theory of sexual selection or an evolutionary perspective of any sort, it was widely appreciated that male animals used traits like horns, tusks, and spurs to win contests for sexual access to females. Not surprisingly, when armaments were reconsidered within the framework of natural and sexual selection, the hypothesis that male-male competition leads to the evolution of armaments was widely accepted and never seriously challenged in the behavior and evolution literature (see Cronin, 1991). In contrast, prior to Darwin, ornamental traits were traditionally viewed as the product of special creation whose purpose was to make the world more interesting and enjoyable for humans (see Hill, 2002). From the outset, framing evolutionary explanations for how ornamental traits evolved proved challenging and contentious. There was a long period of debate in the late nineteenth and early twentieth centuries regarding whether female mate choice played any role in the evolution of ornaments. It is now universally accepted that female mate choice is an important force in the evolution of ornaments, and debate has shifted to the process of ornament evolution. The evolution of ornaments via female choice remains a highly contentious topic in biology with many hypotheses proposed and no clear consensus among researchers (Prum, 2010).

In the following sections, I will outline each of the major hypotheses that have been proposed to explain the evolution of ornaments and armaments and review the evidence in support of each.

THE EVOLUTION OF ARMAMENTS

The utility of armaments in battles among rival animals is easy to observe. Contests over mates are often decided in brutal and even bloody physical struggles. A big, strong body and use of a weapon can be the difference between success and defeat. From the time of Darwin (1859, 1871), it has been easy to conceive how sexual selection could lead to the evolution of traits like armaments that aid in success in contests for mates (Figure 5.3).

A less intuitive idea regarding the outcome of intrasexual selection is that showy traits like bright colors and melodic songs, which might otherwise be taken as ornaments that charm females, could also function as signals of health, vigor, and status in aggressive contests. The basic logic of this hypothesis is that it is only worth engaging in a fight if there is a reasonable chance of winning. If an assessment can be made before a fight, permitting both the stronger and weaker contestant to perceive a mismatch, then the weaker animal benefits by retreating and the stronger animal benefits by allowing the retreat of the weaker (Berglund et al., 1996; West-Eberhard, 1979). But *how can the outcome of a fight be reliably predicted? What stops a weaker fighter from dishonestly signaling superior fighting ability?* Two means to ensure the **honesty** of signals of fighting ability have been proposed: **social mediation** and linking trait production to individual condition (Berglund et al., 1996). I will take up the topic of how production constraints can ensure signal honesty in the

Figure 5.3. The curved horns of a male bighorn sheep (*Ovis canadensis*) serve as battering rams during dramatic and violent contests for harems of females. There are few clearer cases of secondary sexual characteristics serving as armaments.

discussion of **indicator traits** and intersexual selection below. Here I will focus on social mediation.

Both signalers and receivers benefit from honest signals of status because such honest signaling spares them both the costs of unnecessary fighting (Rohwer, 1975). The problem with maintaining signal honesty is that low-status males benefit even more if they can get away with dishonestly signaling high status. The proposed solution is that signals of high status are not always "believed"—they are routinely tested (Rohwer, 1977). So long as the costs of drawing the aggression of dominant males outweigh the benefits of status gained among subordinates through deception, then **cheating** strategies will not succeed. Mathematical models based on **game theory** support the idea that honest signaling can persist by means of such social mediation (Lachmann et al., 2001; Ripoll et al., 2004).

In the past few decades, many studies have provided empirical support for the idea that showy traits can be used in assessment during contests for mates (reviewed in Berglund et al., 1996). For instance, in many species of songbirds, males defend territories that are critical resources for reproduction. Field

> **Box 5.1**
> **Ornamental traits in territorial animals**
>
> Songbirds are among the most ornamented animals. They flash a rainbow of colors; make some of the most complex, loud, and lovely sounds in nature; carry long and elaborate feathers; and engage in dramatic displays. With few exceptions, male songbirds lack an intromittent organ, so forced copulation is typically not possible. Songbirds would seem to be a taxon in which female mate choice has selected for a wide array of ornaments, and indeed, in some songbirds, long tails, colorful feathers, and song are criteria in female mate choice (Hill, 2006). But most songbirds are also highly territorial, and the quality of the territory that a male defends has a large effect on female reproductive success (Emlen & Oring, 1977). Perhaps not surprisingly, in several territorial and highly ornamented songbirds, females show no mating preference for more highly ornamented males, but male ornamentation plays an important role in contests for territories (e.g., eastern bluebird [*Sialia sialis*] [Liu et al., 2007; Siefferman & Hill, 2005]). In these songbirds, apparent ornaments are really armaments. Few species of songbirds have been carefully studied, so at this point there is no way to assess whether colorful feathers and complex songs most frequently function as ornaments or armaments.

studies have shown that song and plumage coloration—two traits that appear to be ornaments—predict male success in acquiring the most preferred territories and procuring access to mates (e.g., Pryke & Andersson, 2003; Siefferman & Hill, 2005) (Box 5.1). Increasing the coloration of males makes potential rivals approach them more carefully and less frequently, and decreasing coloration causes males to be challenged more (examples in Hill, 2010). Hence, armaments can exist either as weapons or as signals of fighting ability.

THE EVOLUTION OF ORNAMENTS

The process by which intersexual selection leads to the evolution of ornamental traits in animals is one of the longest-running and most contentious debates in evolutionary biology. In the nineteenth century Charles Darwin and Alfred Wallace were the key players in the original debate regarding the importance of female choice in ornament evolution. Darwin argued emphatically that female choice was the selective force that gave rise to ornaments while Wallace discounted the possibility of an aesthetic sense in animals and looked for other explanations (Cronin, 1991). Empirical studies in the late twentieth and early twenty-first centuries on a range of animals from flies to primates showed definitively that female choice can select for ornamentation (Andersson, 1994). These studies vindicated Darwin's insistence that female choice was the selective force that drove the evolution of ornaments, but they left unresolved the question of why females used ornaments to choose males. Darwin provided no insight on this fundamental question, and a resolution to how and why female choice for ornamental traits evolves and how such choice leads to the evolution of the range of ornaments seen in nature remains a major unresolved question in behavioral and evolutionary biology. The gist of the discussion regarding the evolution of ornamental traits focuses on whether elaboration of ornaments occurs under selection for arbitrary markers of attractiveness or through female demand for honest signals of the quality of perspective mates. I will consider each of the major hypotheses for the evolution of ornaments via mate choice as well as more specific mechanisms for ornament evolution.

Models of Intersexual Selection Based on Arbitrary Choice by Females

When evolutionary biologists refer to ornamental traits as **arbitrary**, they mean that the expression of ornamental traits is not related to a male's genetic quality or to the quantity or quality of resources controlled by the male. Arbitrary traits hold value exclusively in the realm of sexual attractiveness. The most well-known model for how sexual selection can lead to the evolution of arbitrary traits is the Fisherian or **runaway model** of sexual selection (Figure 5.4). There are, however, at least three other hypotheses for how arbitrary traits might evolve through female mate choice, and these various models of arbitrary mate choice are not mutually exclusive. I will begin by discussing the less prominent models of intersexual selection before focusing on the runaway model of sexual selection.

Species recognition—The first hypothesis that was proposed for the evolution of arbitrary traits is the **species recognition hypothesis**. This hypothesis was proposed in the nineteenth century by Wallace, who argued that for many animals it is vitally important both to recognize individuals of their species and to signal their own species identity (Wallace, 1889). Wallace (1889) discussed the importance of species recognition in social contexts during flocking and other nonsexual congregations, but he focused primarily on species recognition within the context of mate choice. He concluded that mating outside species boundaries is potentially the worst mistake that a female can make, a theme echoed half a century later by Ronald Fisher (Fisher, 1930, 1958). Hence, Wallace proposed that ornaments evolved to serve the very important function of unambiguously signaling species identity primarily in the context of mate selection.

Figure 5.4. The elongated tail of this male paradise flycatcher (*Terpsiphone paradisi*) is the sort of highly elaborated trait that is proposed to have evolved through runaway sexual selection. Alternatively, the ornament may have evolved as a signal of male quality.

For approximately 80 years, from the late nineteenth to the mid-twentieth century, the species recognition hypothesis was the consensus explanation for why ornamental traits evolved (Cronin, 1991). Since the early 1980s, the hypothesis of ornaments as signals of species identity has fallen out of favor in discussions of sexual selection, and many contemporary students of behavioral and evolutionary biology fail to appreciate the influence and success of this idea. For nearly a century, the species recognition hypothesis was the only widely stated explanation for ornamental traits (Cronin, 1991; Hill, 2002). With a surge of interest in and support for runaway and indicator models of sexual selection (both presented below), species recognition is now given little attention. In current discussions of sexual selection it is often not even mentioned. The species recognition hypothesis, however, has not been broadly falsified and remains a viable hypothesis for many ornamental traits of animals.

The demise of the species recognition hypothesis came not from specific tests or empirical observations that disproved it. The hypothesis was eventually deemed inadequate based on theoretical considerations of the sort of traits that should evolve as markers of species identity. Such markers, theorists concluded, should be simple, invariant within a species, and easy to perceive. *Simple* and *invariant* are not accurate descriptors of many ornamental traits, particularly the elaborate ornaments that evolutionary biologists are most anxious to explain. As John Maynard Smith (Maynard Smith, 1991, p. 146) wrote, "it would be absurd to suppose that a male nightingale must sing like that in order for a female to tell that he is not a willow warbler." A consensus quickly emerged in the 1980s that the species recognition hypothesis could not explain elaborate ornaments like a peacock's train or the roar of a red deer (*Cervus elaphus*). The fact that the species recognition hypothesis is not sufficient to explain highly elaborate ornamental traits, however, does not mean that it fails as an explanation for all ornamental traits. A few empirical studies have supported a species recognition function for simple ornaments such as the white forehead spot on species of *Ficedula* flycatchers (Saetre et al., 1997).

Finally, it should be noted that the need to invoke evolutionary forces other than species recognition to explain the evolution of a peacock's tail does not mean that peahens do not use the gaudy trains of males to distinguish peacocks from other pheasant species. Debate regarding the species recognition hypothesis is not about whether ornamental traits can function in species recognition; rather, it is about whether species recognition is sufficient to explain the evolution of the traits.

Sensory exploitation—Another model for the evolution of arbitrary signals of attractiveness is the **sensory exploitation model**. This hypothesis proposes that ornaments evolve because they exploit sensory-response systems that typically function in contexts other than mate choice, such as foraging (Ryan & Keddy-Hector, 1992). By this model, ornamental traits evolve because they stimulate already existing sensory systems and trigger already existing positive behavioral response mechanisms (Ryan & Keddy-Hector, 1992). As an example, consider an animal for which red fruit is a choice dietary item such that the animal is attracted to red objects in its environment. If a red patch of color appeared on a prospective mate in such a species, females may be drawn to and show positive sexual receptivity toward that male because they are predisposed to respond positively to red objects (Rodd et al., 2002). The ornament simply exploits preexisting sensory and behavioral response systems, and the positive response by females leads to the spread of genes for the ornamental trait.

Some of the strongest empirical support for sensory exploitation as an explanation for ornamental traits comes from studies of swordtail fish (genus *Xiphophorus*). As their name implies, swordtails have a sword-like extension from the base of their caudal fins. In all species with swords studied to date, females prefer to mate with males with long swords (Basolo, 1990a). Swordtails are closely related to platyfish, which do not have swords. Through behavioral studies it was discovered that, in some species of platyfish, females also show a sexual preference for males with swords. Reconstruction of the evolutionary history of the swordtails and platyfish showed that the evolution of preference for swords predated the evolution of swords (Basolo, 1990b). The implication was that swords had evolved in some fish lineages because there was a preexisting bias that caused females to favor males with the trait. Researchers found no benefit of the caudal extension other than sexual attractiveness. Studies of the calls of frogs (Ryan et al., 1990) and the colors of fish (Egger et al., 2011) also support the sensory bias models.

The current attitude among research biologists concerning the sensory exploitation model is that it is likely a factor in the evolution of ornamental traits (see the runaway and chase-away models below) but that it is not sufficient as a complete explanation for the evolution of most ornamental traits of animals.

Chase-away model—Another model for the evolution of ornaments that is related to the sensory exploitation model is the ***chase-away model*** of sexual selection. The foundation of this model is the fundamental conflict of interest between males and females in sexually reproducing organisms. Earlier in the chapter I described females as the "choosy sex" and males as the "displaying sex," but under the chase-away model, males would be better described as the "*coercive* sex." Male ornaments are proposed to stimulate females to mate in a manner that benefits males but that is suboptimal for females—overstimulating females to mate too often, at the wrong time, or with a low-quality partner

THE EVOLUTION OF ORNAMENTS AND ARMAMENTS

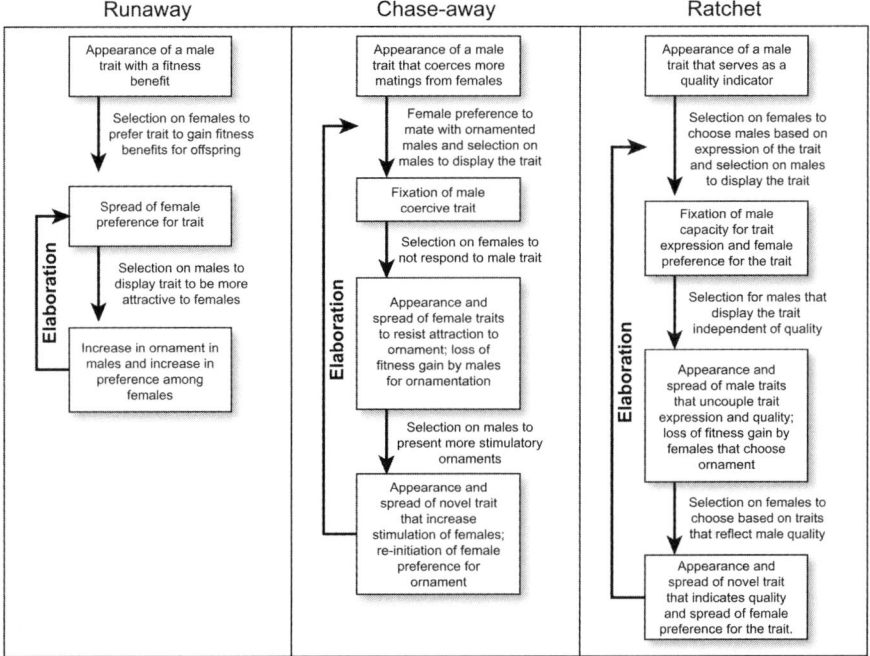

Figure 5.5. Flow charts illustrating the processes proposed by the three models for the elaboration of ornamental traits that can account for elaboration of ornamental traits: runaway model (Fisher, 1958), chase-away model (Holland & Rice, 1998) and ratchet model (Hill, 1994). Under the runaway model, trait elaboration is inevitable once the process ensues, but under the chase-away and ratchet models, elaboration is dependent on the appearance of genetic variants among females that resist attraction to the ornament or on the appearance of shortcut means to trait expression by males, respectively. Evolutionary events are indicated in the boxed portion of the figure, and forces of change are shown by arrows connecting boxes.

(Holland & Rice, 1998). The stimulatory male trait spreads because it enhances male mating success, but the chase-away model proposes that evolutionary escalation ensues. Because being attracted to the ornament is bad for females, they are selected to evolve diminished response systems, thereby resisting the coercion of males. As female responsiveness abates, the advantage to males of having the trait declines such that there is selection on males for any new traits that enhance or expand the stimulatory effect of the ornament. Escalation ensues as males are selected to display more and more elaborate ornamentation to maintain stimulation of females in the face of eroding responsiveness (Figure 5.5).

The chase-away model makes two specific predictions that are different from other models of sexual selection (Holland & Rice, 1998). First, the model proposes declines in female responsiveness to ornamental traits over evolutionary time. Change in female mating preference on an evolutionary time scale is a challenging prediction to test because it requires knowledge of female responsiveness over the evolutionary history of the trait. Evolutionary biologists cannot go back in time to assess female responsiveness (behavior does not typically fossilize), but they can use phylogenetic reconstructions for estimating the past condition of traits, including behaviors (see Chapter 2). A study involving phylogenetic reconstructions of the responsiveness of female wolf spiders (genus *Schizocosa*) to tufts of bristles on males supported predictions of the chase-away model. In species of spiders without tufts, adding tufts to males made them much more sexually attractive to female spiders. In species that had tufts, however, removing or enlarging tufts had no effect on female sexual response. Critically, phylogenetic reconstruction showed that tufts were a derived character and that the species without tufts never had a tufted ancestor (McClintock & Uetz ,1996). One interpretation of these results is that in the lineage with tufts, a chase-away process had selected for reduced response to tufts in females (Holland & Rice, 1998; see also Box 5.2).

A second prediction of the chase-away model is that females should have reduced rather than enhanced fitness if they mate with highly ornamented males (Holland & Rice, 1998). The strongest empirical support for this prediction of reduced female fitness comes not from studies of selection for ornamental traits but rather from studies of mating costs among female fruit flies in experimentally maintained monogamous versus polygynous populations (Rice, 1996; see Chapter 6). In lines of flies that were forced to be monogamous such that the reproductive output of a male was directly tied to its single mate, females had long lives. In lines of flies in which males competed with other males for sexual access to multiple mates, females had shortened lives and lower fecundity. The explanation was that when multiple mating was possible, males added toxic elements to semen to benefit themselves in competition with other males even though it was costly to females (Chapman et al., 1995). When sexual selection was eliminated, the toxic elements in semen declined.

Costs to females of choosing highly ornamented males have proven harder to document. In the runaway model of sexual selection as well as the indicator models described below, females benefit by mating with highly ornamented males. It is universally accepted that mate choice and sexual reproduction impose a cost on females (Andersson, 1994), but a loss of female fitness resulting specifically from a response to male ornamentation is a difficult prediction

> **Box 5.2**
> **The paradox of rooster plumes**
>
> The red jungle fowl (*Gallus gallus*), the wild ancestor of the domestic chicken, is one of the most highly ornamented birds in the world. Hens are brown and drab while roosters have long, elaborate, and colorful feathers covering their bodies. If asked to explain these plumes, most evolutionary biologists would respond that female mate choice or perhaps competition among males drove the evolution of the ornaments and now maintains them. There would be wide consensus that the showy rooster feathers are the result of sexual selection.
>
> Surprisingly, experimental observations do not support either female mate choice or male-male competition as viable explanations for the plumes of roosters. The key experiment involved presenting hens with prospective mates that either had normal ornamental feathers or lacked them entirely. This experiment was possible because there is a gene, called the *hen-feathered* gene, that leaves males bereft of all the ornamental plumes. In every other respect, however, *hen-feathered* males are normal roosters. They have the same fleshy head ornaments as normal males, and they retain their male aggressiveness. J. David Ligon and Patrick Zwartjes (1995) bred red jungle fowl so as to produce brothers that either carried or did not carry the *hen-feather* gene; in other words, they created brothers that had the feathers of hens or that had full ornamental plumes. These sets of brothers were then used in mate-choice experiments. Amazingly, hens showed no preference for males with ornate and colorful plumes—removing all of the long golden and glossy ornamental plumage did not make males any less attractive as mates. In another study, Ligon and colleagues (1990) showed that the plumes have little or no effect on male-male contests. Among proposed sexual selection hypotheses, only the chase-away model can explain why chicken ornaments exist when there is no inter- or intrasexual selection on the trait. According to the chase-away model, the ornamental plumes arose because they coerced females into more mating, and females have escaped male manipulation by evolving lack of preference for the trait.

to test. To date, no specific study as been presented that clearly demonstrates a loss of fitness for females because they were attracted to a male ornament. On the other hand, many studies have proposed to show a net benefit to females for mating with highly ornamented males (Andersson, 1994; Griffith & Pryke, 2006).

Overall, the chase-away model may be the best current explanation for why some ornamental traits like the gaudy plumages of red jungle fowl (*Gallus gallus*) (Box 5.2) have no influence on the mate choices made by females. The importance of the chase-away model as a general explanation for ornamental traits in animals remains unknown.

Runaway sexual selection—The idea that ornamental traits could evolve through a runaway process was first presented in a brief narrative by Fisher (1915) on which he later elaborated (Fisher 1930, 1958). Fisher proposed that female choice and male ornamentation could enter into a self-reinforcing process in which both the ornamental trait and preference for the ornamental trait could become grossly exaggerated. The concept of runaway sexual selection was subsequently formalized in quantitative genetics models, most notably by Russell Lande (1981) and Mark Kirkpatrick (1982), and the runaway model (also known as the Fisherian model and the Lande-Kirkpatrick model) remains one of the most important explanations for the evolution of ornamental traits—particularly fantastic and highly exaggerated traits (Prum, 1997).

The process of runaway sexual selection is proposed to begin with a population of animals in which some males carry genes for a trait that bestows a fitness advantage such as a slightly longer tail that enhances flight ability. (See Arnold [1983] for a lucid and well-illustrated presentation of the runaway model). Some females happen to carry genes that cause them to prefer longer tails—such a preexisting genetic bias invokes sensory exploitation (see above)—and these females benefit because their offspring inherit the genes for longer and functionally better tails. Both longer male tails and female preference for longer tails are beneficial, and hence genes for both greater preference and longer tails spread in the population. As an increasing number of females inherit genes for a preference for longer tails, two important consequences initiate the runaway process. First, the benefits for producing sons that are attractive as mates ("sexy sons") begins to exceed the benefits of producing sons with aerodynamic tails; in other words, the benefit of the trait is no longer determined primarily by natural selection. Moreover, the sexy-son benefit increases with each generation as preference for the trait increases. The stronger the female preference for a large ornament, the greater the selection will be on males to grow even longer tails. In turn, the longer the tails grown by males, the greater are the benefits to females in the form of sons with long and sexy tails.

A second consequence that spurs on a runaway cycle is that the genes for longer tails and the genes for preference for longer tails begin to be inherited together (a process called **linkage disequilibrium**) (O'Donald, 1962). The offspring of a female that has chosen to mate with a long-tailed male passes to her offspring both the genes for longer tails and the genes for preference for longer tails. As a consequence, when a long-tailed male has mating success, it perpetuates both the genes for longer tails and genes for long-tail preference. The results of ever-increasing benefits for both longer tails and preferences for longer tails are that the trait can be elaborated to extreme endpoints and that the direction that the elaboration takes—longer, wider, colored, curled, forked—is constrained only by the esoteric preferences of females and by what variations on the trait arise in the population due to mutation. Eventually, the fitness (natural selection) costs in the form of reduced survival and fecundity from having a burdensome ornament outweigh the benefits of sexy sons and the runaway process is halted.

For about a 10-year period from the early 1980s to the early 1990s, theoretical biologists attempted to test the validity of the runaway model in quantitative genetics models that made various assumptions about the mode of inheritance of both the ornamental trait and the preference behavior, about how rare the trait and preference were at the start of the process, and about the costs that might be associated with mate choice (reviewed in Andersson, 1994; Møller, 1994). These models showed that, theoretically, the runaway sexual selection process could lead to many evolutionary outcomes including ***stable equilibria*** (Figure 5.6).

Despite many attempts over the past four decades to test its validity, the runaway model of sexual selection has proven difficult to test empirically because, like the chase-away model above, true tests require assessments during the process of trait elaboration. Assessment of the end product of a purported runaway process does not allow discrimination from other possible evolutionary models. From the late 1970s into the early 1990s, the runaway model of sexual selection was the model of sexual selection favored by theoreticians, and it was widely presented as a definitive explanation for the evolution of ornamental traits (Andersson 1986, p. 253). Through the 1990s and into the new millennium, however, interest shifted to models of adaptive mate choice (see below). With little success at testing the runaway model of sexual selection, interest in the process declined, and in current evolution and behavior literature, relatively little attention is paid to the runaway model of sexual selection. Nevertheless, the runaway model of sexual selection remains the most likely model to account for extreme forms of ornamental traits (Prum, 2010).

Models of Sexual Selection Based on Adaptive Female Choice for Ornaments

When evolutionary and behavioral biologists discuss sexual selection models based on ***adaptive mate choice***, they are discussing hypotheses under which ornament expression is associated with characteristics of perspective mates that are beneficial outside of a mate-selection context. By choosing highly ornamented males via adaptive mate choice, females receive benefits

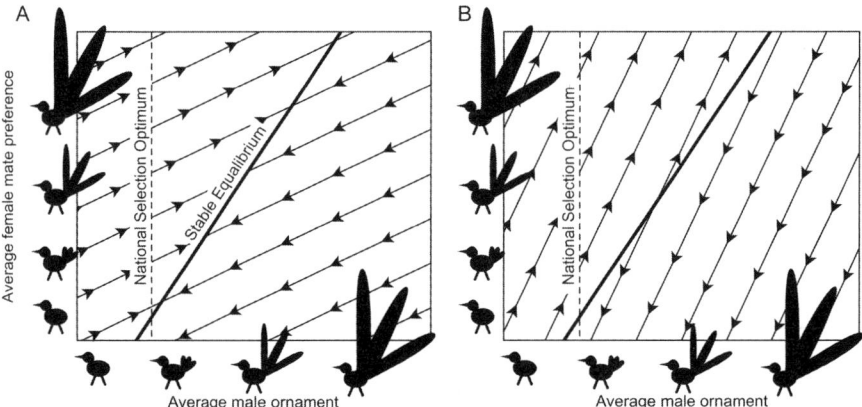

Figure 5.6. Two possible scenarios for runaway sexual selection according to a model created by Lande (1981) and illustrated by Arnold (1983). Plotted on both graphs are the mean ornament expression by males in a population (*x* axis) and mean ornament expression preferred by females (*y* axis). The vertical line near the y axis indicates the trait expression that is optimum under natural selection. The bold line slanting through the figure is the line of equilibrium for expression and preference, which is a function of the intensity of natural versus sexual selection. Once a point on this line of equilibrium is reached, male expression and female preference can remain in a stable state indefinitely. The directional lines running toward or away from the line of equilibrium are evolutionary trajectories predicted for expression and preference depending on the starting, nonequilibrium conditions. The trajectory of the evolving population is a function of the genetic covariance between female trait preference and male trait expression (see Arnold, 1983) (a) When the genetic covariance is relatively weak, then ornamentation reaches a stable expression with moderate elaboration. (b) When the genetic covariance exceeds a certain magnitude, then ornamentation evolves away from the line of equilibrium toward extreme elaboration; in other words, a runaway process ensues. See Arnold (1983) for a detailed description.

in the form of more resources for themselves or their offspring or genes that enhance the fitness of offspring. Females might also avoid costs associated with mating, such as sexually transmitted disease. Models of adaptive mate choice are collectively called ***indicator models*** of sexual selection because the ornamental trait is proposed to indicate important qualities of a perspective mate.

MAINTAINING SIGNAL HONESTY

A central topic related to adaptive mate choice is signal honesty. Indicator models of sexual selection propose that larger, more elaborate, and showier ornaments are associated with high-quality males. The obvious questions related to such a concept is *what stops cheating—what prevents average or low-quality males from sporting big ornaments*? To address the question of signal honesty, in 1975 Amotz Zahavi proposed the **handicap principle** (Zahavi, 1975). Zahavi hypothesized that ornamental traits served as handicaps to survival such that only robust, strong, and healthy individuals could bear the burden of a big ornament. Viewing ornaments as handicaps solved the problem of signal honesty—there is no way to dishonestly bear the burden of a handicapping ornament (Zahavi, 1977). Unfortunately, the handicap model created a new theoretical problem even as it solved the first. Under the handicap model, females who chose males with a big ornament received the benefit of pairing with robust and fit males, but they suffered the cost of passing the handicap to their offspring. Theoretical models indicated that such handicapping traits would not evolve (Maynard Smith, 1976; Bell, 1978; see Pomiankowski & Iwasa, 1998, for an overview) even if they were associated with male quality.

In the decades since the first pioneering papers of Zahavi, the idea of costs associated with ornamentation has become a key concept in behavioral ecology. From the handicap model arose the **condition-dependent model** of ornaments (Kodric-Brown & Brown, 1984). This new thinking shifted the burden of ornamentation away from survival costs onto production costs. Ornaments were hypothesized to be honest signals of individual condition because there was no way for an individual to meet the challenges of trait production without being in good condition (reviewed in Andersson, 1994; Møller, 1994). Once produced, ornaments need pose no survival handicap. Condition has typically been defined as the pool of resources available to the organism such that resources can be allocated either to ornament production or to body maintenance (Rowe & Houle, 1996). Only individuals with large resource pools—and hence high condition—could produce large ornaments. Mathematical models indicated that this form of honest signaling could evolve (Grafen, 1990).

In discussions of allocation trade-offs, energy is generally presented as the limiting resource that must be parsed between ornamentation and body maintenance (Wedekind & Folstad, 1994), but allocation trade-offs could involve molecules such as carotenoid pigments (Figure 5.7) (Lozano, 1994), dietary alkaloid molecules (Iyengar et al., 2001), or any other element needed for both body maintenance and ornamentation. Defining signal honesty in terms of allocation trade-offs, however, discounts a role for genetic contributions to condition. Moreover, several empirical studies have shown that even when resources were not limiting, individuals varied in ornament expression (Bortolotti et al., 1996). To accommodate these problems, condition was

Figure 5.7. The bright yellow coloration of the fleshy ornaments of this wattled lapwing (Vanellus senegallus) result from carotenoid pigments. Such carotenoid coloration may reveal the functionality of vital cellular processes in the displaying male.

defined as the efficiency of vital cellular processes (Hill, 2011), allowing for phenotype and genotype (as well as epigenetic state) to affect individual condition. With ornamentation linked to challenges of trait production, cheating is not possible—only individuals with well-functioning cellular systems can produce elaborate ornaments (Hill, 2011).

THE LEK PARADOX

Whether expression of an ornament is hypothesized to be associated with resource benefits or to be a signal of good genes is extremely important to models of sexual selection. Indicator models founded on direct resource benefits are relatively straightforward: if males with larger ornaments provide more resources, then females who choose such males directly benefit (Hoelzer, 1989). By itself, this model, sometimes called the ***good parent model***, actually provides no mechanism for the elaboration of ornaments (discussed below), but it can explain how such traits function and are maintained. Because males provide no parental care and no resources to females in most animal species, however, resource-based models cannot be universal explanations for ornamental traits. In particular, in many animals with highly conspicuous and elaborate ornaments—which are the traits that the theory of sexual selection was originally developed to explain—males provide nothing to females other than the contents of their ejaculate. In such species, the only explanation for ornamentation founded on adaptive mate choice is that females receive advantageous alleles for their offspring by choosing to mate with highly ornamented males. This ***good genes hypothesis*** requires that males vary in genetic quality and that ornamentation is an honest indicator of these good genes (Williams, 1966).

Ornamentation as a signal of good genes has been an intriguing idea since it was first proposed in the mid-twentieth century. Unfortunately, it is theoretically impossible for a population to maintain standing genetic variation associated with fitness. According to a basic tenet of evolutionary theory, called ***Fisher's fundamental theorem***, at equilibrium the ***heritability*** of fitness will be zero (Fisher, 1930). Stated another way, any alleles that improve fitness will increase in frequency, reaching fixation in just a few generations. For decades, theoreticians have been left with a quandary—the only benefit for female mate choice when males provide no resources is good genes, but good genes are theoretically impossible. This situation is termed the ***lek paradox*** (Kirkpatrick & Ryan, 1991), and it continues to confound good genes explanations of ornament evolution.

The lek paradox remains unresolved but is a rich area of theoretical research in evolutionary biology. One way around the problem is to abandon the concept of adaptive mate choice in cases where males provide no resources—arbitrary mate choice poses no paradox. Alternatively, theoreticians have reconsidered the assumption that most populations are at equilibrium. It has been proposed that mutation can introduce sufficient genetic variation to maintain selection for ornaments that signal good genes (Rice, 1988), but this hypothesis has not been widely supported. As a resolution to the lek paradox, most theoreticians propose that environments are inherently unstable so that equilibrium is never reached (Charlesworth, 1987). In particular, some evolutionary biologists noted that essentially all animals exist in a perpetual evolutionary arms race with their parasites. Hosts constantly evolve novel mechanisms to suppress or defeat parasites, and parasites evolve mechanisms to evade host resistance. Hamilton and Zuk (1982) proposed that genes for parasite resistance will cycle in frequency in response to changes in parasite abundance, which provides perpetual variation in genotype quality, that is, perpetual good genes to be associated with ornamentation (Hamilton & Zuk, 1982). Many studies have shown that parasites affect the expression of ornamental traits, but to date, no study has linked specific genes for parasite resistance to expression of an ornamental trait.

The ***genic capture model*** (Rowe & Houle, 1996), which is the most widely cited resolution to the lek paradox, builds from the observation that in most organisms there is a large amount of genetic variation associated with individual condition and proposes that when ornaments are elaborated, for example through a runaway sexual selection process, they inevitably become associated with condition (i.e., the ornaments become condition dependent). Hence, ornaments become associated with good genes because as they become challenging to produce they become linked to condition, which has substantial genetic variation (Rowe & Houle, 1996). Ornamental traits essentially capture the genetic variation that is an inherent part of individual condition. The genic capture model relies on other models for trait elaboration and only seeks to explain how ornamentation can be associated with good genes.

To date, despite great interest in the idea that ornamental traits might signal the genetic quality of males, no definitive tests of this hypothesis have been conducted. With recent breakthroughs in understanding the genetic architecture of animals, true definitive tests of the good gene hypothesis should soon be possible (Chenoweth & McGuigan, 2010).

A MECHANISM FOR EVOLUTION VIA ADAPTIVE MATE CHOICE

Sexual selection hypotheses that invoke adaptive female choice are generally explanations for the function of ornamental traits or for the evolution of an ornament to a rudimentary level—the degree of elaboration that can be achieved in a single mutational step. Most indicator models provide no mechanisms for the elaboration of ornamental traits from rudimentary states to large and complex structures. All models of ornament elaboration proposed to date rely on a ***feedback mechanism*** through which the ornament and preference for the ornament coevolve (Figure 5.5). Such feedback-based mechanisms are proposed by two models of arbitrary mate choice: the runaway model and the chase-away model. The only model for trait elaboration via adaptive mate choice that invokes a feedback mechanism was not given a name when it was proposed (Hill, 1994), but I will refer to it as the ***ratchet model*** because ornament elaboration is proposed to advance in a ratchet-like manner (Figure 5.5).

The ratchet model begins with a simple trait that is an honest signal of individual quality. Females benefit if they show a preference for the trait because they receive either resources or genetic benefits. Potentially, such an ornament can persist indefinitely as a simple indicator of quality, but there will be selection on males to evolve mechanisms to display the trait independent of their quality. If novel genes arise for such shortcut means to cheaply produce the ornament they will spread rapidly, and the information content of the trait will diminish. Female preference will no longer be beneficial to females. In response to dishonest signaling by males, there will be selection on females to shift preferences to ornamentation that restores the association with male quality—for instance, females will be selected to respond to larger, louder, or more complex ornaments. The coevolutionary cycle that ensues—with males evolving means to produce ornaments without invoking costs and with females selecting for more elaborate and costly forms of ornaments—can result in large, costly, and elaborate traits (Figure 5.5).

The ratchet model is similar to the chase-away model (Holland & Rice, 1998) in that both models focus on a conflict of interest between males and females as the impetus that drives coevolutionary cycles through which ornaments are elaborated (Figure 5.5). In the chase-away model, females are trapped by stimulus-response systems that cause them to make maladaptive choices when mating. The result is ornamentation that is effective at coercing females but that reveals nothing useful about a male. In the ratchet model, in contrast, trait elaboration occurs in response to the demand by females for honest signaling in males. Specific contrasting tests of these two models have not been undertaken, but the fundamental predictions of these models—a loss of fitness through choice for male ornamentation under the chase-away and a gain in fitness through such choice in the ratchet model—should make the hypotheses distinguishable.

INTRASEXUAL SELECTION AS THE FOUNDATION OF INTERSEXUAL SELECTION

Anders Berglund and colleagues (Berglund et al., 1996) proposed that ornamental traits commonly evolve as signals of status and fighting ability either through sexual or social selection. Once ornamental traits exist as signals of status linked to individual condition, they can become the target of female mate choice and begin to function in intersexual selection. As with many other models of sexual selection, this idea can explain the current function of ornamental traits but it does not provide a mechanism for trait elaboration.

SUMMARY

The armaments and ornaments have fascinated biologists since they captured the attention of the original evolutionist, Charles Darwin. It is widely accepted that armaments are the product of aggressive contests for mates, and this hypothesis has not been modified significantly since first proposed by Darwin. The evolution of ornamental traits, however, remains more controversial. After a century and a half of work on the topic, a rich and diverse set of theories has been developed to explain how ornamental traits function and how they evolved. The original hypothesis for the elaboration of traits via sexual selection—the runaway model—remains the most likely explanation for many of the most gaudy and dramatic ornaments. The idea that began as Zahavi's handicap principle has been modified into a more general indicator hypothesis that ornamental traits are linked to the health and vitality of individuals. This indicator model has gained wide support through empirical studies that find links between measures of individual quality and ornamentation. The specific idea from Zahavi that ornaments are handicaps to survival has been generally disproven, and ornaments are now viewed more typically as indicators of condition whose honesty is maintained by challenges posed by trait production. Despite the success of the indicator model in explaining the present function of many ornamental traits, and despite 100 years of thinking about the process, we still lack a well-supported mechanism for how grand and elaborate ornaments evolve.

ACKNOWLEDGMENTS

Over the past two decades the Department of Biological Sciences and College of Science and Mathematics at Auburn University have provided time and resources for me to pursue my studies of sexual selection.

REFERENCES AND SUGGESTED READING

Andersson, M. (1986). Sexual selection and the importance of viability differences: A reply. *Journal of Theoretical Biology*, 120, 251–254.

Andersson, M. (1994). *Sexual Selection*. Princeton, NJ: Princeton University Press.
Arnold, S. J. (1983). Sexual selection: The interface of theory and empiricism. In P. P. G. Bateson (ed.), *Mate Choice* (pp. 67–107). Cambridge, UK: Cambridge University Press.
Basolo, A. L. (1990a). Female preference for male sword length in the green swordtail (*Xiphophorus helleri*). *Animal Behaviour*, 40, 332–338.
Basolo, A. L. (1990b). Female preference predates the evolution of the sword in swordtail fish. *Science*, 250, 808–810.
Bell, G. (1978). The handicap principle in sexual selection. *Evolution*, 32, 872–885.
Berglund, A., A. Bisazza, & A. Pilastro. (1996). Armaments and ornaments: An evolutionary explanation of traits of dual utility. *Biological Journal of the Linnean Society*, 58, 385–399.
Bortolotti, G., J. J. Negro, J. L. Tella, T. A. Marchant, & D. M. Bird. (1996). Sexual dichromatism in birds independent of diet, parasites and androgens. *Proceedings of the Royal Society of London, B*, 263, 1171–1176.
Chapman, T., L. F. Liddle, J. M. Kalb, M. F. Wolfner, & L. Partridge. (1995). Cost of mating in *Drosophila melanogaster* females is mediated by male accessory-gland products. *Nature*, 373, 241–244.
Charlesworth, B. (1987). The heritability of fitness. In J. W. Bradbury & M. B. Andersson (eds.), *Sexual Selection: Testing the Alternatives* (pp. 21–40). London: John Wiley and Sons.
Chenoweth, S. F. & K. McGuigan. (2010). The genetic basis of sexually selected variation. *Annual Review of Ecology, Evolution, and Systematics*, 81–101.
Cronin, H. (1991). *The Ant and the Peacock*. Cambridge, UK: Cambridge University Press.
Darwin, C. (1859). *On the Origin of Species by Natural Selection or the Preservation of Favoured Races in the Struggle for Life*. London: John Murray.
Darwin, C. (1871). *The Descent of Man and Selection in Relation to Sex*. London: Murray.
Davison, G. W. H. (1985). Avian spurs. *Journal of Zoology*, 206, 353–366.
Egger, B., Y. Klaefiger, A. Theis, & W. Salzburger. (2011). A sensory bias has triggered the evolution of egg-spots in cichlid fishes. *PLoS ONE*, 6, e25601. doi:10.1371/journal.pone.0025601.
Emlen, S. T. & L. W. Oring. (1977). Ecology, sexual selection, and the evolution of mating systems. *Science*, 197, 215–223.
Fisher, R. A. (1915). The evolution of sexual preference. *Eugenics Review*, 7, 184–192.
Fisher, R. A. (1930). *The Genetical Theory of Natural Selection*. Oxford, UK: Clarendon Press.
Fisher, R. A. (1958). *The Genetical Theory of Natural Selection*. New York: Dover.
Grafen, A. (1990). Sexual selection unhandicapped by the Fisher process. *Journal of Theoretical Biology*, 144, 473–516.

Griffith, S. C. & S. R. Pryke. (2006). Benefits to females of assessing color displays. In G. E. Hill & K. J. McGraw (eds.), *Bird Coloration*, Vol. 2: *Function and Evolution*. Cambridge, MA: Harvard University Press.

Hamilton, W. D. & M. Zuk. (1982). Heritable true fitness and bright birds: a role for parasites? *Science*, 218, 384–386.

Hill, G. E. (1994). Trait elaboration via adaptive mate choice: Sexual conflict in the evolution of signals of male quality. *Ethology, Ecology and Evolution*, 6, 351–370.

Hill, G. E. (2002). *A Red Bird in a Brown Bag: The Function and Evolution of Ornamental Plumage Coloration in the House Finch*. New York: Oxford University Press.

Hill, G. E. (2006). Female choice for ornamental coloration. In G. E. Hill & K. J. McGraw (eds.), *Bird Coloration*, Vol. 2: *Function and Evolution*. Cambridge, MA: Harvard University Press.

Hill, G. E. (2010). *National Goegraphic Bird Coloration*. Washington, DC: National Geographic Society.

Hill, G. E. (2011). Condition-dependent traits as signals of the functionality of vital cellular processes. *Ecology Letters*, 14, 625–634.

Hill, G. E., R. Montgomerie, C. Roeder, & P. Boag. (1994). Sexual selection and cuckoldry in a monogamous songbird: Implications for sexual selection theory. *Behavioral Ecology and Sociobiology*, 35, 193–199.

Hoelzer, G. A. (1989). The good parent process of sexual selection. *Animal Behaviour*, 38, 1067–1078.

Holland, B. & W. R. Rice. (1998). Perspective: Chase-away sexual selection: Antagonistic seduction versus resistance. *Evolution*, 52, 1–7.

Iyengar, V. K., C. Rossini, & T. Eisner. (2001). Precopulatory assessment of male quality in an arctiid moth (*Utetheisa ornatrix*): Hydroxydanaidal is the only criterion of choice. *Behavioral Ecology and Sociobiology*, 49, 283–288.

Kirkpatrick, M. (1982). Sexual selection and the evolution of female choice. *Evolution*, 36, 1–12.

Kirkpatrick, M. & M. J. Ryan. (1991). The evolution of mating preferences and the paradox of the lek. *Nature*, 350, 33–38.

Kodric-Brown, A. & J. H. Brown. (1984). Truth in advertising: the kinds of traits favored by sexual selection. *American Naturalist*, 124, 309–323.

Lachmann, M., S. Szamado, & C. T. Bergstrom. (2001). Cost and conflict in animal signals and human language. *Proceedings of the National Academy of Sciences, USA*, 98, 13189–13194.

Lande, R. (1981). Models of speciation by sexual selection on polygenic traits. *Proceedings of the National Academy of Sciences, USA*, 78, 3721–3725.

Ligon, D. J. (1999). *The Evolution of Avian Breeding Systems*. Oxford, UK: Oxford University Press.

Ligon, J. D., R. Thornhill, M. Zuk, & K. Johnson. (1990). Male-male competition, ornamentation, and the role of testosterone in sexual selection in red jungle fowl. *Animal Behaviour*, 40, 367–373.

Ligon, J. D. & P. W. Zwartjes. (1995). Ornate plumage of male red junglefowl does not influence mate choice by females. *Animal Behaviour*, 49, 117–125.

Liu, M., L. Siefferman, & G. E. Hill. (2007). An experimental test of female choice relative to male structural coloration in eastern bluebirds. *Behavioral Ecology and Sociobiology*, 61, 623–630.

Lozano, G. A. (1994). Carotenoids, parasites, and sexual selection. *Oikos*, 70, 309–311.

Maynard Smith, J. (1976). Sexual selection and the handicap principle. *Journal of Theoretical Biology*, 57, 239–242.

Maynard Smith, J. (1991). Theories of sexual selection. *Trends in Ecology and Evolution*, 6, 146–151.

McClintock, W. J. & G. W. Uetz. (1996). Female choice and pre-existing bias: Visual cues during courtship in two *Schizocosa* wolf spiders (Araneae: Lycosidae). *Animal Behaviour*, 52, 167–181.

Møller, A. P. (1994). *Sexual Selection and the Barn Swallow*. Oxford, UK: Oxford University Press.

O'Donald, P. (1962). The theory of sexual selection. *Heredity*, 17, 541–552.

Petrie, M., T. Halliday, & C. Sanders. (1991). Peahens prefer peacocks with elaborate trains. *Animal Behaviour*, 41, 323–331.

Pomiankowski, A. & Y. Iwasa. (1998). Handicap signaling: Loud and true? *Evolution*, 52, 928–932.

Prum, R. O. (1997). Phylogenetic tests of alternative intersexual selection mechanisms: Trait macroevolution in a polygynous clade (Aves: Pipridae). *American Naturalist*, 149, 668–692.

Prum, R. O. (2010). The Lande-Kirkpatrick mechanism is the null model of evolution by intersexual selection: Implications for meaning, honesty, and design in intersexual signals. *Evolution*, 64, 3085–3100.

Pryke, S. R. & S. Andersson. (2003). Carotenoid-based status signalling in red-shouldered widowbirds (*Euplectes axillaris*): Epaulet size and redness affect captive and territorial competition. *Behavioral Ecology and Sociobiology*, 53, 393–401.

Rice, W. R. (1988). Heritable variation in fitness as a prerequisite for adaptive female choice—the effect of mutation-selection balance. *Evolution*, 42, 817–820.

Rice, W. R. (1996). Sexually antagonistic male adaptation triggered by experimental arrest of female evolution. *Nature*, 381, 232–234.

Ripoll, J., J. Saldana, & J. C. Senar. (2004). Evolutionarily stable transition rates in a stage-structured model: An application to the analysis of size distributions of badges of social status. *Mathematical Biosciences*, 190, 145–181.

Rodd, F. H., K. A. Hughes, G. F. Grether, & C. T. Baril. (2002). A possible non-sexual origin of mate preference: Are male guppies mimicking fruit? *Proceedings of the Royal Society of London, B*, 269, 475–481.

Rohwer, S. (1975). The social significance of avian winter plumage variability. *Evolution*, 29, 593–610.

Rohwer, S. (1977). Status signaling in Harris' sparrows: Some experiments in deception. *Behaviour*, 61, 107–128.

Rowe, L. & D. Houle. (1996). The lek paradox and the capture of genetic variance by condition dependent traits. *Proceedings of the Royal Society of London, B*, 263, 1415–1421.

Ryan, M. J., J. H. Fox, W. Wilczynski, & A. S. Rand. (1990). Sexual selection for sensory exploitation in the frog *Physalaemus pustulosus*. *Nature*, 343, 66–67.

Ryan, M. J. & A. Keddy-Hector. (1992). Directional patterns of female mate choice and the role of sensory biases. *American Naturalist*, 139, S4–S35.

Saetre, G. P., T. Moum, S. Bures, M. Kral, M. Adamjan, & J. Moreno. (1997). A sexually selected character displacement reinforces premating isolation. *Nature*, 387, 589–592.

Siefferman, L. & G. E. Hill. (2005). UV-blue structural coloration and competition for nestboxes in male eastern bluebirds. *Animal Behaviour*, 69, 67–72.

Skogsmyr, I. & A. Lankinen.(2002). Sexual selection: An evolutionary force in plants. *Biological Reviews*, 77, 537–562.

Stutchbury, B. J. & E. S. Morton. (1995). The effect of breeding synchrony on extra-pair mating systems in songbirds. *Behaviour* 132, 675–690.

Trivers, R. L. (1972). Parental investment and sexual selection. In B. Campbell (ed.), *Sexual Selection and the Descent of Man, 1871–1971* (pp. 136–179). Chicago: Aldine.

von Schantz, T., G. Goransson, G. Andersson, I. Froberg, M. Grahn, A. Helgee, et al. (1989). Female choice selects for a viability-based trait in pheasants. *Nature*, 337, 166–169.

Wallace, A. R. (1889). *Darwinism*. London: Macmillian.

Wedekind, C. & I. Folstad. (1994). Adaptive or nonadaptive immunosuppression by sex hormones? *American Naturalist*, 143, 936–938.

West-Eberhard, M. J. (1979). Sexual selection, social competition, and evolution. *Proceedings of the American Philosophical Society*, 123, 222–234.

Westneat, D. F. & I. R. K. Stewart. (2003). Extra-pair paternity in birds: Causes, correlates, and conflict. *Annual Review of Ecology and Evolution*, 34, 365–396.

Williams, G. C. (1966). Natural selection, the costs of reproduction, and a refinement of Lack's principle. *American Naturalist*, 100, 687–690.

Zahavi, A. (1975). Mate selection—a selection for a handicap. *Journal of Theoretical Biology*. 53, 205–214.

Zahavi, A. (1977). The cost of honesty (further remarks on the handicap principle). *Journal of Theoretical Biology*, 67, 603–605.

6

Sexual Conflict: All's Not Fair in Love—It's War!

Zenobia Lewis

INTRODUCTION

The male lion (*Panthera leo*) is feared by all. But it is the female lion that should perhaps be most cautious. The penis of the male lion is covered in tiny barbs, made of the hard protein keratin, usually found in claws and nails. These barbs tear the lining of the uterus during copulation, which is painful for the female, but which allows the male to remove any previously inseminated rival males' sperm and thereby minimize sperm competition. Stimulation by the barbs also sends signals to the female's brain, which help induce the release of eggs. Despite the unpleasant nature of copulation in this species, during their short fertile period each month female lions will mate up to 100 times a day. It is not only female lions that are subjected to this somewhat harsh treatment; across diverse animal species males exhibit spiky, barbed penises and other adaptations that result in the males harming their female mates over or during reproduction.

Until the 1970s reproduction was viewed as a venture between males and females, a view derived from overestimation of the occurrence of **lifelong monogamy** (see Table 6.1 for definitions of the types of mating systems found in the animal world) in the wild and maintained by outdated Victorian values with regards to sex (Birkhead, 2010). We now know differently—males and females often have conflicting interests when it comes to reproduction, and

Table 6.1. Types of animal mating systems.

Type	Description
Monogamy	Male and female within a pair mate only once.
Monandry	Female typically mates with only one male.
Polyandry	Female typically mates with multiple males.
Monogyny	Male typically mates with only one female.
Polynygy	Male typically mates with multiple females.
Promiscuity	Both males and females mate with multiple partners.

these conflicts of interest can have profound evolutionary effects on both sexes. It was Geoff Parker (1979, p. 124) who first coined the term ***sexual conflict***, defining it as the "the conflict in evolutionary interests between males and females." He noted that sexual conflict is rooted in the differential investment in reproduction by the two sexes, as described in Chapter 5. Today, two forms of sexual conflict are recognized: ***intralocus sexual conflict*** is a form of conflict occurring when the same genes are expressed in both males and females but are selected in opposite directions. A classic example is that of the human hip; it has been suggested that in our evolutionary past, there was strong selection for wide hips in women, due to the risk of injury during childbirth. However, opposing selection in males favoured smaller hips due to improved locomotor function. Thus, there may have been an ***evolutionary tug-of-war*** between males and females over the optimal hip width (Rice & Chippindale, 2001). Intralocus sexual conflict is thought to be a powerful and pervasive evolutionary process, yet to date surprisingly few concrete empirical examples exist (Bonduriansky & Chenoweth, 2009; Lewis et al., 2011). In this chapter we will focus on the second form of sexual conflict, ***interlocus sexual conflict***. This is where the conflict between males and females is mediated by different genes and traits.

In many cases, sexual conflict occurs because both males and females in the majority of species mate with more than one individual. If there is true monogamy between a male and female, the reproductive interests of the pair—mate, reproduce, and care for offspring where appropriate—are perfectly aligned, and there is no sexual conflict (Rice, 2000). However, where one or both of the pair remate with one or more other individuals, there is the potential for sexual conflict. For example, if a female has mated with a poor-quality mate the first time around, it would benefit her to "trade up" should a better-quality mate arrive on the scene (reviewed in Andersson, 1994; Jennions & Petrie, 2000); this would be detrimental to the first male in a number of ways. If the female mates with both males within the same reproductive cycle, this would likely lead to competition between the sperm of the two males within the female reproductive tract and could therefore decrease the number of offspring sired by the first male. Alternatively, in species that exhibit joint parental care, the first male may end up providing costly paternal care to offspring that he has not sired. Thus in many species where females mate with multiple males, males have evolved traits such as ***mate guarding*** (reviewed in, for example, Alcock, 1994), and ***plugs*** that block the female reproductive tract (e.g., Orr & Rutowski, 1991; Shine, Olsson, et al., 2000) to deny access to other males until after the female has laid her eggs or given birth. This is an example of sexual conflict over mating rate. However, conflict between the male and female can, and does, occur over any aspect of reproduction, from courtship, through mating and fertilization, to parental care of offspring (see Figure 6.1).

A UNIVERSAL TRUTH?

Sexual conflict seems to be universal in nature, with examples seen across all animal groups. In social mammals such as lions and gorillas, where the females care for their offspring for long periods of time, the females are often rendered nonreceptive to further reproduction until their current offspring become independent or die. In such species, new males will often kill the offspring, thereby bringing the female into estrus sooner (reviewed in Hrdy, 1979). ***Infanticide*** is a rather extreme example of premating sexual conflict; another example is seen in red-sided garter snakes (*Thamnophis sirtalis parietalis*), where it is the females themselves that are under threat during the premating period. This species typically exhibits populations with highly male-biased sex ratios, resulting in strong competition between males for mates (Shine, O'Connor, et al., 2000). On emerging from their communal overwintering dens, the snakes form mass, writhing, mating aggregations (see Figure 6.2), containing hundreds of males and very few females. The consequences for the females include forced copulation, exhaustion, physiological stress, and in some cases death (Shine, O'Connor, et al., 2000; Shine et al., 2004). Perhaps unsurprisingly, even humans are thought to exhibit sexual conflict over, for example, family size, fidelity within relationships, and provisioning to offspring (reviewed in Mulder & Rauch, 2009).

Recent studies have begun examining what appears to be a fascinating example of sexual conflict over fertilization in waterfowl. It was first documented a century ago that "rape" (***forced copulation***) is extremely common in ducks, with the males attempting to coerce the females into mating so

1. **Pre-mating and courtship**
   - Expression of sexual characters
   - Perception of courtship stimuli
   - Courtship gifts
   - Body size
   - Sexual cannibalism
   - Infanticide
   - Pre-copula duration

2. **Mating**
   - Identity of partner
   - Inbreeding / incest avoidance
   - Mating frequency
   - Duration of mating
   - Duration of mate guarding
   - Number of mates

3. **Fertilization**
   - Sperm use
   - Sperm selection
   - Rate of fertilization / avoidance of polyspermy
   - Sex allocation
   - Allocation to male or female function in hermaphrodites
   - Sperm digestion

4. **Parental investment in gametes**
   - Clutch size
   - Hatching date

5. **Parental investment per offspring**
   - Care per offspring
   - Probability of desertion
   - Provisioning via genomic imprinting

Figure 6.1. Sexual conflict can occur over any aspect of, and during any episode of, the reproductive process. (Adapted from Chapman, 2006, with permission from Elsevier)

violently they will on occasion drown them (Huxley, 1912). Ducks are unusual for birds in that the males have penises; even more remarkable is the structure and functioning of the duck phallus. The duck penis is a highly complex, coiled structure, resembling a corkscrew (Brennan et al., 2010). At rest within the male, the penis is essentially inside out. Prior to copulation, the

Figure 6.2. A red-sided garter snake (*Thamnophis sirtalis parietalis*) mating aggregation; hundreds of male compete for access to only a few females. (Photo by Jon Webb)

male does not develop an erection; instead, as copulation commences, the penis quickly fills with fluid and is everted explosively into the female reproductive tract. Penis eversion, followed by ejaculation and reversion back into the male, all take place in a matter of seconds (Brennan et al., 2010). The structure and functioning of the duck penis is likely to facilitate forced copulations in this group; however, it seems females have evolved to counteract this. The female reproductive tract is also highly complex, with many blind-ending pouches and a coiled structure; however, interestingly, it coils in the opposite direction to this male penis, making it somewhat challenging for the male to inseminate her (Brennan et al., 2007). Although as yet untested, Birkhead (2010) has suggested that during forced copulations, females tighten their reproductive tract, making intromission more difficult and directing the phallus into one of the blind-ended pouches; however, when mating with desired mates, the female relaxes her tract, permitting intromission to occur.

The insects exhibit some of the most interesting examples of conflict over reproduction. Perhaps the most famous is seen in the mantids, where the female eats the male during copulation and thereby derives nutrients, which increase her fecundity. Lelito and Brown (2006) have shown that the male praying mantis *Tenodera aridifolia sinensis* has adapted to recognize the level of potential danger presented by the female and adopts risk-averse behavior as appropriate. Female water striders, or "pond skaters" as they are more commonly known, struggle during copulation and attempt to kick the mounted male from their back; the males have responded in evolutionary terms by evolving spines and claspers on their legs, which assist them in gripping the female (Arnqvist & Rowe, 2002). Closely related to the water striders, the Zeus bugs—tiny water bugs found in New Guinea and tropical Australia—are relatively new to science, having only been discovered in the past 10 years (Polhemus & Polhemus, 2000; Andersen & Weir, 2001). The group exhibits a remarkable mating system whereby the male, diminutive in size, remains almost permanently attached to the back of the female, feeding from a pair of glands on her back that produce a wax-like secretion (Arnqvist et al., 2007). It has been suggested that the male behavior is an extreme form of mate guarding; the adult sex ratio in wild populations is strongly male-biased, and thus single males have little opportunity to find unoccupied females. Rather than risk losing their mate by dismounting to forage for food, males stay put and feed from the female glands. *Why does the female permit this*? Males of this group take further liberties with the females, stealing prey items when they are feeding and apparently even wounding them, evidenced by the presence of melanized scarring on the backs of most of the females. When blocking the female glands experimentally, Arnqvist and colleagues (2006) observed that males increased the extent to which they steal prey items from the female, suggesting that the female glands have evolved to minimize the costly behaviors exhibited by the males.

One of the more bizarre examples of sexual conflict in insects is exhibited by the bed bugs (see Figure 6.3, reviewed in Siva-Jothy, 2006). Bed bugs engage in traumatic insemination; rather than employing more traditional routes, males inject sperm directly into the female body cavity through the abdominal wall via their syringe-like penises. Unsurprisingly, this is costly for the female; eventually, after multiple matings, it results in death. However, to combat this, females have evolved a specialized organ, the ***ectospermalege***. This structure acts essentially as a mating guide for the male, directing his penis to a specific region of the female abdomen containing cells involved in the immune response, which help minimize the long-term damage to the female (Reinhardt et al., 2003). From the ectospermalege, the male's sperm migrates to the ovaries to inseminate the female's eggs. It seems the male submits to this female control over the site of insemination as even low rates of piercing outside of the ectospermalege can reduce female offspring production

Figure 6.3. Bedbug males engage in traumatic insemination, where they ejaculate their sperm directly into the female body cavity, one of the most bizarre mating behaviors in the animal world. (Photo by Richard Naylor)

by 50 percent, thus proving highly costly for both the female and the male (Morrow & Arnqvist, 2003).

THE SEXUAL ARMS RACE

How does sexual conflict operate? Imagine a mutation arises in a male, which results in an ***adaptation*** that causes harm or generates a cost to his female mate but is beneficial to him. This would result in selection acting on the female to evolve a ***counteradaptation*** that reduces or prevents the male-induced harm. In turn, selection then acts on the male to evolve another adaptation that overcomes or reduces the action of the female counteradaptation, and so on (Rice, 2000). This results in continual, cyclical coevolution between the male and female, known as ***sexually antagonistic coevolution (SAC)***. The ensuing "arms race" is akin to the host-parasite relationship, whereby the host and parasite are locked in a continual battle to gain one-upmanship over the other.

Chapter 5 described a number of mechanisms by which male traits can evolve under ***sexual selection*** in the absence of direct benefits to females such as food and parental care. Sexual conflict is inherently linked to sexual selection; females may mate with some males but not others as a result of sexual conflict, thereby leading to variation in reproductive success among males and contributing to the evolution and maintenance of exaggerated male traits. Thus sexual conflict is often invoked as a further mechanism by which sexual selection can operate. The difference with traditional models of sexual selection is that these tend to focus on direct and indirect benefits of mating preference to the female; under sexual selection via sexual conflict, female choice is an attempt to minimize the costs of mating rather than maximize benefits (Rice & Holland, 1997; Holland & Rice, 1998). The relative importance of sexual conflict as an instrument of sexual selection, compared to more traditional mechanisms, is still hotly debated (e.g., Chapman et al., 2003; Cordero & Eberhard, 2003; Córdoba-Aguilar & Contreras-Garduño, 2003; Eberhard & Cordero, 2003; Eberhard, 2004).

For behavioral scientists, one unfortunate consequence of the manner in which sexual conflict operates, in addition to lack of consensus over terminology and manner of operation, is that it is difficult to study. As a result of the nature of the sexual arms race, past adaptations are hidden, and at any given moment it is difficult to measure the relative costs and benefits of the conflict to each sex. However, behavioral scientists have found ways to get around this. Currently one of the most fruitful methods of investigating sexual conflict, as seen in the *Drosophila melanogaster* case study below, is through the use of **experimental evolution**, where experimental treatments are imposed on replicate populations of organisms in the laboratory; the organisms are then measured for traits of interest over time to see how these traits evolve. In addition, comparative studies across species, and population crosses within species, have proven fruitful as, again, we shall see in some of the examples given below.

THE LINK WITH MATING SYSTEM

The degree of sexual conflict is often invoked as being related to the level of promiscuity within a species or population, and in the past it has been suggested that sexual conflict does not occur where pairs are monogamous (e.g., Rice, 2000). However, it is becoming increasingly recognized that this is not the case; in fact, sexual conflict can promote monogamy in polygamous species (reviewed in Hosken et al., 2009).

As noted above, in some species where females mate with multiple mates, a male will guard the female after copulation or block her reproductive tract with a mating plug, thereby protecting his reproductive success to some degree. In some cases, the female is permanently prevented from remating.

For example, in the housefly, *Musca domestica*, the male transfers compounds to the female in his ejaculate that permanently switch off her receptivity to other males (Andrés & Arnqvist, 2001). However, females will remate with another male if given the opportunity, and there is evidence that multiple mating could increase the reproductive success of **polyandrous** females through nutrients transferred to females via the seminal fluids (Arnqvist & Andrés, 2006). Thus, in this species, males enforce **monandry** in females despite the fact it may be costly to female reproductive success.

It is not only males that exert control over the remating of their partners. The burying beetles, *Nicrophorus* spp., are a fascinating group of organisms for a number of reasons, one of which is the fact that it is one of the few insect groups in which both males and females engage in parental care (reviewed in Scott, 1998). Mated pairs seek out, bury, and then defend the carcass of a small vertebrate, for example a rodent or bird. The females lay their eggs in the carcass, and one or both of the parents care for the developing offspring. If the pair manages to find a carcass large enough to support more than one brood of offspring, it would increase the reproductive success of the male to rear a brood with a second female on the same carcass. This would, however, be costly for the resident female, as her offspring would be subject to competition for the food resource. In one species, *Nicrophorus defodiens*, males that are defending a large carcass will attempt to emit pheromones to attract a second mate. However, their efforts are often thwarted by the resident female, who will push, undercut, and bite the male to minimize the time he spends signalling to other prospective mates (Eggert & Sakaluk, 1995) and thereby decrease the chance of his obtaining a second mate. Thus, in this species, sexual conflict over remating can result in females enforcing **monogyny** on males.

Sexual conflict over remating can also result in both sexes within a pair being forced to mate only once. In many spider species, males exhibit so-called one-shot genitalia, inseminating the female by breaking off parts of their genitalia containing their sperm and leaving them in the female reproductive tract (e.g., Fromhage & Schneider, 2006). This blocks the female reproductive tract, thereby preventing her from remating with other males, but also renders the male unable to remate despite the fact that he may survive the ordeal. Thus, in theory, conflict over remating in such cases could result in monogamy. However, this is not necessarily the end of the conflict; two recent studies in different spider species suggest that both the female and the male can, in evolutionary terms, eventually override the stalemate, in the case of the female by adapting to remove the plug (Kuntner et al., 2009), and in the case of the male by refraining from damaging his genitalia and instead partitioning sperm across several copulations (Schneider & Michalik, 2011).

A CASE STUDY: SEXUAL CONFLICT IN A FRUIT FLY

One of the best-understood examples of sexual conflict is over mating rate in the fruit fly model organism, *D. melanogaster* (Figure 6.4). In 1989, a team at Edinburgh University showed that for female *D. melanogaster*, mating has an unfortunate consequence—they die sooner (Fowler & Partridge, 1989). Kevin Fowler and Linda Partridge compared the adult longevity of females housed with, and allowed to mate freely with, wild-type males (the "high-mating" group), and females alternately housed with wild-type males and males that had had their genitalia cauterised (the "low-mating" group). Thus the low-mating group females were constantly exposed to males, as were the high-mating group, but they had fewer opportunities to actually mate. It was found that high-mating-group females had significantly shorter lifespans than low-mating-group females, and as they died sooner they exhibited a correlated decrease in reproductive success. But *why should males "wish" their mates to die quickly, particularly when their decreased lifespan results in their producing fewer offspring*? And *how is the conflict mediated*? In 1995 the same team, now based at University College London, took further steps towards answering these questions. They essentially repeated the experiment, but this time they exposed females to experimental, genetically modified males of two types.

Figure 6.4. The fruit fly model organism *Drosophila melanogaster* presents one of the best-understood examples of sexual conflict. (Photo by Tracey Chapman)

One group was unable to produce sperm or the major components of seminal fluid ("DTA males"), while the other produced normal seminal fluid but not sperm ("*tudor*" males) (Chapman et al., 1995). As before, cauterized males from each of the three stocks were also used to control for nonmating exposure of females to males. They found that females that mated to the *tudor* males, and thus received the full complement of seminal fluid, died sooner than females mated to the DTA males, suggesting that it was something in the seminal fluid that was reducing female lifespan.

In most species, seminal fluid is a veritable cocktail of components produced by the male accessory glands, seminal vesicles, ejaculatory duct, ejaculatory bulb, and testes and transferred to the female during mating. These so-called **seminal fluid proteins** (SFPs) have wide-ranging effects on the female, some good and some bad (reviewed in Avila et al., 2011). For example, in *D. melanogaster*, certain SFPs are essential to the female for processes such as sperm storage in, and release from, the sperm-storage organ (e.g., Neubaum & Wolfner, 1999; Tram & Wolfner, 1999; Avila et al., 2010). Of benefit to the male, SFPs also increase the oviposition or egg-laying rate of the female (e.g., Herndon & Wolfner, 1995; Heifetz et al., 2001; Ravi Ram & Wolfner, 2007), thus ensuring that she produces many eggs fertilized with his sperm. However, this is where the conflict occurs. Although the female may now have sufficient sperm to fertilize her eggs, she may not wish to fertilize all of them using the male's sperm; as we saw in Chapter 5 on sexual selection, there are numerous reasons why the female may wish to remate with one or more other males. If the female were to remate, the male's sperm would likely be subject to **sperm competition**. In *D. melanogaster*, subsequent males sire a greater proportion of a female's offspring (Gromko et al., 1984), and the first male's reproductive success would therefore decrease. However, males have weapons to counter this. The wide-ranging effects of male SFPs on females include manipulation of the female to decrease her interest in other males; after receiving SFPs, mated females will actively reject courting males (e.g., Aigaki et al., 1991).

What of the impact on female lifespan with which we began this story? To our knowledge, the SFP-induced reduction in female longevity is an unfortunate side effect of the overall actions of SFP; it is unlikely that male *D. melanogaster* "intentionally" set out to harm their mates. As stated previously, sexual conflict is predicted to result in a cyclical arms race, with males and females continuously evolving to counter the selection imposed on them by the opposite sex. Indeed, it seems female *D. melanogaster* have tricks up their proverbial sleeves to counter the effects of male SFPs. A more recent study, again by Tracey Chapman's group, utilized experimental evolution to examine female responses to male-induced harm in *D. melanogaster* (Wigby & Chapman, 2004). They created replicate populations subject to differing levels of sexual conflict by manipulating the adult sex ratios of each line; it was predicted that in populations with male-biased sex ratios, females would be subject to more frequent mating attempts from males and would therefore mate at a higher rate than females from populations with equal-sex, and, in turn, female-biased sex ratios. Thus, sexual conflict was predicted to be highest in male-biased populations, followed by equal-sex-ratio populations, followed by female-biased populations. After 31 generations of these treatments, Stuart Wigby and Chapman (2004) measured the frequencies at which females were copulating in the different populations. As predicted, females were courted and mated most frequently in male-based populations, followed by the equal-sex-ratio populations, followed by the female-biased populations. They also compared the adult longevity of females by housing them with wild-type (i.e., nonselected) males. Remarkably, females from the male-biased or high-sexual-conflict populations survived longer than females from the other populations, suggesting that after generations of being subject to high levels of sexual conflict, the females had evolved a counteradaptation to minimize the costs induced by males. The nature of this counteradaptation is as yet unknown, although it is thought to be a modification in the female receptors that are manipulated by the male SFPs.

The story does not quite end there. Another study has shown that *D. melanogaster* females seem to prefer the males that induce the most harm; it is known that females of this species prefer larger males, and Urban Friberg and Göran Arnqvist (2003) showed that larger males induce a greater longevity cost to females. It is currently unknown how the cost and body-size traits are linked, or whether females actually prefer larger males or whether the target of their preference is something linked to body size. However, this finding provides support to models of sexual selection by sexual conflict that suggest female preference can result in the evolution and maintenance of exaggerated male traits despite the fact that they may impose costs on females.

It is only more recently, with the advent of the molecular age, that we have come closer to understanding the biochemical processes underlying the sexual conflict in *D. melanogaster*. We now know that one of the chief culprits mediating male-induced harm to females is the SFP Acp70a, more commonly known as sex peptide (SP) (Wigby & Chapman, 2005). Interestingly, experimental application of SP to female *Helicoverpa armigera* moths has been found to induce similar effects to those in *D. melanogaster* (Fan et al., 2000), and a homologue of SFP—a highly genetically similar molecule—has recently been found in mosquitoes (Dottorini et al., 2007; reviewed in Avila et al., 2011).

These studies suggest that the insect female physiological pathway, which is targeted by males via SFPs, may be an ancient and highly conserved one.

CONSEQUENCES OF SEXUAL CONFLICT

Sexual conflict is thought to have the potential to drive reproductive isolation and hence *speciation* (reviewed in Ritchie, 2007). Numerous theoretical models predict that cycles of SAC between males and females can to lead to rapid evolutionary change within species and, in turn, promote the development of reproductive barriers between populations that can eventually result in speciation (e.g., Parker & Partridge, 1998; Gavrilets & Waxman, 2002; Hayashi et al., 2007). Until recently, the best evidence we had that sexual conflict could lead to speciation came from comparative studies across different species. For example, Arnqvist and colleagues (2000) compared speciation rate across pairs of groups of closely related insects that differed in their potential for sexual conflict. Their reasoning was that in insect species where females mate with many males, there should be greater potential for sexual conflict than in species in which females mated only once, and thus groups of insects with higher levels of sexual conflict should contain more species if sexual conflict can result in speciation. They found this was indeed the case; groups where females mated with many males exhibited speciation rates four times as high as in groups where females only once. Subsequent studies have produced contrasting results; for example, similar studies across butterflies, mammals, and spiders (Gage et al., 2002) and another in birds (Morrow et al., 2003) found no evidence for a relationship between female mating rate and speciosity. However, a more recent comparative analysis of all the data of speciosity versus sexual selection, and hence sexual conflict, collected to date suggests that there is a small but significant association between the two (Kraaijeveld et al., 2011).

There is some experimental evidence that sexual conflict could lead to reproductive isolation. Oliver Martin and David Hosken (2003) created replicate populations of the dung fly (*Sepsis cynipsea*) under experimental evolution where they were subject to varying levels of sexual conflict. They did this by manipulating the densities of the populations, as sexual conflict is predicted to be stronger in larger populations, where there are more frequent interactions between males and females (Gavrilets, 2000). After 35 generations of evolution, Martin and Hosken (2003) found that in the populations under potentially high levels of sexual conflict, females had evolved to be better at resisting male attempts to mate compared to females from lines with little or no sexual conflict. In theory this could result in reproductive barriers to mating, as females that evolve increased resistance to males of their population due to sexual conflict may refuse to mate with males from other populations altogether. Over time, this barrier to gene flow between the populations could eventually result in ***divergence*** and finally speciation. A recent study in a grasshopper potentially provides the first evidence of this happening in wild populations. Yoshikazu Sugano and Shin-ichi Akimoto (2007) had previously shown that crossing different populations of the Japanese grasshopper *Podisma sapporensis* revealed similar asymmetries in their mating behaviors to the experimental lines described above; males of a given population mated more frequently with females from a different population than with females of their own population, yet females tended to prefer males of their own population. Further analysis, utilizing additional populations of the grasshopper, supports the hypothesis that there is asymmetric premating reproductive isolation within this species as a result of geographic differences in male and female mating propensity that could over time result in divergence (Sugano & Akimoto, 2011).

It is not yet clear whether divergence under sexual conflict is a general phenomenon; indeed, several other experimental studies have found weak or no evidence that sexual conflict promotes reproductive isolation (e.g., Wigby & Chapman, 2006; Gay et al., 2009). Increasingly, genetics and molecular studies are informing our knowledge with regard to how speciation operates. It has been found that proteins involved in reproduction, which we would predict to be crucial players in the operation of sexual conflict, are the most fast-evolving group of proteins (e.g., Swanson et al., 2001; reviewed in Swanson & Vacquier, 2002), lending some support to predications that sexual conflict can lead to reproductive isolation.

It has also been suggested that sexual conflict could lead to the extinction of populations. A number of theoretical models suggest that, for example, male-induced costs to females could lower the overall fitness of a population to the extent that the population may become inviable and go extinct (reviewed in Kokko & Brooks, 2003). For example, Rankin and colleagues (2011) theorized that the evolution of male-induced "harassment" (cost) to females could result in a positive feedback cycle whereby fewer females survived each generation as a result of the harassment, and those that did had higher mortality as a result of the consequent high ratio of males to females in the population and thus increased levels of harassment. Eventually, according to their model, the population collapsed. However, further to this, they found that the evolution of a female countertrait could ameliorate this and prevent population extinction. A comparative analysis of sexual selection and the risk of extinction in mammals found no support for the hypothesis (Morrow & Fricke, 2004), and to date we have no direct empirical evidence that sexual conflict has or could drive a population to extinction.

However, there are conditions where sexual conflict could exacerbate the population decline to extinction in conjunction with other factors. Some populations of the fruit fly *Drosophila pseudoobscura* are infected with a **selfish genetic element (SGE)** called sex ratio (SR) meiotic drive; SGEs are pieces of DNA that defy the usual rules of inheritance and, through a variety of mechanisms, promote their own transfer to the next generation at the expense of the other genes in an organism (reviewed in Hurst & Werren, 2001). In *D. pseudoobscura*, males infected with SR cannot sire sons and thus only produce daughters, which can lead to populations with female-biased sex ratios. This can be costly to a female that mates with an SR male for several reasons; for example, not only does she only produce daughters, but her offspring may have reduced fitness due to inheriting the SGE (Jaenike, 2001). Price and colleagues (2008) showed that the presence of SR in a population results in females evolving increased mating rates; it was suggested that this promoted sperm competition and thus reduced the numbers of offspring that SR males sired, thereby minimizing to some degree the costs associated with mating with such males. However, a subsequent study showed that males countered this by evolving an increased ability to suppress this female mating response, suggesting that SR can promote sexual conflict in this species (Price et al., 2009). The final twist to this tale is that if females within an SR-infected population are prevented from mating multiply, and thereby decreasing the number of their offspring sired by SR males, the population can go extinct due to a lack of males (Price et al., 2010). Although not directly shown, this suggests that in a population infected with SR, if sexual conflict prevents females from resisting the costs associated with SR, it could lead to extinction.

CONCLUSIONS AND FUTURE DIRECTIONS

The term *sexual conflict* was first defined by Parker in 1979, yet it is only in the last decade that interest in this field of research has gained momentum. Increasingly, as with many topics within science, new technologies and innovative use of preexisting technologies are allowing us to delve deeper into the mechanisms underlying the operation of sexual conflict. For example, a team at Syracuse University recently produced **transgenic** *D. melanogaster* males expressing proteins in their sperm that fluoresced red or green under specific wavelengths of light (Manier et al., 2010). As a result, they could follow the sperm on their journey through the female reproductive tract and all the while unambiguously discriminate between the sperm of two males. This has allowed them to resolve some of the long-standing questions about mechanisms of postcopulatory sexual selection. One of their findings, which had previously only been inferred (Snook & Hosken, 2004), was that following copulation and before egg-laying commences, females eject sperm from their body, presumably allowing them to excise some control over who, out of their multiple mates, fertilizes their eggs. Michal Polak's team at the University of Cincinnati has been using laser surgery to study the function of the microscopic structures used in reproduction by insects. For example, using high-precision lasers to remove individual spines—in some cases less than 1 μm in size—from the genitals of *D. melanogaster* males, and then measuring the reproductive success of the males, they found that the spines aided males in obtaining copulations and in competing with rival males but had no effect on insemination and fertilization success (Polak & Rashed, 2010). This has led Polak and Arash Rashed to suggest that the broad morphological diversity seen in *Drosophila* genitalia is the result of sexual conflict prior to insemination, supporting the idea that if the spines are an adaptation to overcome female resistance to mating, differences in spine morphology across species may represent differences in the form and intensity of female resistance (Arnqvist & Rowe, 2005).

Many suspected examples of sexual conflict, some of which have been described here, we actually know very little about. Until we can measure the actual relative costs and benefits to males and females of their reproductive behaviors, something that is surprisingly difficult to do, we are in danger of being left only with anecdotal or speculative evidence. For example, it may appear that a male trait is costly to females, yet it might be the case that, in fact, the costs of female resistance to the trait actually outweigh the cost imposed by the trait itself. In addition, we still have little understanding of how conflicts between males and females might be resolved if indeed resolution occurs. *Who, if anyone, wins?* Or *Is there no resolution, only continuing cycles of adaptation and counteradaptation?* Quantifying animal behavior in general, including sexual conflict, is particularly difficult under field conditions; however, it is important not to become too focused on the laboratory, despite the diverse avenues that new techniques open up for us. Although such laboratory studies undoubtedly increase our understanding of how sexual selection operates, as with all studies of evolution and behavior, what happens in natural conditions in the field is of paramount importance. Sexual conflict is a fascinating aspect of animal behavior, but we still have a great deal to learn.

ACKNOWLEDGMENTS

The author would like to thank the Japan Society for the Promotion of Science for funding the research trip to Japan where much of this chapter was conceived. She would also like to thank Professor Takahisa Miyatake and Doctor Tom Price for valuable discussion on the topic, and her Animal Behaviour and Sociobiology students for providing inspiration.

REFERENCES AND SUGGESTED READINGS

Aigaki, T., I. Fleischmann, P. Chen, & E. Kubli. (1991). Ectopic expression of sex peptide alters reproductive behaviour of female *Drosophila melanogaster*. *Neuron*, 7, 557–563.

Alcock, J. (1994). Postinsemination associations between males and females in insects: The mate-guarding hypothesis. *Annual Review of Entomology*, 39, 1–21.

Andersen, N. M. & A. T. Weir. (2001). New genera of Veliidae (Hemiptera: Heteroptera) from Australia, with notes on the generic classification of the subfamily Microveliinae. *Invertebrate Taxonomy*, 15, 217–258.

Andersson, M. (1994). *Sexual Selection*. Princeton, NJ: Princeton University Press.

Andrés, J. A. & G. Arnqvist. (2001). Genetic divergence of the seminal signal-receptor system in houseflies: The footprint of sexually antagonistic coevolution. *Proceedings of the Royal Society of London, B*, 268, 399–405.

Arnqvist, G. & J. A. Andrés. (2006). The effects of experimentally induced polyandry on female reproduction in a monandrous mating system. *Ethology*, 112, 748–756.

Arnqvist, G., M. Edvardsson, U. Friberg, & T. Nilsson. (2000). Sexual conflict promotes speciation in insects. *Proceedings of the National Academy of Sciences, USA*, 97, 10460–10464.

Arnqvist, G., T. M. Jones, & M. A. Elgar. (2006). Sex-role reversed nuptial feeding reduces kleptoparasitism of females in Zeus bugs (Heteroptera; Veliidae). *Biology Letters*, 2, 491–493.

Arnqvist, G., T. M. Jones, & M. A. Elgar. (2007). The extraordinary mating system of Zeus bugs (Heteroptera: Veliidae: *Phoreticovelia* sp.). *Australian Journal of Zoology*, 55, 131–137.

Arnqvist, G. & L. Rowe. (2002). Antagonistic coevolution between the sexes in a group of insects. *Nature*, 415, 787–789.

Arnqvist, G. & L. Rowe. (2005). *Sexual Conflict*. Princeton, NJ: Princeton University Press.

Avila, F. W., K. Ravi Ram, M. C. Bloch Qazi, & M. F. Wolfner. (2010). Sex peptide is required for the efficient release of stored sperm in mated *Drosophila* females. *Genetics*, 186, 595–600.

Avila, F. W., L. K. Sirot, B. A. LaFlamme, C. D. Rubinstein, & M. F. Wolfner. (2011). Insect seminal fluid proteins: Identification and function. *Annual Review of Entomology*, 56, 21–40.

Birkhead, T. R. (2010). How stupid to have thought of that: Post-copulatory sexual selection. 2010. *Journal of Zoology*, 281, 78–93.

Bondurianksy, R. & S. F. Chenoweth. (2009). Intralocus sexual conflict. *Trends in Ecology and Evolution*, 24, 280–288.

Brennan, P. L. R., C. J. Clark, & R. O. Prum. (2010). Explosive eversion and functional morphology of the duck penis supports sexual conflict in waterfowl genitalia. *Proceedings of the Royal Society of London, B*, 277, 1309–1314.

Brennan, P. L. R., R. O. Prum, K. G. McCracken, M. D. Sorensen, R. E. Wilson, & T. R. Birkhead. (2007). Coevolution of male and female genital morphology in waterfowl. *PLoS ONE*, 5, e418.

Chapman, T. (2006). Evolutionary conflicts of interest between males and females. *Current Biology*, 16, R744–R754.

Chapman, T., G. Arnqvist, J. Bangham, & L. Rowe. (2003). Response to Eberhard and Cordero, and Córdoba-Aguilar and Contreras-Garduño: Sexual conflict and female choice. *Trends in Ecology and Evolution*, 18, 440–441.

Chapman, T., L. F. Liddle, J. M. Kalb, M. F. Wolfner, & L. Partridge. (1995). Cost of mating in *Drosophila melanogaster* females is mediated by male accessory gland products. *Nature*, 373, 241–244.

Cordero, C. & W. G. Eberhard. (2003). Female choice of sexually antagonistic male adaptations: A critical review of some current research. *Journal of Evolutionary Biology*, 16, 1–6.

Córdoba-Aguilar, A. & J. Contreras-Garduño. (2003). Sexual conflict. *Trends in Ecology and Evolution*, 18, 439–440.

Dottorini, T., L. Nicolaides, H. Ranson, D. W. Rogers, A. Crisanti, & F. Catteruccia. (2007). A genome-wide analysis in *Anopheles gambiae* mosquitoes reveals 46 male accessory gland genes, possible modulators of female behaviour. *Proceedings of the National Academy of Sciences, USA*, 104, 16215–16220.

Eberhard, W. G. (2004). Rapid divergent evolution of sexual morphology: Comparative tests of antagonistic coevolution and traditional female choice. *Evolution*, 58, 1947–1970.

Eberhard, W. G. & C. Cordero. (2003). Sexual conflict and female choice. *Trends in Ecology and Evolution*, 18, 438–439.

Eggert, A. K. & S. K. Sakaluk. (1995). Female-coerced monogamy in burying beetles. *Behavioral Ecology and Sociobiology*, 37, 147–153.

Fan, Y. L., A. Rafaeli, P. Moshitzky, E. Kubli, Y. Choffat, & S. W. Applebaum. (2000). Common functional elements of *Drosophila melanogaster* seminal peptides involved in reproduction of *Drosophila melanogaster* and *Helicoverpa armigera* females. *Insect Biochemistry and Molecular Biology*, 30, 805–812.

Fowler, K. & L. Partridge. (1989). A cost of mating in female fruitflies. *Nature*, 338, 760–761.

Friberg, U. & G. Arnqvist. (2003). Fitness effects of female mate choice: Preferred males are detrimental for *Drosophila melanogaster* females. *Journal of Evolutionary Biology*, 16, 797–811.

Fromhage, L. & J. M. Schneider. (2006). Emasculation to plug up the females: The significance of pedipalp damage in *Nephila fenestrata*. *Behavioral Ecology*, 17, 353–357.

Gage, M. J. G., G. A. Parker, S. Nylin, & C. Wiklund. (2002). Sexual selection and speciation in mammals, butterflies and spiders. *Proceedings of the Royal Society of London, B*, 269, 2309–2316.

Gavrilets, S. (2000). Rapid evolution of reproductive barriers driven by sexual conflict. *Nature*, 403, 886–889.

Gavrilets, S. & D. Waxman. (2002). Sympatric speciation by sexual conflict. *Proceedings of the National Academy of Sciences, USA*, 99, 10533–10538.

Gay, L., P. E. Eady, R. Vasudev, D. J. Hosken, & T. Tregenza. (2009). Does reproductive isolation evolve faster in larger populations via sexually antagonistic coevolution? *Biology Letters*, 5, 693–696.

Gromko, M. H., D. G. Gilbert, & R. C. Richmond. (1984). Sperm transfer and use in multiple mating system of *Drosophila*. In R. L. Smith (ed.), *Sperm Competition and the Evolution of Animal Mating Systems* (pp. 372–427). New York: Academic Press.

Hayashi, T. I., M. Vose, & S. Gavrilets. (2007). Genetic differentiation by sexual conflict. *Evolution*, 61, 516–529.

Heifetz, Y., U. Tram, & M. F. Wolfner. (2001). Male contributions to egg production: The role of accessory gland products and sperm in *Drosophila melanogaster*. *Proceedings of the Royal Society of London, B*, 268, 175–180.

Herndon, L. A. & M. F. Wolfner. (1995). A *Drosophila* seminal fluid protein, Acp26a, stimulates egg laying in females for 1 day after mating. *Proceedings of the National Academy of Sciences, USA*, 92, 10114–10118.

Holland, B. & W. R. Rice. (1998). Chase-away sexual selection: Antagonistic seduction versus resistance. *Evolution*, 52, 1–7.

Hosken, D. J., P. Stockley, T. Tregenza, & N. Wedell. (2009). Monogamy and the battle of the sexes. *Annual Review of Entomology*, 54, 361–378.

Hrdy, S. B. (1979). Infanticide among animals: A review, classification, and examination of the implications for the reproductive strategies of females. *Ethology and Sociobiology*, 1, 13–40.

Hurst, G. D. D. & J. H. Werren. (2001). The role of selfish genetic elements in eukaryotic evolution. *Nature Reviews*, 2, 597–606.

Huxley, J. S. (1912). A "disharmony" in the reproductive habits of the wild duck (*Anas boschas L.*). *Biologisches Zentralblatt*, 32, 621–623.

Jaenike, J. (2001). Sex chromosome meiotic drive. *Annual Review of Ecology, Evolution and Systematics*, 32, 25–49.

Jennions, M. & M. Petrie. (2000). Why do females mate multiply? A review of the genetic benefits. *Biological Reviews*, 75, 21–64.

Kokko, H. & R. Brooks. (2003). Sexy to die for? Sexual selection and the risk of extinction. *Annales Zoologici Fennici*, 40, 207–219.

Kraaijeveld, K., F. J. L. Kraaijeveld-Smit, & M. E. Maan. (2011). Sexual selection and speciation: The comparative evidence revisited. *Biological Reviews*, 86, 67–377.

Kuntner, M., S. Kralj-Fisher, J. M. Schneider, & D. Li. (2009). Mate plugging via genital mutilation in nephilid spiders: An evolutionary hypothesis. *Journal of Zoology*, 277, 257–266.

Lelito, J. P. & W. D. Brown. (2006). Complicity over sexual cannibalism? Male risk taking in the praying mantis *Tenodera aridifolia sinensis*. *American Naturalist*, 168, 263–269.

Lewis, Z., N. Wedell, & J. Hunt. (2011). Evidence for strong intralocus sexual conflict in the Indian meal moth, *Plodia interpunctella*. *Evolution*, 65, 2085–2097.

Manier, M. K., J. M. Belote, K. S. Berben, D. Novikov, W. T. Stuart, & S. Pitnick. (2010). Resolving mechanisms of competitive fertilization success in *Drosophila melanogaster*. *Science*, 328, 354–357.

Martin, O. Y. & D. J. Hosken. (2003). The evolution of reproductive isolation through sexual conflict. *Nature*, 423, 979–982.

Morrow, E. H. & G. Arnqvist. (2003). Costly traumatic insemination and a female counter-adaptation in bed bugs. *Proceedings of the Royal Society of London, B*, 270, 2377–2381.

Morrow, E. H. & C. Fricke. (2004). Sexual selection and the risk of extinction in mammals. *Proceedings of the Royal Society of London, B*, 271, 2395–2401.

Morrow, E. H., T. E. Pitcher, & G. Arnqvist. (2003). No evidence that sexual selection is an "engine of speciation" in birds. *Ecology Letters*, 6, 228–234.

Mulder, M. B. & K. L. Rauch. (2009). Sexual conflict in humans: Variations and solutions. *Evolutionary Anthropology*, 18, 201–214.

Neubaum, D. M. & M. F. Wolfner. (1999). Mated *Drosophila melanogaster* females require a seminal fluid protein, Acp 36DE, to store sperm efficiently. *Genetics*, 153, 845–857.

Orr, A. G. & R. L. Rutowski. (1991). The function of the sphragis in *Cressida cressida* (Fab) (Lepidoptera, Papilionidae): A visual deterrent to copulation attempts. *Journal of Natural History*, 25, 703–710.

Parker, G. A. (1979). Sexual selection and sexual conflict. In M. S. Blum & N. A. Blum (eds.), *Selection and Reproductive Competition in Insects* (pp. 123–166). New York: Academic Press.

Parker, G. A. & L. Partridge. (1998). Sexual conflict and speciation. *Philosophical Transactions of the Royal Society of London, B*, 353, 261–274.

Polak, M. & A. Rashed. (2010). Microscale laser surgery reveals adaptive function of male intromittant genitalia. *Proceedings of the Royal Society of London, B*, 277, 1371–1376.

Polhemus, D. A. & T. Polhemus. (2000). Additional new genera and species of Microveliinae (Heteroptera: Veliidae) from New Guinea and adjacent regions. *Tijdschrift voor Entomologie*, 143, 91–123.

Price, T. A. R., D. J. Hodgson, Z. Lewis, G. D. D. Hurst, & N. Wedell. (2008). Selfish genetic elements promote polyandry in a fly. *Science*, 322, 1241–1243.

Price, T. A. R., G. D. D. Hurst, & N. Wedell. (2010). Polyandry prevents extinction. *Current Biology*, 20, 471–475.

Price, T. A. R., Z. Lewis, D. T. Smith, G. D. D. Hurst, & N. Wedell. (2009). Sex ratio drive promotes sexual conflict and sexual coevolution in the fly *Drosophila pseudoobscura*. *Evolution*, 64, 1504–1509.

Rankin, D. J., U. Dieckmann, & H. Kokko. (2011). Sexual conflict and the tragedy of the commons. *American Naturalist*, 177, 780–791.

Ravi Ram, K. & M. F. Wolfner. (2007). Sustained post-mating response in *Drosophila melanogaster* requires multiple seminal fluid proteins. *PLoS Genetics*, 3, 2428–2438.

Reinhardt, K., R. Naylor, & M. T. Siva-Jothy. (2003). Reducing a cost of traumatic insemination: Female bedbugs evolve a unique organ. *Proceedings of the Royal Society of London, B*, 270, 2371–2375.

Rice, W. R. (2000). Dangerous liaisons. *Proceedings of the National Academy of Sciences, USA*, 97, 12593–12955.

Rice, W. R. & K. Chippindale. (2001). Intersexual ontogenetic conflict. *Journal of Evolutionary Biology*, 14, 685–693.

Rice, W. R. & B. Holland. (1997). The enemies within: Intergenomic conflict, interlocus contest evolution (ICE), and the intraspecific red queen. *Behavioral Ecology and Sociobiology*, 41, 1–10.

Ritchie, M. G. (2007). Sexual selection and speciation. *Annual Review of Ecology, Evolution and Systematics*, 38, 79–102.

Schneider, J. M. & P. Michalik. (2011). One-shot genitalia are not an evolutionary dead end—regained male polygamy in a sperm limited spider species. *BMC Evolutionary Biology*, 11, 197.

Scott, M. P. (1998). The ecology and behaviour of burying beetles. *Annual Review of Entomology*, 43, 595–618.

Shine, R., D. O'Connor, & R. T. Mason. (2000). Sexual conflict in the snake den. *Behavioral Ecology and Sociobiology*, 48, 392–401.

Shine, R., M. M. Olsson, & R. T. Mason. (2000). Chastity belts in gartersnakes: The functional significance of mating plugs. *Biological Journal of the Linnaean Society*, 70, 377–390.

Shine, R., B. Philips, T. Langkilde, D. I. Lutterschmidt, H. Waye, & W. T. Mason. (2004). Mechanisms and consequences of sexual conflict in garter snakes (*Thamnophis sirtalis*, Colubridae). *Behavioral Ecology*, 15, 654–660.

Siva-Jothy, M. T. (2006). Trauma, disease and collateral damage: Conflict in cimicids. *Philosophical Transactions of the Royal Society of London, B*, 361, 269–275.

Snook, R. R. & D. J. Hosken. (2004). Sperm death and dumping in *Drosophila*. *Nature*, 428, 939–941.

Sugano, Y. C. & S. Akimoto. (2007). Asymmetric mating in the brachypterous grasshopper *Podisma sapporensis*. *Ethology*, 113, 301–311.

Sugano, Y. C. & S. Akimoto. (2011). Mating asymmetry resulting from sexual conflict in the brachypterous grasshopper *Podisma sapporensis*. *Behavioral Ecology*, 22, 701–709.

Swanson, W. J., A. G. Clark, H. M. Waldrip-Dail, M. F. Wolfner, & C. F. Aquadro. (2001). Evolutionary EST analysis identifies rapidly evolving male reproductive proteins in *Drosophila. Proceedings of the National Academy of Sciences, USA*, 98, 7375–7379.

Swanson, W. J. & V. D. Vacquier. (2002). The rapid evolution of reproductive proteins. *Nature Reviews Genetics*, 3, 137–144.

Tram, U. & M. F. Wolfner. (1999). Male seminal fluid proteins are essential for sperm storage in *Drosophila melanogaster. Genetics*, 153, 837–844.

Wigby, S. & T. Chapman. (2004). Female resistance to male harm evolves in response to manipulation of sexual conflict. *Evolution*, 58, 1028–1037.

Wigby, S. & T. Chapman. (2005). Sex peptide causes mating costs in *Drosophila melanogaster. Current Biology*, 15, 16–21.

Wigby, S. & T. Chapman. (2006). No evidence that experimental manipulation of sexual conflict drives premating reproductive isolation in *Drosophila melanogaster. Journal of Evolutionary Biology*, 19, 1033–1039.

7

A Nest of Vipers: Conflict and Cooperation in Families

Michelle Pellissier Scott

How sharper than a serpent's tooth it is
To have a thankless child!
—Shakespeare, *King Lear*

Happy families are all alike; every unhappy family is unhappy in its own way.
—Tolstoy, *Anna Karenina*

INTRODUCTION

While it may be true that happy human families are all alike, the diversity of happy families in the nonhuman world is substantial. Evolution has sculpted a wide variety of solutions for launching the next generation. In many species, parental behavior may be manifested only in the choice of a place to leave the eggs where they might have the best chance of survival. In addition, eggs may be camouflaged or infused with a distasteful substance to protect them from predators. Beyond that, there is a wide range of types of parental care: incubation and nest guarding, live birth, and female and/or male care of offspring. Each increase in care places the emphasis on rearing fewer, higher-quality young.

Once obligate parental care beyond leaving sheltered eggs to fend for themselves has evolved, the trail gets twisted. Mostly family members have each other's best interests at heart, but there are basic conflicts of interests between parents, between parents and offspring, and among offspring. There may be an evolutionary benefit for fathers who desert; offspring may want more resources than parents are prepared to give; and each offspring would prefer to be first in its parents' heart. There are questions of the optimal number of young that will result in the greatest number of reproducing heirs, and there are some dark ways parents have to achieve their goal.

This chapter will put parental behaviors in the perspective of natural selection. Traits have evolved because individuals who acted in that way in a given environment, with given resources, competitors, and predators, left more young. Topics will include an evaluation of the circumstances under which providing extended care should be beneficial to the parent, why females are usually the caregivers in some taxa but biparental care is common in others, and why some offspring forgo their own reproduction to help parents rear the next generation. This chapter will also explore what determines the optimal clutch or brood size and how that is achieved, discuss how offspring communicate their needs and parents respond, and finally conclude with the nature of the conflict between parents and offspring as members seek to optimize their own future reproduction. The diversity and complexity of family life across the animal kingdom not only provides interesting stories but also offers us a way to further appreciate the workings of natural selection and adaptation.

WHY PROVIDE PARENTAL CARE?

Parental care is any form of behavior that increases the survival and ultimate reproductive success of offspring. In the most basic form of care, females provide a nutrient-rich egg. However, males in some species may contribute to this type of care as well in the form of a food offering during courtship, common in birds, or in the form of a nutritious mass packaged with the sperm, commonly seen in insects. This ***nuptial gift***, which can be substantial (the record may be a sperm-packet "gift" weighing 25 percent of the male katydid's body mass), can offset some of the energetic cost to the female and may increase the number or mass of her eggs (Gwynne, 1986, 1988a, 1988b). Nest construction, which in birds is frequently a joint effort, is another form of prefertilization care. After fertilization, the forms and benefits of parental care are usually obvious: eggs and live young need to be protected and fed. Even after young are nutritionally independent, parents may facilitate social interactions. In spotted hyenas (*Crocuta crocuta*) and some Old World primates, for example, daughters inherit their mothers' rank (Eng et al., 2000; Bergstrom & Fedigan, 2010), which is facilitated through maternal intervention, aggression towards low-ranking individuals, and coalition support. These high-ranking daughters generally have better access to food and can better protect their young than low-ranking females.

There is a *trade-off* between the amount of parental care provided and the number of young that can be reared, and there is a continuum from leaving many young to fend for themselves at one extreme and rearing fewer, more competitive young at the other. Mammals, with their reproductive mode of gestation, lactation, and long period of dependence, represent the latter extreme. But *why don't all animals provide so well for their young*? This question can be addressed by examining the costs and benefits of providing care. **Parental investment**, any behavior that increases offspring fitness but decreases the parents' ability to invest in other offspring, is a concept that lends itself better to testing hypotheses, because it incorporates costs, than does parental care. In a ***cost/benefit analysis***, the potential outcomes of the different options are scored in terms of *fitness*, which refers specifically to ***reproductive success***. It describes the ability both to survive and to reproduce. This approach allows for a comparison of mothers and fathers using the same ***currency*** (current and future offspring). It is the costs and benefits of any behavior that are the grist for the mill of natural selection.

Specific costs and benefits have very much to do with ***life history*** (life span, reproductive schedule, etc.) of each animal. Large animals with a long life would be predicted to have fewer offspring each time they breed and to provide more care to each. They might also be predicted to be more conservative (***risk averse***) when they are younger and have greater reproductive potential in the future; older individuals might be more ***risk prone*** and willing to pay higher energetic costs for late-born young. Jane Goodall's chimpanzee "Flo" was a wonderful mother and reared many successful young, but she overprotected and overindulged her last, to his detriment (Goodall, 1971). In addition to the influence of body size and life span, ***life history traits*** evolve in response to specific environmental characteristics, such as predictability. The effect of environmental predictability on parental behaviors is nicely illustrated with a comparison of house mice (*Mus musculus*) and the similarly sized marsupial mouse, *Antechinus stuartii*. House mice can have five to 10 litters of six to eight young each in a year. Juvenile mortality is high. Females are sexually mature in six weeks, and males are mature in eight weeks. In the wild, they can, but usually do not, live for a year. *Antechinus* (Figure 7.1), on the other hand, have a single, highly synchronized, short breeding season each year, and within a day of female ovulation all males in the population die of

Figure 7.1. *Antechinus stuartii* mother with youngsters that are almost old enough to be left in the nest. (Photo by Diana Fisher and Andras Kazei)

stress-related symptoms. Females have six to ten young, and juvenile mortality is lower than that of mice. There is a period of four months of gestation and lactation, and when they are weaned, the combined mass of the litter is three and a half times that of the mother. Mice are opportunistic breeders and live in a somewhat unpredictable environment, so females breed often and fast before conditions change. *Antechinus* lives in a predictable environment; females make a very large energetic investment in young, and males put all their energy into a single reproductive event as the probability of living until the next year would have been remote. Indeed, males that put the most energy into finding mating partners are the most successful (Braithwaite, 1979; Scott & Tan, 1985).

Birds and mammals are striking examples of parental care; offspring would not survive without a huge expenditure of energy by the mother and, for most birds, by the father as well. Most other taxa contain some species for which parental care is **obligatory** (Clutton-Brock, 1991). Although they might not spring to mind, some insects also provide extensive care to young. Male and female burying beetles (*Nichrophorus* spp.) cooperate to bury small vertebrate carcasses that will become food for their young (Figure 7.2). Parents remain in the burial chamber provisioning and guarding young. Carcasses (a **bonanza resource**) are rare, and finding one is unpredictable. In this case, this environmental factor has selected for life history traits and behavior on the far end of the continuum of extensive care. Once they have the opportunity (a carcass),

Figure 7.2. Burying beetle (*Nichrophorus orbicollis*) female feeding four-day-old larvae. (Photo by M. W. Moffett)

females spend more than half of their reproductive life spans rearing a single brood. Males help because two adults can successfully defend the resource and brood from infanticidal intruders who seek to rear their own young on this rare resource (Scott, 1990; Trumbo, 1991).

Burying beetles may be an example of particularly elaborate biparental care in an insect, but parental care of one sort or another is widespread and diverse (Table 7.1). It is thought to be confined to species for which eggs or young are clumped in time and space, which in turn is dependent on the temporal and spatial distribution of resources. The dispersion, quality, and persistence of resources influence how *sedentary* or *nomadic* a species is and what *competitors*, *predators* and *parasites* it must deal with. These in turn determine if parental care can be an effective strategy, for instance when there is something that parents are able to do to increase the survival of young. Because burying beetles use a small, discrete resource, behaviors to protect it and the young have evolved; and with burial and extended care, a suite of *adaptations* (behavioral, physiological, and anatomical) coevolved from characteristics of nonparental ancestors (Trumbo, 1996). For example, once burying beetle larvae were protected underground by parents, they no longer had to be heavily

Table 7.1. Types of parental care by insects and numbers of orders in which parents exhibit the behaviors listed.

Behavior	Maternal care	Paternal care	Biparental care
Eggs covered, then abandoned	6	0	0
Parents remain and guard eggs	12	2	4
Parents remain and care for young	12	2	5
Females extend development internally and give birth to live young	10	–	–
Males provide nutrition with sperm	–	9	–

protected by body armor as other (nonparental) carrion beetle larvae are. Parents feed from the carcass, as do nonparental carrion beetles, but they must also feed their young with predigested regurgitant.

WHO PROVIDES CARE?

Care can be provided by females, males, or both. Because this trait most likely evolved from a condition of no care, the benefits of increased survival and quality are assumed to outweigh the costs of providing the care. These costs may differ for males and females, even for the same behavior. For example, pair-bonded male and female birds may both bring food to the brood, but the cost to future fitness may be greater for the male if he is giving up the opportunity to search for and mate with another female who will rear his young. The female rarely has the opportunity to find a care-giving male to rear her young. On the other hand, if the mother ate the food she collected and left the father to feed the chicks, she would have more energy to produce more eggs; the energy gain for the same behavior of the father would have less direct fitness consequences.

Each offspring has one mother and one father; thus the benefits from rearing it are shared equally by the mother and father regardless of which provides the care. However, if one parent can do the job, either parent would gain by leaving the other to do it. We can therefore expect conflict between the sexes over the amount or duration of care the other sex provides. This conflict drives an unusual mating system in the Eurasian penduline tit (*Remiz pendulinus*). These small birds have very short pair bonds lasting only up to a week, and both sexes practice ***serial polygamy***. In this species one parent alone can incubate and rear the brood, but 30 percent of the clutches are deserted. Both sexes gain if they desert first and leave the other to do it—provided they can find an unmated and willing new partner. Desertion by the male is costly to the female because she can either stay, which take substantial time and energy, or she can desert the eggs and leave them to die. If the female deserts, leaving the father to incubate, she can produce a new clutch (Szentirmai et al., 2007).

Out-and-out desertion is extreme; various factors can influence the level at which males and females work. When one parent deserts or is experimentally removed, the remaining parent usually increases its workload, if it is able. Burying beetle females do not increase their level of care when their mates are removed or experimentally handicapped with extra weights, whereas males do work harder under these circumstances. This suggests that females already work as hard as they can; if need be, males can step up, but they usually work less hard than their mate (Seizi & Masahiro, 2009). Parents may also adjust their level of care based on the perceived quality of their mate. For example, female zebra finches (*Taeniopygia guttata*) paired with attractive males (in this case, males with orange or red leg bands) feed their chicks more than those with unattractive mates (those with green or blue leg bands) (Burley, 1986).

Because females produce a comparatively large, nutrient-rich egg relative to the male's small, mobile sperm, their lifetime reproductive success depends mostly on their energy intake, whereas a male's lifetime reproductive success depends more on the number of matings he can get. In some species the consequence of this is extreme; a few males in the population mate with many females and most mate with none. In this case, when the ***variance*** in male success is greater than the variance in female success (i.e., most females have closer to the average number of young in their lifetimes), there is strong selection (called ***sexual selection***) for competition among males for access to receptive females. This is not to say that females do not also undergo sexual selection. Females may compete for high-quality males, territory, or social status, which will increase their fitness. In some species like the ***polyandrous*** jaçana or the ***sex-role-reversed*** spotted hyena, sexual selection is stronger on females than on males (Clutton-Brock, 2007). But generally the greater competition among males pushes for ***secondary sexual traits*** such as weapons and ornaments. This potential for some males to do better than others in the population such that only a subset breed also selects for exclusive female parental care because the costs for the breeding male to provide care is greater than that for females (Kokko & Jennions, 2003).

There are additional reasons why female care has evolved more often than male care. Females can be more certain that they are the mothers of their young than males can be that they are the fathers. If females mate with multiple males before eggs are fertilized, or, as with some fish, if a nonparental male sneaks in and releases sperm just as the nest owner is spawning, males are less certain of paternity and benefits of parental care are devalued, theoretically. ***Extra-pair copulations*** are quite common with pair-bonded birds, but this does not necessarily devalue paternal care. On average, males in a population have equal success ***cuckolding*** other males; thus a male may be caring for step offspring while his are being cared for by another male.

Reduced parental effort rather than desertion is a more common effect of reduced confidence in paternity. Results from theoretical modeling and experiments examining the effect of paternity on parental effort are mixed (Sheldon, 2002). When either male or female collared flycatchers (*Ficedula albicollis*) were temporarily removed when the female was supposed to be fertile, males seemed to use that as an indication of cuckoldry and reduced paternity, and they did reduce their share of provisioning the youngsters (Sheldon, 2002). For this response to evolve, we assume that there must be a benefit in future fitness for the males; otherwise, they would just go ahead and feed these potential step offspring. These male flycatchers had the same probability of survival if they worked harder or less hard, but those that reduced their care had larger forehead patches the following year. This is a sexually selected trait, and those with larger patches sired a larger proportion of the offspring in their nests.

The ***association hypothesis*** (or order-of-gamete-release hypothesis) suggests another factor that makes maternal care more likely than paternal care (Gross & Shine, 1981). With internal fertilization, males release their gametes first, and this puts the female in a cruel bind. The male is free to desert immediately, and if he has, the female has the choice of providing care alone or abandoning the breeding attempt (assuming that some care is necessary for the survival of the young).

Male care and biparental care are common in some groups, especially in fish and frogs. The association hypothesis offers an explanation for the frequency of male care by nest-building fish. Females are induced to release their eggs in the nest that males guard and are more attracted if there are already eggs in the nest, advertising what attentive fathers the owners are. (In some species, males are even known to steal eggs from the nests of other males [Rohwer, 1978].) Males release sperm after females spawn, so the female is free to desert first, but the stronger force selecting for paternal care is that the cost to the male of remaining is relatively low compared to that of the female. He remains on his territory after the female leaves, guarding the eggs from predators and attracting additional females. Furthermore, care is sharable, meaning that it does not increase as more females deposit eggs in his nest.

A second reason for the commonness of paternal care in fish is that growth is indeterminate. As females grow their fecundity increases; thus, if caring reduces growth, they suffer a high cost to future reproduction. In species for which size in males does not have a strong effect on fitness, the cost of care, and consequential reduced growth, is less for them than it is for females.

Models for cost-benefit analyses have been developed using **game theory** borrowed from economics. To provide care or not is seen as a trade-off. The major factors input into these models are offspring survival with zero, one, or two parents; the availability of additional mates if the male or female deserts; certainty of parentage; and the increase in fecundity if the female deserts. The potential payoff is calculated for each option, given that the other parent will or will not provide care. Thus if offspring survival is much better with two parents and few females are available as additional mates, males are expected to stay and help. Imagine the case of emperor penguins (*Aptenodytes forsteri*). Their harsh breeding environment has selected for biparental care. Without the father returning to relieve the mother, the chick would have no chance of survival. Furthermore, most females are already paired.

EXTENDED FAMILIES

The African saying that it takes a village to raise a child may be true, at least as a metaphor, for some nonhuman species as well. Pair-bonded parents are assisted by older siblings or even less related group members in numerous species of birds and mammals. **Cooperative breeding** is good for the extended family because groups can be more efficient than individuals through division of labor and specialization, and there is opportunity for **social learning**. But *why should grown offspring give up their own chance to reproduce to help their parents?* Benefits are varied and complicated. Especially if the new brood is comprised of full siblings, they are helping to promote the success of their genes. (This type of selection is called **kin selection**.) There is a net fitness gain if the number of young that a first-time breeder can expect is low and its assistance to parents results in much better survival of the new brood. Although the close relatedness of parents and helpers certainly facilitates the evolution of helping, it is perhaps not the most important factor. In many species, nonrelated individuals help, and in various ways this increases their personal reproductive success. Helping may be the "dues" they pay to be allowed to stay. Many recent studies show that there can be direct benefits to delaying dispersal from the family territory, even if they do not help with the new young. By staying, they may increase their survival and chance of gaining a territory and breeding in the future. They may also gain skills that increase their future fitness (Clutton-Brock, 2002; Komdeur & Ekman, 2010).

Nothing matches social insects for extended families! Individuals in a family of ants, termites, and many species of bees and wasps do not reproduce themselves but help the queen to rear their younger siblings. Honey bee (*Apis mellifera*) colonies, for example, consist of tens of thousands of individuals, and the workers do all the housework and care unstintingly for the eggs and larvae before they become the hive's foragers when they are about two weeks old. Sterile worker ants are often morphologically specialized for their specific and clearly defined roles. Leafcutter ants (*Atta* spp.) have highly specialized castes. The **nurse ants** are relatively small, and their job, as their name implies, is to care for the brood. They keep them clean and feed the begging larvae the fruits from their fungal garden and even with sterile eggs that they produce for just for nutrition. Armored **soldier** termites defend the family with weaponry of huge pincers or nozzle-like heads that spray caustic substances on enemies. The soft-bodied **workers** perform the mundane household duties and tend to the youngsters' needs. These **altruistic** workers in most, but not all, cases are helping to rear closely related kin and thus pass their genes on indirectly. However, selection at the colony level has also been important in the evolution of this division of labor. Larger colonies that are more efficient and have grown fast are better competitors and more likely to ultimately give rise to daughter colonies (Wilson & Wilson, 2007). If there is a persistent food supply and individuals have built a defensible nest, the groundwork is laid for social living. Division of labor with some individuals specialized for reproduction and others for caring for the young can then follow.

FAMILY PLANNING

Just as the amount of care is a trade-off between current and future offspring, the number of young to have and when to have them is a similar trade-off. Large, long-lived species take a conservative strategy and save energy for future young. Even shorter-lived species face the same "decision" of how many young to have and how much care to provide. Parents are typically optimistic or at least make room for the possibility of the best outcome. Nonetheless, mothers do not produce the maximum number of young that they could. Birds tend to have an optimal clutch size (Lack, 1947; see Williams, 1966). Mothers that lay too many eggs are overly optimistic and bring fewer young to independence; those that lay too few may rear all but produce fewer than the female that has chosen the intermediate. At least, that is what common sense and theory would predict. However, this has been difficult to demonstrate with experimental manipulation of clutches because parents differ in quality and much is dependent on the availability of resources at the time. Trade-offs are ultimately mediated by a number of factors including both chance events like predation and storms, biotic factors like parasite and prey abundance, and competition from conspecifics.

Blue tits (*Parus caeruleus*) are an exceptionally well-studied bird. The effects of ***reproductive effort*** (clutch size) on offspring quantity and quality and on parental future reproduction have been studied in populations from Scandinavia to Corsica. An experimental manipulation of clutches, in which eggs were removed or added to random nests, in an English population supported the hypothesis that individuals optimize their clutch size. The number of young surviving was best in nests that were not manipulated (Pettifor, 1993a). This same study also found no overall negative effect of brood size on the survival or future reproduction of the parents (Pettifor, 1993b), although there did seem to be a significant negative effect of brood size on survival of males or females in some individual years. The key to the variability between years was most likely due to differences in the caterpillar crop (the main food source for parental tits). A population of blue tits in Corsica differs significantly from more northern ones: there is less food, less water, and higher temperatures, and birds suffer a higher parasite load (Blondel et al., 1998). These tits have smaller clutches and breed later than the northern ones. In this study, clutch manipulation did not significantly affect the number of young reared, but it did affect their quality. Young from larger broods were smaller and lighter, and females raised in these larger broods bred later themselves and laid fewer eggs than those from control or reduced clutches. As with the previously mentioned study, there was no consistent effect of clutch manipulation on adult survival or reproduction the following year. However, in this study there was a big effect of year-to-year variation in environmental conditions, especially in caterpillar abundance. Not only in some years was there better overall fledgling success, but also the effect of the manipulation differed. In one year, clutches reduced by two eggs did relatively well, but those increased by two eggs did exceptionally poorly, compared to the same manipulations in other years (Blondel et al., 1998).

In addition to the number of young to produce, the frequency at which to produce them and even the sex of the young are all components of family planning. Long-lived species often do not have young every year. Large mammals, with their very high costs of reproduction, have young infrequently. African elephant females (*Loxodonta africana*) give birth every four to nine years, but humpback whales (*Megaptera novaeangliae*), in spite of their large size and slow growth, reproduce on average every other year. It is sometimes the case for mammals that the sex of one offspring affects future offspring. Rearing a larger male to independence is metabolically more costly than rearing a smaller female to independence. Female red deer (*Cervus elaphus*) that have a son are more likely not to breed the following year than those that have a daughter (Clutton-Brock et al., 1982). The effect of raising sons is even more subtle in humans. Data from Lutheran church records of Finnish farmers and fishermen in the eighteenth and nineteenth centuries show that having a son takes more out of the mother than having a daughter, and succeeding offspring have a lower birth weight. Furthermore, surviving children born after a son were less likely to have offspring of their own than those born after a daughter (Rickard et al., 2007), presumably because of their poor start in life.

HOW PARENTS PLAY FAVORITES

The future is uncertain, and while parents of some species produce what nature tells them will give them the best outcome of healthy, viable offspring, others have more alarming practices for achieving the optimal brood or clutch size. They rely on *infanticide* and *siblicide*! The optimal clutch size for burying beetles depends entirely on the size of the carcass they have buried. For example, *Nicrophorus orbicollis* can successfully rear young on a tiny 10 g mouse or a small 60 g chipmunk, but only a few larvae survive on the mouse, and 50 or so can be raised on the chipmunk. Because she has plenty of food to convert to eggs and because some may not hatch, the female overproduces. There seems to be a minimum of about 25 eggs that she oviposits (Scott, 1997). She and the male have assessed the carcass size by walking around it and measuring (Trumbo & Fernandez, 1995). With this information both male and female kill excess young as eggs hatch and larvae arrive on the carcass (Bartlett, 1987; Trumbo, 1990). Although we might imagine a cartoon image of parents counting off arriving youngsters and reaching the cut-off point after which they eat the rest, it has been hard to assess the basis of their mathematical skills. We do know that parents are hormonally and behaviorally sensitive to the number of begging larvae (Panaitof et al., 2004); perhaps the frequency of interactions provides all the information they need.

Antechinus, the marsupial mice with the annual male die-off, also produce more offspring than can be accommodated. They give birth to such underdeveloped young that the cost of overproduction is low. Females have a limited number of teats (that differs by population, perhaps depending on resource availability), and the trek from the genital opening to the pouch is treacherous. Because each young fastens onto its own teat and stays there for six weeks or so, once the teats are occupied, no more young can survive. Mothers may ignore or simply eat the extras (Lee & Cockburn, 1985).

It may seem both gruesome and wasteful that parents should resort to cannibalism to achieve the best brood size, but it is not that uncommon, occurring in mammals, birds, fish, and reptiles as well as insects (Klug & Bonsall, 2007). Paternal care is common in fish. While often this consists of males guarding a nest in which the female has left her eggs, in some cases the male's mouth serves as the "nest." Mouth-brooding male cardinalfish (*Pterapogon kauderni*) pick up the egg mass into their mouths immediately after spawning. As long as things go well, eggs stay there protected from predators and well aerated until embryos hatch. But this comes with a cost to the fathers, as they must fast and, as the breeding season progresses, they lose condition. Partial brood cannibalism is common, and the entire brood disappears into the father's stomach not infrequently. This ***filial cannibalism*** by male cardinalfish becomes much more prevalent later in the season as the father's body condition decreases (Okuda & Yanagisawa, 1996) and tends to happen early in the mouth-brood phase as young eggs have greater nutritional value and less reproductive value than younger eggs.

An energy-based explanation for filial cannibalism in teleost fish seems to be valid in most cases (Manica, 2007), but other factors, such as brood size or age, availability of additional mates, and the probability of paternity can also influence this "decision." Male threespined sticklebacks (*Gasterosteus aculeatus*), a fish that makes and guards a nest in which females deposit their eggs, suffer from "sneaky" males that wiggle into nest of the territorial male and deposit sperm that fertilize some of the eggs. Thus the nest owner cannot be 100 percent sure that he is the father of all the young. Clutches containing a high proportion of foreign eggs are much more likely to be entirely cannibalized. But *how can the male make an informed judgment?* In an experiment in which from 0 to 100 percent of one male's eggs were exchanged with those of another male's, all the eggs were destroyed and presumably eaten significantly more often when the male was caring for fewer than 50 percent of his own young. Moreover, the risk of total brood cannibalism was greater early in the breeding season, but only by males with many of their own eggs; males with few of their own eggs cannibalized the brood at a high rate throughout the season. Some males only cannibalized part of the brood, and in these cases, the greater the proportion of their own eggs that were in the nest, the more foreign eggs were consumed (Mehlis et al., 2010). *What cues can they use to recognize alien eggs?* This experiment ruled out all but cues from the eggs themselves (e.g., no rival males were ever present), and olfactory cues associated with the eggs are the most likely.

While parents sometimes do the dirty work to eliminate excess offspring themselves, in some cases they leave it to the siblings. Sand tiger sharks (*Carcharias taurus*) kill and eat brothers and sisters *in utero* (Gilmore et al., 1983). Embryo pronghorn antelope (*Antilocapra americana*) have a spear of

Figure 7.3. A Nazca booby mother (*Sula granti*) looks on while the first-born chick evicts the second born from the nest, leaving it to certain death. (Photo by Dave Anderson)

dead tissue that they use to kill their twins in the narrow uterus (Forbes, 2005). Workers of many social insects eat eggs oviposited by other workers because they are less related to these future offspring that they are to those of the queen. But siblicide is best studied, and is most poignant, in birds, especially in large sea-going species that normally have small clutches. Nazca booby (*Sula granti*) mothers lay two eggs but always raise only one. The older and larger chick forces its weaker sibling out from under its mother, where it dies a pitiful death in the beating sun (Humphries et al., 2006).

Leaving one offspring to eliminate the "extras" is a very practical evolutionary solution. Parents face multiple uncertainties: critical resources may vary such that the parent cannot predict the optimal brood size, or some offspring may not hatch, die prematurely, or be intrinsically flawed. To evaluate the effects of these uncertainties, members of the brood are thought of as ***core*** or ***marginal*** offspring. The core is the number of young that are normally raised to independence, and the marginal offspring are those that make up the overproduction—those that might make it in the best of times (Mock & Forbes, 1995). The marginal member(s) of the brood are usually handicapped, often by being hatched later, which make them easier to be killed. In Nazca boobies siblicide is ***obligate*** (Figure 7.3)—99.95 percent of broods fledged only one chick, and the most common cause of death was the action of the other chick (Humphries et al., 2006); in other species, the blue-footed booby (*Sula nebouxii*) for example, siblicide is *facultative*.

There are several hypotheses that might predict why parents overproduce and for which species siblicide is facultative and which obligate (Mock & Forbes, 1995). The **resource-tracking hypothesis** suggests that parents produce the number of young that can be raised in a good year and are prepared to sacrifice them if resources turn out to be insufficient. **Hatching asynchrony** in red-winged blackbirds (*Agelaius phoeniceus*) creates core and marginal offspring, and in this way they keep the costs of sacrificing marginal offspring lower by creating weaker brood members. In an experiment in which the core or marginal brood was enlarged, the growth of the core youngsters was not affected by additional marginal ones, but alteration of the core brood significantly reduced the growth of the marginal ones (Forbes & Glassy, 2000). Parents feed core offspring what they need to develop properly, and marginal ones just suffer more with extra core. Similarly, cattle egrets (*Bubulcus ibis*) produce two core offspring and one marginal one, the consequence of being last hatched. As long as there are three young in the nest the core chicks, especially the second hatched, attack the third, but it is the nutritional condition of this third that determines the outcome; if it is weak and malnourished already, it will succumb, but in a good year it becomes strong enough to survive (Mock & Lamey, 1991).

A second hypothesis that might explain the overproduction and consequential siblicide has been called the **insurance hypothesis**. This predicts that the junior chick would be killed even if food were plentiful, but parents produce a back-up chick in case something happens to the first. Whereas Nazca boobies virtually never raise more than one chick, their cousin the blue-footed booby often raises two or more. Nazca booby parents forage farther from land than do the blue-footed parents. If these longer commutes limit the amount of food that can be delivered, the resource-tracking hypothesis would be supported. However, Nazca and blue-footed boobies have similar foraging efficiency, and both could fledge two chicks (Anderson & Ricklefs, 1992), but the hatching success of the Nazca booby is much lower than that of the blue-footed booby, apparently due to exceptionally high infertility or early embryonic death (Anderson, 1990). On average, the benefit of having the back-up chick is greater than the cost of producing the wasteful second youngster that will just be killed. Thus fatal sibling rivalry is in the best interests of the parents (as well of as the core chicks), and their lack of interference is expected.

There is a third hypothesis for the evolution of siblicide, and that is as a **back-up food supply** for the dominant offspring (Mock & Forbes, 1995).

Parents thus convert relatively low-cost nutrients (marginal young) for emergency rations for the more-likely-to-succeed youngsters. This probably accounts for the sand tiger sharks' siblicidal acts, but birds and mammals seldom eat their dead siblings. Usually siblicide for nutrients is not specifically orchestrated by the parents as it is when it occurs for insurance or resource-tracking reasons. By practicing siblicide, the individual is eliminating some of its own genes that could pass to the next generation; thus cannibalism for the food value should be practiced on nonkin. The cannibalistic morph of spadefoot toads (*Spea multiplicata*) is characterized by a large head and jaw muscles. The development of this morphology is environmentally cued by the type of food that is available when the cannibals have an advantage over the omnivore morph. In fact, the cannibalistic toads show some ability to recognize siblings and refrain from eating them; the omnivores have no such skill in recognizing kin (Pfenning, 1999).

"JUST TELL ME WHAT YOU NEED"

Family dynamics include complicated trade-offs between family members and elaborate **communication signals**; getting and giving food is the central arena on which these dynamics are played out. Human babies, nestling birds, and even insect larvae are far from the helpless little creatures they may seem. They can run their parents ragged with demands to get more for themselves and strive to seem more needy than their siblings. In most species where parents provision, youngsters **beg**. Burying beetle larvae, like baby birds, are relatively quiet when no parents are around. But as soon as the larvae detect a parent or the nest shakes and the parent bird casts a shadow, the young beg energetically. Young beetles rise up, clamor over parental feet, and push their heads into the mouth of the parent (Rauter & Moore, 1999); nestlings raise their necks, extend their legs, open their beaks, call loudly, and jostle for position to be the closest mouth to receive the caterpillar. Begging behavior has even been reported in a frog. Taiwanese tree frog (*Chirixalus eiffingeri*) tadpoles immediately aggregate around the mother when she returns to the nest. They stiffen their tails, vibrate vigorously, and nip at her skin around her genital opening and thighs. This induces her to lay unfertilized eggs for food, which the tadpoles eagerly consume (Kam & Yang, 2002). *How should parents respond to begging? Do begging youngsters honestly communicate their needs (as opposed to their wants), and do parents respond to differences in begging vigor by feeding the most energetic first?*

This extravagant and often loud begging is assumed to be metabolically expensive and even dangerous if it attracts predators. Offspring pay this price to outcompete their siblings for resources. Parents are equally related to each of their young and might favor an equitable distribution, but each individual values its own welfare twice as much as it values it siblings' because they share only 50 percent of the same genes at best. So as long as parents deliver individual packets of food, it behooves each youngster to try to get a disproportionately large share of it. This suggests that begging is a means to manipulate parents to deliver more resources than is optimal for the parent because it is more than that particular individual needs. However, begging is a communication signal, and as such it is thought to have evolved because is confers a benefit to both the sender and the receiver; that is, parental response plays a role in shaping the begging signal. If the young differ in condition, parents might benefit if they can tell which youngster is in the greatest need to preferentially feed it because it would get a greater benefit than its better-fed siblings. Thus parents gain if greater need is signaled by more vigorous begging. The costly nature of begging suggests that it should be an honest signal of need (Kilner & Johnstone, 1997); that is, if it did not come with a cost, each would demand everything it could get, and parents would gain no information and would ignore differences.

Many studies have shown that begging intensity is an honest indication of need. For example, yellow-headed blackbird (*Xanthocephalus xanthocephalus*) nestlings in an experimentally altered brood of four consistently begged longer than those in a brood of three (Price, 1996) and begged longer if they were smaller than their competitors or in poor condition. Nestlings that were deprived of food for 30, 60, or 90 minutes begged harder the longer they had gone without. Male yellow-headed blackbird youngsters grow faster and must reach a greater weight before fledging than their sisters. Males begged louder and longer and ate more than females for any hunger level (Price et al. 1996). In this species at least, begging can communicate both long-term (males need more and begged more) and short-term (longer deprivation led to longer begging) need. Furthermore begging worked. When begging by the yellow-headed blackbird nestling was enhanced by adding recorded vocalizations, parents doubled their feeding visits, suggesting that they do use the information provided by begging (Price, 1998).

Not surprisingly, most studies on begging have been conducted on birds. Studies on burying beetles have also supported the prediction that begging can be an **honest** signal of need. Food-deprived larvae and smaller, younger larvae begged longer than their better-fed and larger brood mates. Begging increased with brood size up to a point, and the time spent feeding the larvae by both parents together was strongly associated with the time spent begging by each larva. When the hunger levels of individuals within the brood was experimentally manipulated, parents fed hungry larvae more than well-fed ones, but this bias occurred because hungry larvae spent more time begging (Smiseth & Moore, 2002, 2007, 2008).

Parental response to the honest signal of begging may not always be to feed the hungriest, smallest, or neediest. Earwig (*Forficula auricularia*) mothers regurgitate food individually to their nymphs, and chemical cues communicate offspring need. Older offspring and those in better physiological condition have higher value to parents because the expected parental investment needed to successfully rear them to independence is lower. In this species, rather than preferentially feeding the neediest nymphs, mothers exposed to the chemical cues of well-fed youngsters foraged longer and gave more food to nymphs than mothers exposed to cues from poorly fed nymphs (Mas et al., 2009).

In spite of the above examples, results of studies on the relation of begging intensity and need and the effect on parental provisioning have been mixed. Offspring begging signals are often complex, with multiple components (begging loudness and duration, stretch height, etc.). Older siblings in a brood that has hatched asynchronously may be able to stretch higher and therefore get fed more just because they block the access of other chicks. Older chicks can beg harder just because they have more energy to spare. Lastly, the interpretation of begging as an honest signal of need depends on begging being costly. Experimental results have been very mixed, however. In some experiments, recorded begging calls near artificial nest have attracted predators; in other studies, there is no difference between attraction to begging calls and white noise. Some studies have measured the energetic cost of begging, which ranges from only a little more to about 25 percent more than the resting metabolic rate, but usually there seems to be no cost in either growth or survival (Kilner & Johnstone, 1997).

PARENT-OFFSPRING CONFLICT

Although most of the time parents "want" what is best for their offspring and vice versa, each share only half of their genes with the other. In the eyes of the offspring, this devalues parental sacrifices, and in the eyes of the parents, future offspring could benefit from energy saved. Thus the level of care that maximally benefits the young is more than the parents should give if they maximize their lifetime reproductive success. Furthermore, offspring care more about their own welfare that than of a sibling, future or current (Trivers, 1974; see Figure 7.4). The greater the discrepancy between offspring and parental optima, the greater the conflict. Often conflict flares towards the

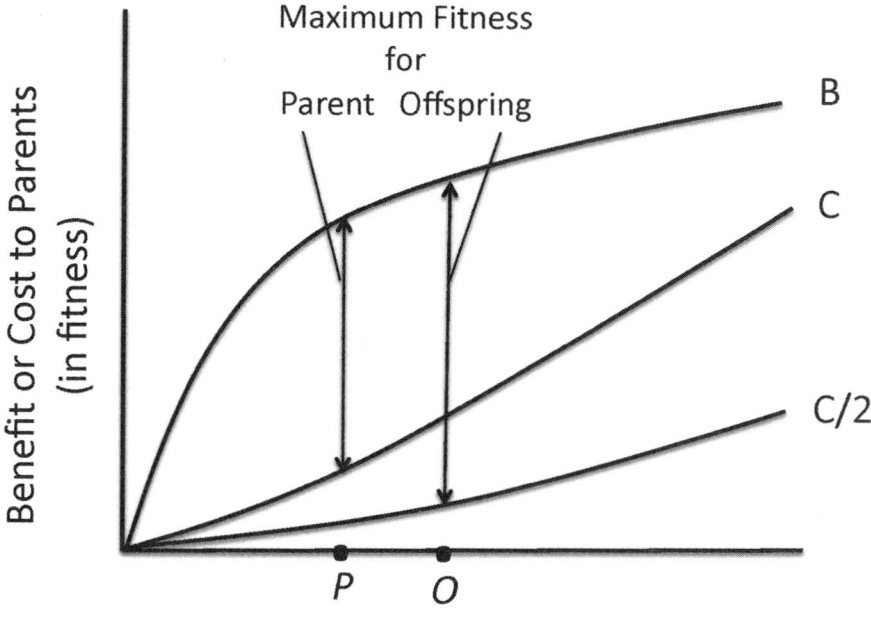

Figure 7.4. The basis of parent-offspring conflict. The benefit to the offspring increases quickly with the first investment because a little does a lot of good, but as the young get more, additional investment makes less of a difference. Another way to think of it is that as the offspring grow and become more independent, benefit levels off as more food or care does not increase survival or quality very much. The cost to the parent (in terms of future reproduction) in providing this investment is expected to be more or less linear, each bit of increase in effort being the same cost at a high or low level. Thus for the parent the maximum net fitness gain (B–C) is at parental investment P. However, from the offspring's perspective, the cost to the parent is only half as much because the parent and future offspring only share half of the same genes. Thus the maximum net fitness gain (B–1/2C) is at parental investment level O. Between time period or investment level P and O, offspring want more than parents want to give, and there is a conflict. (Redrawn from Trivers, 1974)

end of the stage when young are dependent on parental feeding. Fledglings harass their parents for more food when they are already fatter than their parents, and young mammals can have temper tantrums when denied milk during weaning.

One of the most startling examples of ***parent-offspring conflict*** has been proposed to manifest itself during human pregnancy (Haig, 1993). Some pregnancies progress smoothly; baby and mother appear to be in sync, and the baby is born at a reasonable size with minimal fuss. But more often than not, the mother has endured months of pregnancy sickness and possibly dangerously high blood pressure and gestational diabetes. The latter two problems can be attributed to the demands of the fetus to get more nourishment through the placenta than is good for the mother. ***Preeclampsia***, or dangerously high blood pressure, can benefit the fetus because it increases the blood flow through the placenta, and more blood means more nutrients. This condition is common when the placental growth is poor and the placenta is embedded shallowly in the uterus. It appears to be the fetus's way of compensating, but the mechanism is not yet known. In ***gestational diabetes***, the fetus secrets a hormone (human placental lactogen) that reduces the mother's sensitivity to insulin. The result is increased blood glucose levels that the fetus uses to good advantage.

Offspring begging can be seen as manipulation to get from the parents more than they want to give, or, if it is costly, it may be the mediator to resolve the conflict; parents can know what they need, not just what they want. If parents adjust the rate of feeding in response to begging and offspring adjust their begging in response to the amount of food received, parent-offspring conflict can also be understood as an evolutionary process between supply and demand, each affecting the other. Thus the level of begging and the level of feeding have some genetic basis and have evolved together. This relationship can be explored with ***cross-fostering experiments*** in which some members of one brood are exchanged with those of another, leaving a mix of genetic and foster offspring in each brood. Some experiments, but not others, have shown a positive relation between begging and feeding; that is, in families where offspring are genetically inclined to beg more, parents are genetically inclined to feed more (Smiseth et al., 2008). This experimental approach can often reveal if parents ultimately control feeding or if offspring begging manipulates parents to the detriment of their future reproductive success. Species differ.

CONCLUSIONS—NEW HORIZONS

Through exploring parental behavior, this chapter has sought to elucidate some of the many ways that family members cooperate or are in conflict with other members. The discipline of ***behavioral ecology*** is well developed, and hypotheses to predict the behavioral outcome of interactions have been proposed; this has encouraged tests of these hypotheses in a wide diversity of species from invertebrates to mammals. Much less is known about the mechanisms that regulate parental and offspring behavior. Arguably the most exciting are hormonal and genetic mechanisms. Two examples follow for which these mechanisms have been uncovered as a suggestion for how fruitful further exploration should be.

Many birds lay eggs that hatch asynchronously. Later-hatched chicks have a disadvantage, and this may or may not be in the best interest of the mother. Mothers may vary the amount of the hormone **testosterone** that they put into the eggs. The general consequences are that chicks from eggs with higher testosterone beg more vigorously. The consequences of increased begging on offspring fitness are complicated: It can have either a negative or positive effect on growth and survival. Most likely this is related to available resources. In lean years, the extra effort put into begging reduces growth, and in good years offspring get more food. One would predict that mothers would apply more testosterone to either level the playing field for late-born youngsters or to increase the competitive asymmetry, depending on the desired outcome to raise all chicks or to allow siblicide to proceed at a low cost to the older chick. However, there is not strong evidence for this prediction (Smiseth et al., 2011). We need more information on the commonness of maternal hormonal manipulation of offspring condition and on how endogenous offspring hormones (rather than experimentally altered titers) affect begging and whether this might be a counterweapon to help regulate their begging.

More and more of the genetic basis for behavior is being discovered each month, and among the most interesting new research is the genetic basis for monogamy and paternal care in mammals. Monogamy is rare in mammals, and even for these species, individuals can be "unfaithful." Rodent mating systems vary from **monogamous** to **promiscuous**. The monogamous prairie vole's (*Microtus ochrogaster*) social behavior is characterized by pair bonding, biparental care, and increased defense by males, whereas in the promiscuous meadow vole (*Microtus pennsylvanicus*) males seek out new females after each mating and provide little assistance rearing the litter. The dramatic increase in the receptors for the neurotransmitter **vasopressin** in the appropriate brain centers facilitates fidelity and good parenting in the monogamous males. (This brain center, the ventral pallidum, is generally involved with addiction and pleasure perception.) The genetic basis for this increase in vasopressin receptors has been identified as a stretch of repeats in the section of DNA that regulates the gene *Avpr1a*; the more repeats, the more receptors, the greater sensitivity to vasopressin, and the stronger the partner preference. When the extra bit of DNA was introduced into the genome of the promiscuous meadow vole, these males provided more paternal care, retrieving and licking the pups (Nair & Young, 2006). There is increasing evidence that this "monogamy gene" may be linked to pair-bonding behavior in many mammal species, including humans (Walum et al., 2008). Although the genetic basis of behavior will no doubt prove to be very complex, the more we know about the underpinnings of the differences in social behavior, the more exciting the revelations become and the more we can expand our understanding into human behavior and the evolution of behavior in general.

ACKNOWLEDGMENTS

I would like to thank my colleagues, Jessica Bolker, David Berlinsky and Adrienne Kovach, for years of good conversations and the National Science Foundation and the U.S. Department of Agriculture for financial support.

REFERENCES AND SUGGESTED READING

Anderson, D. J. (1990). Evolution of obligate siblicide in boobies. 1. A test of the insurance-egg hypothesis. *American Naturalist*, 135, 334–350.

Anderson D. J. & R. E. Ricklefs. (1992). Brood size and food provisioning in masked and blue-footed boobies (*Sula* spp.). *Ecology*, 73, 1363–1374.

Bartlett, J. (1987). Filial cannibalism in burying beetles. *Behavioral Ecology and Sociobiology*, 21, 179–183.

Bergstrom, M. L. & L. M. Fedigan. (2010). Dominance among female white-faced capuchin monkeys (*Cebus capucinus*): Hierarchical linearity, nepotism, strength and stability. *Behaviour*, 147, 899–931.

Blondel, J., M. Maistre, P. Perret, S. Hurtrez-Boussès, & M. M. Lambrechts. (1998). Is the small clutch size of a Corsican blue tit population optimal? *Oecologia*, 117, 80–89.

Braithwaite, R. W. & A. K. Lee. (1979). A mammalian example of semelparity. *American Naturalist*, 113, 151.

Burley, N. (1986). Sexual selection for aesthetic traits in species with biparental care. *American Naturalist*, 127, 415–445.

Clutton-Brock, T. H. (1991). *The Evolution of Parental Care*. Princeton, NJ: Princeton University Press.

Clutton-Brock, T. H. (2002). Breeding together: Kin selection and mutualism in cooperative vertebrates. *Science*, 296, 69–72.

Clutton-Brock, T. H. (2007). Sexual defection in males and females. *Science*, 318, 1882–1885.

Clutton-Brock, T. H., F. E. Guinness, & S. D. Albon. (1982). *Red Deer: Behavior and Ecology of Two Sexes*. Edinburgh, UK: Edinburgh University Press.

Eng, A. L., K. Esch, L. Smale, & K. Holekamp. (2000). Mechanisms of maternal rank inheritance in the spotted hyena, *Crocuta crocuta*. *Animal Behavior*, 60, 323–332.

Forbes, S. (2005). *A Natural History of Families*. Princeton, NJ: Princeton University Press.

Forbes, S. & B. Glassy. (2000). Asymmetric sibling rivalry and nestling growth in red-winged blackbirds (*Agelaius phoeniceus*). *Behavioral Ecology and Sociobiology*, 48, 413–417.

Gilmore, R. G., J. W. Dodrill, & P. A. Linley. (1983). Reproduction and embryonic development of the sand tiger shark, *Odontaspis taurus* (Rafinesque). *Fishery Bulletin*, 81, 201–225.

Goodall, J. v. L. & H. van Lawick. (1971). *In the Shadow of Man*. New York: Houghton Mifflin.

Gross, M. R. & R. Shine. (1981). Parental care and mode of fertilization in ectothermic vertebrates. *Evolution*, 35, 775–793.

Gwynne, D. T. (1986). Courtship feeding in katydids (Orthoptera: Tettigoniidae): Investment in offspring or in obtaining fertilizations? *American Naturalist*, 128, 342–352.

Gwynne, D. T. (1988a). Courtship feeding and the fitness of female katydids (Orthoptera: Tettigoniidae). *Evolution*, 42, 545–555.

Gwynne, D. T. (1988b). Courtship feeding in katydids benefits the mating male's offspring. *Behavioral Ecology and Sociobiology*, 23, 373–377.

Haig, D. (1993). Genetic conflicts in human pregnancy. *Quarterly Review of Biology*, 68, 495–532.

Humphries, C. A., V. D. Arevalo, K. N. Fischer, & D. J. Anderson. (2006). Contributions of marginal offspring to reproductive success of Nazca booby (*Sula granti*) parents: Tests of multiple hypotheses. *Oecologia*, 147, 379–390.

Kam, Y.-C. & H.-W. Yang. (2002). Female-offspring communication in a Taiwanese tree frog, *Chirixalus eiffingeri* (Anura: Rhacophoridae). *Animal Behaviour*, 64, 881–886.

Kilner, R. & R. A. Johnstone. (1997). Begging the question: Are offspring solicitation behaviours signals of need? *Trends in Ecology and Evolution*, 12, 11–15.

Klug, H. & M. B. Bonsall. (2007). When to care for, abandon, or eat your offspring: The evolution of parental care and filial cannibalism. *American Naturalist*, 170, 886–901.

Kokko, H. & M. Jennions. (2003). It takes two to tango. *Trends in Ecology and Evolution*, 18, 103–104.

Komdeur, J. & J. Ekman. (2010). Adaptation and constraints in the evolution of delayed dispersal: Implications for cooperation. In T. Szekely, A. Moore, & J. Komdeur (eds.), *Social Behavior: Genes, Ecology and Evolution* (pp. 306–327). Cambridge, UK: Cambridge University Press.

Kvarnemo, C. (2010). Parental care. In D. F. Westneat & C. W. Fox (eds.), *Evolutionary Behavioral Ecology* (pp. 451–467). Oxford, UK: Oxford University Press.

Lack, D. (1947). The significance of clutch size. *Ibis*, 89, 302–352.

Lee, A. K. & A. Cockburn. (1985). *Evolutionary Ecology of Marsupials*. Cambridge, UK: Cambridge University Press.

Manica, A. (2007). Filial cannibalism in teleost fish. *Biological Reviews*, 77, 261–277.

Mas, F., K. F. Haynes, & M. Kölliker. (2009). A chemical signal of offspring quality affects maternal care in a social insect. *Proceedings of the Royal Society of London, B*, 276, 2847–2853.

Mehlis, M., T. C. M. Bakker, L. Engqvist, & J. G. Frommen. (2010). To eat or not to eat: Egg-based assessment of paternity triggers fine-tuned decisions about filial cannibalism. *Proceedings of the Royal Society of London, B*, 277, 2627–2635.

Mock, D. W. & L. S. Forbes. (1995). The evolution of parental optimism. *Trends in Ecology and Evolution*, 10, 130–133.

Mock, D. W. & T. C. Lamey. (1991). The role of brood size in regulating egret sibling aggression. *American Naturalist*, 138, 1015–1026.

Nair, H. P. & L. J. Young. (2006). Vasopressin and pair-bond formation: Genes to brain to behavior. *Physiology*, 21, 146–152.

Okuda, N. & Y. Yanagisawa. (1996). Filial cannibalism by mouthbrooding males of the cardinal fish, *Apogon doederleini*, in relation to their physical condition. *Environmental Biology of Fishes*, 45, 397–404.

Panaitof, S. C., M. P. Scott, & D. W. Borst. (2004). Plasticity in juvenile hormone in male burying beetles during breeding: Physiological consequences of the loss of a mate. *Journal of Insect Physiology*, 50, 715–724.

Pettifor, R. A. (1993a). Brood-manipulation experiments. I. The number of offspring surviving per nest in blue tits (*Parus caeruleus*). *Journal of Animal Ecology*, 62, 131–144.

Pettifor, R. A. (1993b). Brood-manipulation experiments. II. A cost of reproduction in blue tits (*Parus caeruleus*). *Journal of Animal Ecology*, 62, 145–159.

Pfennig, D. W. (1999). Cannibalistic tadpoles that pose the greatest threat to kin are the most likely to discriminate kin. *Proceedings of the Royal Society of London, B*, 266, 57–61.

Price, K. (1996). Begging as competition for food in yellow-headed blackbirds. *Auk*, 113, 963–967.

Price, K. (1998). Benefits of begging for yellow-headed blackbird nestlings. *Animal Behaviour*, 56, 571–577.

Price, K., H. Harvey, & R. Ydenberg. (1996). Begging tactics of nestling yellow-headed blackbirds, *Xanthocephalus, xanthocephalus*, in relation to need. *Animal Behaviour*, 51, 421–435.

Rauter, C. M. & A. J. Moore. (1999). Do honest signalling models of spring solicitation apply to insects? *Proceedings of the Royal Society of London, B*, 266, 1691–1696.

Rickard, I. J., A. F. Russell, & V. Lummas. (2007). Producing sons reduces lifetime reproductive success of subsequent offspring in pre-industrial Finns. *Proceedings of the Royal Society of London, B*, 274, 2981–2988.

Rohwer, S. (1978). Parental cannibalism of offspring and egg raiding as a courtship strategy. *American Naturalist*, 112, 429–440.

Scott, M. P. (1990). Brood guarding and the evolution of male parental care in burying beetles. *Behavioural Ecology and Sociobiology*, 26, 31–39.

Scott, M. P. (1997). Reproductive dominance and differential ovicide in the communally breeding burying beetle *Nicrophorus tomentosus*. *Behavioral Ecology and Sociobiology*, 40, 313–320.

Scott, M. P. & T. N. Tan. (1985). Radionuclide determination of male mating success in natural populations. *Behavioural Ecology and Sociobiology*, 17, 29–33.

Seizi, S. & N. Masahiro. (2009). To compensate of not? Caring parents respond differently to mate removal and mate handicapping in the burying beetle, *Nicrophorus quadripunctatus*. *Ethology*, 115, 1–6.

Sheldon, B. C. (2002). Relating paternity to paternal care. *Philosophical Transactions of the Royal Society of London, B*, 357, 341–350.

Smiseth, P. T. & A. J. Moore. (2002). Does resource availability affect offspring begging and parental provisioning in a partially begging species? *Animal Behaviour*, 63, 577–585.

Smiseth, P. T. & A. J. Moore. (2007). Signaling of hunger by senior and junior larvae in asynchronous broods of a burying beetle. *Animal Behaviour*, 74, 699–705.

Smiseth, P. T. & A. J. Moore. (2008). Parental distribution of resources in relation to larval hunger and size rank in the burying beetle *Nicrophorus vespilloides*. *Ethology*, 114, 789–796.

Smiseth, P. T., M. P. Scott, & C. Andrews. (2011). Hormonal regulation of offspring begging and mediation of parent-offspring conflict. *Animal Behaviour*, 81, 507–517.

Szentirmai, I., T. Székely, & J. Komdeur. (2007). Sexual conflict over care: Antagonistic effects of clutch desertion on reproductive success of male and female penduline tits. *Journal of Evolutionary Biology*. 20, 1739–1744.

Trivers, R. L. (1974). Parent-offspring conflict. *American Zoologist*, 14, 249–264.

Trumbo, S. T. (1990). Reproductive benefits of infanticide in a biparental burying beetle, *Nicrophorus orbicollis*. *Behavioral Ecology and Sociobiology*, 27, 269–273

Trumbo, S. T. (1991). Reproductive benefits and the duration of paternal care in a biparental burying beetle, *Nicrophorus orbicollis*. *Behaviour*, 117, 82–105.

Trumbo, S. T. (1996). Parental care in invertebrates. In J. S. Rosenblatt & C. T. Snowdon (eds.), *Parental Care: Evolution, Mechanisms and Adaptive Significance* (pp. 3–51). New York: Academic Press.

Trumbo, S. T. & A. G. Fernandez. (1995). Regulation of brood size by male parents and cues employed to assess resource size by burying beetles. *Ethology, Ecology and Evolution*. 7, 313–322.

Walum, H., L. Westberg, S. Henningsson, et al. (2008). Genetic variation in the vasopressin receptor 1a gene (AVPR1A) associates with pair-bonding behavior in humans. *Proceedings of the National Academy of Sciences, USA*, 105, 14153–14156.

Williams, G. C. (1966). Natural selection, the cost of reproduction, and a refinement of Lack's principle. *American Naturalist*, 100, 687–690.

Wilson, D. S. & E. O. Wilson. (2007). Rethinking the theoretical foundation of sociobiology. *Quarterly Review of Biology*, 82, 327–348.

8
Make Space Enough between You*: Intraspecific Variation in Animal Spacing

Nancy G. Solomon and Brian Keane

INTRODUCTION

The ways in which animals occupy and share space have intrigued biologists for decades, and there is an extensive body of literature describing the spacing patterns of many animal species from diverse taxa (Allee, 1931; Brown & Orians, 1970; Maher & Lott, 2000). Many of the initial investigations of the mechanisms causing individuals to exhibit a particular type of spacing behavior (***proximate factors***) and why a specific type of spacing system evolved (***ultimate factors***) focused on interspecific differences in space use. These patterns of space use reflect the average or most common pattern of spacing among individuals of a specific species, and all individuals of the same species were thought to display the same species-specific pattern of spatial dispersion and space use (Brown, 1975).

Evidence began to appear suggesting that a ***phylogenetic*** explanation for spacing patterns was not always correct (Brown, 1975). Different spacing patterns were found within the same taxa, and similarities in spacing patterns were noted in unrelated species living under similar environmental conditions.

---

*Shakespeare, *Antony and Cleopatra*, Act 2, Scene 3

If spacing patterns were not related to the evolutionary history of a particular species, then it is likely that ecological factors in the present-day environment, such as finding food resources or avoiding predators, which are critical for survival and reproduction, are influencing spacing patterns.

Dale Lott (1984) presented considerable evidence that variation in space use occurs within and between numerous geographically distinct populations of the same species. Individuals of many species exhibit flexibility in space use depending on environmental (e.g., distribution of critical resources, predation, population density) and individual (e.g., genotype, age) conditions (for more details, see Lott, 1991). There are now hundreds of publications documenting intraspecific variation in spacing systems across a broad spectrum of animal taxa (for reviews see Lott, 1991; Maher & Lott, 2000; Adams, 2001; Nilsen et al., 2005). These studies demonstrate that intraspecific variation in space use between geographically distinct populations as well as within populations is common instead of being rare instances of atypical behavior as they were previously thought to be.

Intraspecific variation in spacing patterns is typically viewed as an adaptive response to differences in environmental conditions (e.g., ***conspecifics***, ***heterospecifics***, and habitat structure) or to different characteristics of individuals, such as developmental stage or genotype. Because it appears that the same types of proximate factors that influence interspecific spacing patterns also affect intraspecific space use in a number of species, intraspecific variation in spacing patterns provides an excellent tool for examining the determinants of spacing patterns without the variation being confounded by phylogenetic influences.

Progress in understanding the proximate factors that determine the manner in which individuals distribute themselves in space will greatly enrich studies in ***socioecology*** and ***population biology*** because the spacing behavior of individuals can have consequences for social interactions, patterns of reproduction, and population structure and dynamics (Figure 8.1). Furthermore, a better understanding of intraspecific variation in space use should have practical implications for the management and conservation of animal populations.

Rather than attempting to review the extensive body of literature on intraspecific variation in animal spacing systems in this chapter, our aim here is to highlight several main points regarding intraspecific variation in spacing using examples from diverse taxa. We first summarize some of the common types of animal spacing patterns seen in nature. Next, we review some of the empirical support for the proximate mechanisms underlying particular patterns of spatial dispersion. Finally, we discuss the consequences of intraspecific variation

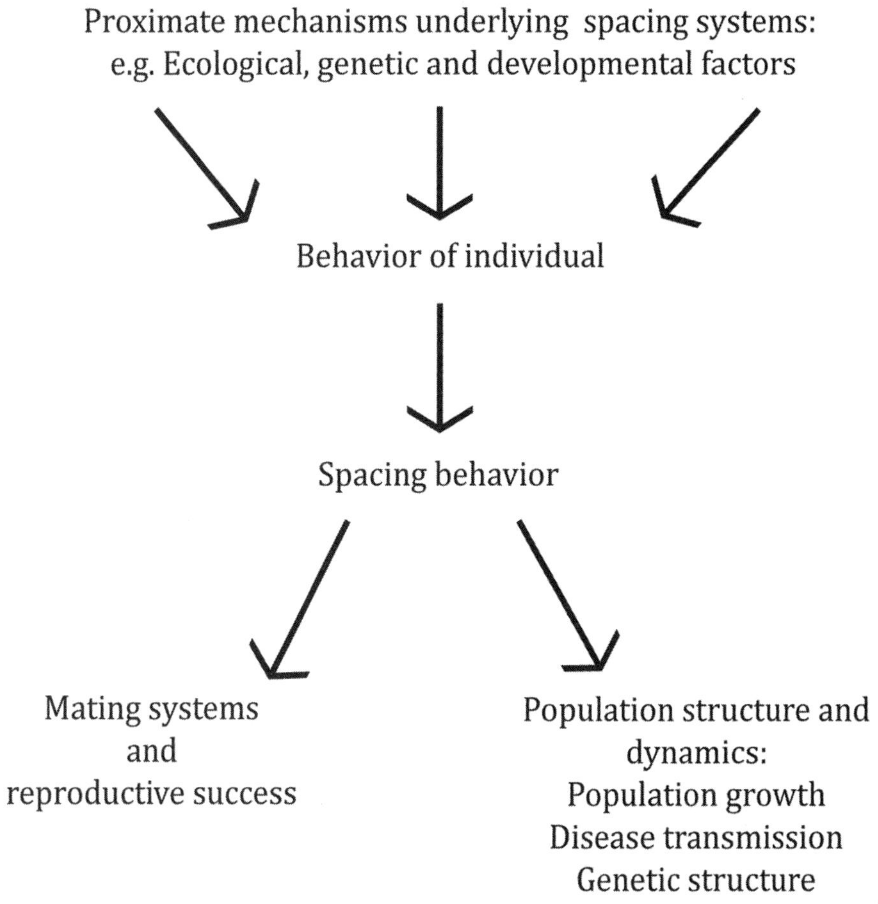

Proximate mechanisms underlying spacing systems:
e.g. Ecological, genetic and developmental factors

↓

Behavior of individual

↓

Spacing behavior

↙ ↘

Mating systems
and
reproductive success

Population structure and
dynamics:
Population growth
Disease transmission
Genetic structure

Figure 8.1. Proximate factors (e.g., ecological or genetic) can result in intraspecific variation in space use by differentially influencing the behavior of individuals. The manner in which individuals are spatially distributed can have substantial consequences for reproductive behavior as well as population structure and dynamics.

in spacing patterns on mating behavior, population-level processes such as population growth, and processes at even a larger scale like conservation.

TYPES OF SPACING SYSTEMS

The spacing pattern of individuals is typically viewed as a continuum ranging from discrete aggregates of individuals, to a lack of spatial pattern due to random spacing, to very uniformly spaced individuals. Random spacing is very unusual among animals, and most patterns of spatial dispersion tend towards either an aggregated or uniform distribution. Although animal space use is really a continuum, the spatial distribution of individuals relative to each other can be categorized into a small number of commonly recognized discrete spacing systems.

Random Distribution of Individuals

A *random distribution* of individuals is only seen in a few animals, such as Poli's stellate barnacle (*Chthamalus stellatus*) on rocky shores in southwestern England and southern Europe (Yahner, 2012). The position of one individual is independent of another individual; that is, individuals do not attract or repel each other. This pattern of distribution is thought to occur in habitats where resources are spatially and temporally consistent.

Uniform Distribution of Individuals

Uniform spacing is often due to direct agonistic interactions between individuals that repel each other (e.g., *territorial behavior*) but can also result from a uniform distribution of resources within a habitat.

Territoriality—In 1920, British ornithologist Henry E. Howard, in his book *Territory in Bird Life*, first described the concept of territoriality, although naturalists as far back as the seventeenth century described behavior encompassed by Howard's definition. More recently Christine Maher and Lott (1995) reviewed definitions of territoriality that have been applied to vertebrates and suggested that a practical definition should include both behavioral and ecological elements. They proposed that a useful definition of *territory* would be a fixed area or space containing specific resources such as food, nest sites, or mates from which one or more individuals (territory residents) exclude other members of the same species (conspecifics). Territorial behavior may include overt aggressive behavior, visual displays, vocalizations, or scent marking. These behaviors typically require an increased energetic investment and can decrease survival through injury due to fighting or a greater risk of predation while engaged in territorial behavior. Therefore, the evolution and maintenance of territoriality requires that the average overall fitness of territorial individuals exceed that of nonterritorial individuals.

There may be *sexual dimorphism* in territorial behavior. In some species males are territorial, and in others only females defend exclusive areas. Except when paternal care is critical for offspring growth, development, or survival, the resource that males are expected to defend to maximize their fitness is opposite-sex conspecifics. In contrast, females would maximize their fitness by defending critical resources necessary to prevent *infanticide* and successfully rear offspring. For this reason, territorial males may often benefit from accepting additional females into their territories. However, they should repel other males from their territories because the latter represent a threat to the resident males' fitness (***mate guarding*** hypothesis; Wittenberger & Tilson, 1980; Back et al., 2002). In contrast, females should secure an area encompassing sufficient critical resources, such as nest sites and food, against competitors of both sexes (***resource defense*** hypothesis; Ostfeld, 1985; Back et al., 2002). Consistent with this latter hypothesis are observations that increased food abundance often decreases the level of female territoriality (Ostfeld, 1985, 1986; Ylönen et al., 1988). Additionally, female territoriality can be a mechanism to prevent infanticide through territorial behavior directed at males or other females (Wolff, 1993).

Male-female pairs—Male-female pairs are found in a number of invertebrates and vertebrates (see Table 1 in Matthews, 2002; Martin et al., 2007). In these species, a single adult male and female have extensively overlapping ***home ranges*** from which they exclude all other conspecifics. This type of spacing system can occur when a male and conspecific female spend most of their time together and coordinate behavior (***associated social monogamy***). One example of this type of social monogamy (a male and female living together) is seen in snapping shrimp (genus *Alpheus*) in which a male-female pair jointly defends a territory. Lauren Matthews (2002) suggested that advantages accrued from territorial cooperation (sharing in defense and food) may be one factor that may have selected for social monogamy because both partners are likely to benefit by sharing the time and energetic costs of defending and maintaining their territory. In some of these male-female associations, both members of the pair also participate in parental care of offspring (Kleiman, 1977; Wittenberger & Tilson, 1980; Clutton-Brock, 1991; Woodroffe & Vincent, 1994).

In another type of social monogamy, a male and female may share a territory but forage and sleep alone (***dispersed social monogamy***). Dispersed social monogamy is much more rare than associated social monogamy and is found in some small mammals like the Cape porcupine (*Hystrix africaeaustralis*; Corbet & van Aarde, 1996), tree shrews (e.g., *Tupaia tana*; Munshi-South et al., 2007), nocturnal prosimians (e.g., fat-tailed dwarf lemur, *Cheirogaleus medius*; Fietz, 1999), and the Japanese serow (*Capricornis crispus*; Kishimoto, 2003), a bovid. In these species, males and females are territorial and solitary foragers. There is overlap between the territory of one adult male and one female, allowing a male to defend resources necessary for himself and a mate. Further, female dispersion in the Japanese serow typically prevents males from defending the home ranges of more than one female, although a small proportion of males (19–36 percent) manage to do so (Kishimoto, 2003).

Many hypotheses have been proposed to explain the evolution and maintenance of social monogamy. When females are widely dispersed, often thought to be the result of resource dispersion or female-female aggression, each male home range may only be able to overlap the home range of primarily one female, resulting in social monogamy. Territories defended by one male and one female may occur where males do not differ much in body mass so one male cannot prevent other males from gaining access to females and the population sex ratio is 1:1. Benefits due to joint territorial defense and mate guarding also have been proposed to explain social monogamy. These hypotheses are not necessarily mutually exclusive, and both joint territorial defense and mate guarding have been thought to explain social monogamy in pair-living snapping shrimp of the genus *Alpheus*, for example (Matthews, 2002). Finally, the need for biparental care can explain social monogamy in some species (see Table 3.2 in Brotherton & Komers, 2003), but not others.

Aggregations

Aggregated patterns of individuals may be due to social tendencies (e.g., cooperative care of offspring or predator defense) or to a clumped distribution of resources (e.g., food or mates). Some types of groups form when multiple individuals are simply attracted to some feature of the environment (Brown & Orians, 1970; Wittenberger, 1981) in contrast to the social groups formed by other species such as communal nesters or **eusocial** animals such as hymenoptera species (see Table 9.1 in Bourke, 1997) or naked mole-rats (*Heterocephalus glaber*; Jarvis, 1981; Lacey & Sherman, 1991). In eusocial species, there is a much more complex social structure (e.g., division of labor). Avian flocks, fish schools, and mammalian herds would be intermediate in this continuum of sociality. In flocks, schools, and herds, social factors hold individuals together, but there is no division of labor and the groups are without a dominant breeding pair or leader as seen in eusocial insects and mammals. Many classification schemes have been proposed to define different types of groups, but we will not discuss classification schemes or all forms of animal spatial grouping in this chapter. Rather, we will restrict ourselves to only a few types of space use here.

Fish schools, bird flocks, or mammalian herds—Two important criteria used in defining a group of this type are density and behavior. Density abruptly increases when the edge of this type of group is encountered, and there is often some type(s) of complex, coordinated behavior (such as coordinated movement) resulting from social interactions between individuals (Beiswenger, 1975).

Aggregations such as fish schools, bird flocks, or mammalian herds appear to result from the interplay between long-range attraction and short-range repulsion among individuals (Herbert-Read et al., 2011; Katz et al., 2011). Groups of this type are proposed to form when predator pressure is high (Hamilton, 1971). Flocks or schools may also form when high-quality feeding areas are controlled by territorial conspecifics. The occurrence of a flock or school increases intruder pressure to the point that territorial defense becomes more costly than the benefit received, and thus territoriality will disappear (Robertson et al., 1976). Thus, fish schools, bird flocks, and mammal herds can also provide foraging benefits (Lazarus, 1979; Partridge, 1982).

Coloniality—When individuals breed in territories that are densely clustered in space and these territories do not contain resources other than breeding sites, we refer to these animals as **colonial** (Danchin et al., 2008). Coloniality is seen in about 13 percent of avian species and almost all seabirds (Chiozzi et al., 2011). Coloniality has also been reported in some mammalian species (e.g., elephant seals, *Mirounga leonine*, Baldi et al., 1996; prairie dogs, *Cynomys leucurus* and *C. ludovicianus*, Hoogland, 1981; numerous species of bats, Davis et al., 1962; Bradbury, 1977) and some fish (e.g., damselfish, *Eupomacentrus parties*, Myrberg, 1972; bluegill sunfish, *Lepomis macrochirus*, Gross & MacMillan, 1981).

Coloniality is thought to have arisen multiple times so selection pressures favoring its evolution may differ between species (Rolland et al., 1998). Coloniality may have evolved and be maintained due to limited favorable breeding sites (e.g., seabirds, Veen, 1977; and elephant seals, Baldi et al., 1996), enhanced foraging efficiency (e.g., a colony acting as an information center, Ward & Zahavi, 1973; Brown, 1986), or allowing individuals more time to forage if less time is needed for vigilance or protection against predation through a number of mechanisms such as many eyes to watch for predators, the **confusion effect**, or collective defense (reviewed in Wittenberger & Hunt, 1985; Siegel-Causey & Kharitonov, 1990; Danchin et al., 2008). Results from studies testing these hypotheses are not consistent. In some studies (e.g., Olsthoorn & Nelson, 1990; Chiozzi et al., 2011) no nest-site limitation was found. Furthermore, a negative relationship between per capita nest predation risk and colony size or nearest-neighbor distance was found in some studies, but no relationship between these variables or the opposite pattern has been found in other studies (Møller, 1987; Anderson & Hodum, 1993). Some investigators also have suggested that there are problems with the hypothesis that colonies function as information centers to enhance food finding (Mock et al., 1988; Richner & Heeb, 1995). Heinz Richner and Philipp Heeb (1996) proposed that, in contrast to the information center hypothesis, colonies might serve as centers for recruiting individuals for group foraging. Additionally, it has been proposed that coloniality may be a side effect of conspecific attraction in habitat selection or sexual selection (Morton et al., 1990; Danchin & Wagner, 1997). Conspecific cues can lead to the formation of invertebrate colonies (Meadows & Campbell, 1972), and evidence suggests that this can occur in some vertebrates. For example, Jan Veen (1977) was able to attract sandwich terns (*Sterna sandvicencis*) to an area using decoys. These individuals settled in between the decoys initially and then settled on territories around the newly established colony. Finally, Gene Morton and colleagues (1990) proposed that colonies might form due to benefits to older males from extra-pair matings with females whose partners were younger males. Monogamous females may benefit from having aggregated territories like those seen in a colony for the same reason because females with male social partners also could obtain extra-pair matings (Wagner, 1993; Wagner et al., 1996).

Communal nesting—**Communal nesting** occurs when more than one breeding female occupies a single nest or chamber within a burrow. Communal nesting is fairly widespread among many species of insects (Wilson, 1971) and refers to a relatively simple form of social organization where individuals share a nest and each individual produces and provisions her own offspring. There is no reproductive division of labor and no overlap of generations such as is seen in eusocial insects (Michener, 1974). Communal nesting also has been reported in numerous species of birds and mammals such as rodents and banded mongooses (*Mungos mungo*; Gilchrist et al., 2004). In these species, females often share in care of offspring (Hayes, 2000; Vehrencamp & Quinn, 2004), and offspring may benefit from an increase in care (Lewis & Pusey, 1997; Branchi, 2009). Despite these benefits, communal nesting is often costly. In some species, such as greater anis (*Crotophaga major*), in which multiple monogamous pairs nest together, females eject each other's eggs (see also Vehrencamp & Quinn, 2004). In this species, though, benefits may outweigh costs because pairs in larger groups have greater reproductive success (Riehl, 2011).

PROXIMATE MECHANISMS UNDERLYING SPACING SYSTEMS

Ecological and Social Determinants of Spacing Systems

The most common ecological factor cited for its influence on spacing patterns is food resources (Maher & Lott, 2000). This is not surprising as a number of theoretical models predict that variation in food resources could influence space use (Carpenter & McMillen, 1976; Hixon, 1980; Maher & Lott, 2000; McLoughlin et al., 2000; Adams, 2001). Food resources can influence spacing behavior in numerous ways, through their abundance, distribution, accessibility (the degree to which food can be acquired by individuals), density, predictability, quality, rate of replenishment, and type (e.g., fruit, insect prey). Because food is relatively easy to quantify and study compared to other ecological factors, such as predation pressure, the importance of food as a factor determining spacing systems may have received more emphasis than its importance would warrant (Stamps, 1994). It is also likely that more than one variable can influence spacing behavior (Maher & Lott, 2000).

The abundance of food has been one of the most frequently studied effects of food resources, and numerous studies have found a relationship between food abundance and territoriality. However, the nature of this association has not been consistent among studies. For example, Hawaiian honeycreepers (*Vestiaria coccinea*; Carpenter, 1987), California voles (*Microtus californicus*; Ostfeld, 1986), and grey-sided voles (*Clethrionomys rufocanus*; Ims, 1987) exhibited decreased territoriality at higher food abundance. In contrast, lower food abundance was correlated with a lack of territoriality in other species such as the Townsend solitaire (*Myadestes townsendi*; Lederer, 1981), acorn woodpecker (*Melanerpes formicivorus*; Hannon et al., 1987), and pronghorn (*Antilocapra Americana*; Maher, 1994). While these findings regarding food abundance and territorial behavior may seem irreconcilable, they are consistent with several cost-benefit models that predict that the relationship between food abundance and territoriality is an inverted U-shaped function (Carpenter & MacMillen, 1976; Grant, 1993; Maher & Lott, 2000). These models predict that territoriality occurs when the quantity of food is moderate, but when food is scarce or extremely abundant then territories are more energetically costly to defend than the benefits obtained from defense (i.e., not economically defendable *sensu* Brown, 1964).

Food abundance can also affect the size of an individual's territory. A negative correlation between food abundance and home range (an area that an animal moves through in search of food and mates) or territory size also has been observed in numerous species (Powell, 1987; Sandell, 1989; Maher & Lott, 2000; Nilsen et al., 2005), and models of optimal territory size predict such a relationship should often be expected (Hixon, 1980; Schoener, 1983). The results of a natural experiment indicate a causal relationship between food abundance and territory size. Home-range size and overlap of adult male desert iguanas (*Dipsosaurus dorsalis*) were compared before and after a sandstorm that substantially reduced food resources (e.g., leaves, flowers) of the herbivorous lizards (Krekorian, 1976). When food resources were less abundant after the storm, home-range size was significantly greater than before the storm. Furthermore, although male home ranges did not overlap prior to the storm, there was extensive overlap of home ranges afterwards. Controlled field manipulations of the abundance of natural food sources also have shown decreases in territory size in response to higher food levels in mollusks (e.g., owl limpet, *Lottia gigantean*; Stimson, 1973), insects (e.g., *Leucotrichia pictipes*; Hart, 1985), fish (e.g., rainbow trout, *Oncorhynchus mykiss*; Keeley & McPhail, 1998), lizards (e.g., spiny lizard, *Sceloporus jarrovi*; Simon, 1975), birds (rufus hummingbird, *Selasphorus rufus*; Hixon et al., 1983), and mammals (e.g., eastern chipmunk, *Tamias striatus*; Mares et al., 1982). The importance of these experimental studies is that they demonstrate a direct causal link between food abundance and territory size.

The distribution of food resources is another commonly investigated variable proposed to influence spacing patterns. Some experimental manipulations showed that territoriality occurred when food was clumped but not when it was evenly distributed (Zahavi, 1971; Grant & Guha, 1993). In addition, differences in the distribution of food are correlated with the distribution of individuals such as prairie voles (*Microtus ochrogaster*). Dicots, which are critical in the diet of prairie voles, were more clumped at a study site in Indiana than at a geographically distinct study site in Kansas. Adult females, males, and nest sites also were more clumped in Indiana than in Kansas, resulting in considerable home-range overlap among neighboring females compared to females in the Kansas population (Streatfeild et al., 2011).

As with food abundance, results from observational studies examining food distribution and spacing patterns are inconsistent, which may be partially due to the fact that there are a number of differences in methodologies used in previous studies. It is also possible that the relationship between the food distribution and territoriality may be best explained, as was the relationship between food quantity and territoriality, by an inverted U-shaped curve where moderately clumped resources are the only ones that are worth defending (Craig & Douglas, 1986).

Investigations of both food abundance and distribution generally showed that limited, patchy food resulted in increased territoriality. Consistent with these findings, when food was abundant and patchy, animals did not defend territories (e.g., black bears, *Ursus americanus*; Rogers, 1987).

Some studies also showed that types of prey could influence space use. In Great Britain, badgers (*Meles meles*) feed primarily on earthworms, which are very abundant, but in continental Europe earthworms are typically much less abundant and badgers tend to specialize on foods such as rabbits (*Oryctolagus cuniculus*), insects, and fruit (Johnson et al., 2000). Throughout continental Europe, badgers are typically solitary with little home-range overlap among conspecifics, whereas in Great Britain badgers commonly live in large social groups of up to 30 individuals sharing a home range.

The ease of finding and killing prey can also result in differences in spacing patterns. Coyotes (*Canis latrans*) are typically solitary predators, but several adults may hunt together sharing a common home range when prey items that are difficult for single individuals to kill are abundant (Bowyer, 1987). A study of coyotes in Alberta, Canada, showed that variation in the average size of prey eaten by coyotes was positively correlated with group size, and coyotes were more likely to be solitary when their food base was primarily small prey items such as rodents and fawns instead of adult deer (Bowen, 1981). Finally, nesting Arctic skuas (*Steercorarius parasiticus*) that feed primarily on lemmings defend territories while conspecifics in the same area that feed mainly on fish taken from other birds nest colonially (Andersson & Gotmark, 1980).

Changes in space use also can occur in response to meeting energetic demands associated with the seasonal change in food resources. The lesser spotted woodpecker (*Dendrocopos minor*) inhabits deciduous forests in Europe. During the summer, individuals feed on surface-living adult insects, but from autumn to spring they feed almost exclusively on beetle larvae in dead branches of deciduous trees (Wiktander et al., 2001). Home-range size increases from about 100 hectares in late spring to over 700 hectares in winter for males and females. During the winter, these birds also become solitary and nonterritorial. The pattern of space use in **scatter-hoarding** Eurasian red squirrels (*Sciurus vulgaris*) also changes seasonally in response to food abundance. Home ranges of males and females are largest in spring and summer when seed supplies are depleted and the squirrels feed more on less preferred food items such as flowers and buds (Wauters & Dhondt, 1992). Home ranges are smallest in autumn and winter when seeds are abundant and comprise 70 percent of the diet.

Spatial organization and space use is also correlated with predation risk (see Krause & Ruxton, 2002, for a detailed overview). In general, individuals benefit in a number of ways from aggregation and group formation in the presence of predation (e.g., **many-eyes effect**, **selfish herd** effect, predator confusion, communal defense). Evidence from a number of taxa suggests that larger groups can detect the approach of predators more effectively (i.e., the many-eyes hypothesis, reviewed by Elgar, 1989), allowing each group member to spend less time being vigilant and more time foraging (fish, Godin et al., 1988; birds, Lazarus, 1979; Cresswell, 1994; mammals, Fairbanks & Dobson, 2007; van Schaik et al., 1983, but see Smith & Cain, 2009). This hypothesis assumes that information on predator presence or approach is transmitted clearly among group members (Krause & Ruxton, 2002) and is supported by some (e.g., Magurran & Higham, 1988; Treherne & Foster, 1981) but not all experimental manipulations that have been conducted (Lima, 1995).

The many-eyes hypothesis is not necessarily distinct from the ***dilution effect***, which states that as group size increases, the risk of predation for an individual decreases because the predator can only catch or eat a certain number of prey. The benefit from the dilution effect may be offset by the cost of larger groups becoming more conspicuous. Thus, these two factors need to be examined together to determine if there is a net benefit from grouping (see also Turner & Pitcher, 1986). These ideas are also related to the selfish herd effect (Hamilton, 1971), where individuals not only benefit from being in a group versus being solitary but also benefit by keeping other group members between themselves and a potential predator. Finally, the predator confusion effect describes a reduction in the ability of a predator to single out and kill an individual when it is in a group versus when it is solitary. This effect would have the biggest impact when group members look very similar and any individual that looks quite different would enable the predator to focus upon it more easily (Krause & Ruxton, 2002).

Evidence for particular effects of grouping is limited because it is difficult to distinguish among them. Additionally, if one of these particular benefits is not sufficient for an individual to become part of a group, it may be that individuals gain multiple benefits from grouping. In general, studies have provided support for the effect of predation on spacing behavior, particularly for group formation under conditions of higher predation, but it is important to also consider the costs of group formation, such as increased conspicuousness to a predator, when attempting to determine net benefits to grouping.

Population density is another ecological variable that has often been cited as correlated with intraspecific variation in spacing behavior. Although prairie voles are considered socially monogamous, within a population they may live as single individuals, male-female pairs, or in groups containing at least two adults of the same sex but frequently three or more voles. The proportion of groups increases with increasing density in natural populations (Getz et al., 1987) as well as in seminatural populations (Cochran & Solomon, 2000; Lucia et al., 2008). Home-range size is also smaller at higher densities (Solomon et al., 2009). As found in studies of food abundance and distribution, the results of studies in which the relationship between population density and territoriality was examined are inconsistent. Many investigators report the occurrence of territoriality only at low or moderate population densities and that territorial defense disappears at high density, although this not a universal result (reviewed by Maher & Lott, 2000). It may be that different investigators use the terms *low*, *moderate*, and *high* differently or that the relationship between population density and territoriality also is an inverted U-shaped function where the benefit-cost ratio is greatest when population density is moderate. In studies where food quantity and population density were both examined, population density appeared to have a stronger influence on behavior, with high density resulting in a lack of territoriality (reviewed in Maher & Lott, 2000). When food distribution and density were examined concurrently, territoriality occurred when food was clumped. Territoriality was also found when food was uniformly distributed and population density was moderate but not when density was low or high (Rubenstein, 1981). The latter result supports the hypothesis that the relationship between density and territoriality is nonlinear.

Some investigators have hypothesized that the relationship between population density (or food distribution) and territoriality occurs because individuals can assess intruder pressure (Ferron & Ouellet, 1989). The causal relationship that was proposed is that increased intruder pressure resulted in decreased territoriality due to the increased costs of defense (reviewed by Maher & Lott, 2000). This result is seen in some species of coral reef fishes (Norman & Jones, 1984; Tricas, 1989) and birds (Myers et al., 1979; Eberhard & Ewald, 1994) where the primary ecological correlate of territory size is intrusion by conspecifics.

Aside from food resources, predation, and population density, numerous other ecological variables have been implicated as contributing to intraspecific variation in space use. Specific features of the abiotic habitat have been correlated with patterns of space use. For example, rainbow trout (*Salmo gairdneri*) are more territorial in fast-flowing water relative to still water (Cole & Noakes. 1980). Additionally, soil characteristics are correlated with home-range size in tuco-tucos (*Ctenomys talarum*), suggesting that the energetics of digging may influence home-range size (Cutrera et al., 2006). Individuals in several species of rodents that are solitary in the summer may share nest sites in the winter (red-backed vole, *Clethrionomys rutilus*, West, 1977; meadow voles, *Microtus pennsylvanicus*, Madison, 1984; white-footed mice, *Peromyscus leucopus*; Madison et al., 1984). Nest sharing in these species may save energy by lowering the cost of thermoregulation. For comprehensive reviews of the potential ecological factors related to intraspecific variation in space use see Lott (1991) and Maher and Lott (2000).

Although most studies focus on just one ecological variable, it is likely that spacing behavior in nature is affected by multiple variables operating simultaneously. Greater insight into the ecological determinants of variation in space use will come from studies examining multiple ecological variables concurrently and then determining which variables best explain spacing systems.

For example, Ernest Keeley and J. Donald McPhail (1998) showed that territory size in juvenile steelhead trout (*Oncorhynchus mykiss*) was significantly inversely related to food abundance, but there was no significant affect of intruder pressure on territory size.

Individual Characteristics and Spacing

Intrapopulation differences in spacing behavior may be affected by characteristics of individuals (Steury & Murray, 2003). Larger male sphecid wasps (*Philanthus basilaris*) were significantly more likely to be territorial than small males (O'Neill, 1983). In the spiny lizard (*Sceloporus jarrovi*), territory size was influenced by sex and body size. Male territories were twice the size of female territories after controlling for body size. Within each sex, larger lizards had larger territories (Simon, 1975). In juvenile Atlantic salmon (*Salmo salar*), 88 percent of the variation in territory size was explained by differences in body size and age, whereas food abundance only explained an additional 2 percent of the variation (Keeley & Grant, 1995).

Genetic Basis for Spacing Behavior

There are not many studies that have been conducted on the genetic influences on spacing behavior, but existing studies suggest that intraspecific differences in spatial organization may also reflect genetic differences among individuals within or among populations (spiders, Uetz & Cangialosi, 1986; fish, Seghers, 1974; mammals, Hammock & Young, 2005). Dominic Wright and colleagues (2003) conducted a well-controlled breeding experiment to determine if there were differences in schooling behavior in zebrafish (*Danio rerio*) within and between populations. There was no significant difference in schooling between populations, but there was a within-population difference, indicating that schooling had a genetic component (Wright et al., 2003). Geographically isolated populations of guppies (*Poecilia reticulata*) also show differences in schooling behavior, even within the same stream. Guppies from populations in areas with predatory fish have a much greater tendency to school than guppies from areas without predatory fish. Guppies in streams without predators tend to be solitary (Seghers, 1974). Laboratory breeding experiments confirmed that differences in the tendency to school had an underlying genetic basis. In a small number of species, **alternative reproductive tactics** (different behavior patterns shown by one sex within the same species to enhance their reproductive success), with their concomitant differences in space use (territorial versus nonterritorial), are genetically based, and males do not switch tactics (e.g., bluegill sunfish, *Lepomis macrochirus*, and Pacific salmon, *Oncorhynchus* spp., Gross, 1984; marine isopods, *Paracerceis sculpta*, Shuster & Wade, 1991; ruffs, *Philomachus pugnax*, Lank et al., 1995; tree lizards, *Urosaurus ornatus*, Moore et al., 1998).

The **neuropeptide** arginine vasopressin has been shown to be important in the mediation of social behavior in male prairie voles through its action on the vasopressin 1a receptor (V1aR; Insel et al., 1994; Wang et al., 1998; Phelps, 2010). Individual variation in V1aR neural expression is correlated with length polymorphisms of microsatellite DNA within the regulatory region of the gene (*avpr1a*) encoding the V1aR (Hammock & Young, 2005). In voles maintained in outdoor rodent enclosures, neural V1aR expression in areas of the brain associated with spatial memory strongly covaried with space use, suggesting a possible link between male space use and *avpr1a* genotype (Ophir et al., 2008, but see Solomon et al., 2009).

Neurobiological Basis for Spacing Behavior

During the past few decades, evidence has accumulated showing that gonadal hormones, particularly testosterone, regulate territoriality in males at the beginning of the breeding season by apparently affecting levels of aggression (Nelson, 2000; Sinervo et al., 2000). Intrasexual competition during the breeding season results in transient increases in testosterone that, in turn, induce aggression toward same-sex conspecifics (Wingfield et al., 1987, 1990). In direct contrast, during the nonbreeding season when testosterone levels are low and the testes are regressed (Wingfield, 1994), numerous species of male birds and some species of mammals and reptiles still display territoriality (Caldwell et al., 1984; Logan, 1992; Moore & Marler, 1987; Gwinner et al., 1994). Some evidence suggests that territoriality during the nonbreeding season is modulated by conversion of testosterone to estrogen and the interaction of estrogen with the estrogen receptor (Soma et al., 1999; Wingfield et al., 2001). Because testosterone is typically low during the nonbreeding season, Kiran Soma and John Wingfield (2001) suggested that a testosterone precursor, dehydroepiandrosterone (DHEA), might be the substrate that is converted to estrogen (see Vanson et al., 1996). DHEA is also present in the blood plasma of temperate songbirds, some tropical birds (Hau et al., 2004), and mammals such as red squirrels (Boonstra et al., 2008) during the nonbreeding season at significantly higher levels than testosterone or estrogen. The seasonal pattern of DHEA parallels that of territorial aggression. Soma and Wingfield (2001) and Rudy Boonstra and colleagues (2008) suggest that the possible involvement of DHEA in aggression during the nonbreeding season may be adaptive because it allows delivery of sex steroids to specific brain regions without the cost of maintaining high levels of plasma testosterone. In the nonbreeding season, typically winter, many birds and small mammals increase their immune and thermogenic (heat-producing) activity so they can survive through the winter. Because high levels of testosterone suppress the immune system and increase energetic costs, maintenance of high levels of testosterone would increase winter mortality (Wingfield et al., 2001).

Previous studies also suggest that neuropeptides, such as arginine vasopressin and serotonin, modulate aggression in a number of species including crustaceans, amphibians, birds, and mammals. For example, neural infusions of vasopressin inhibited aggressive chases and songs in two species of territorial birds (Goodson, 1998a, 1998b). Similarly, some species of birds and mammals also show an inverse relationship between serotonin levels and aggression (Edwards & Krawitz, 1997; Sperry et al., 2003; Ten Eyck, 2008).

There is much less known about the hormonal responses to a territorial intrusion in females. The few studies that exist suggest that hormonal control of territorial aggression in females is different from that in males. Female song sparrows (*Melospiza melodia*) that experienced simulated territorial intrusions had significantly lower levels of testosterone and dihydrotestosterone than controls (Elekonich & Wingfield, 2000). There were no differences between groups with or without simulated intrusions in levels of estradiol, progesterone, or corticosterone. In a similar study with the monogamous California mouse (*Peromyscus californicus*), a decrease in progesterone and the progesterone/testosterone ratio was the only hormonal change following simulated territorial intrusion (Davis & Marler, 2003), again suggesting that territorial aggression by females is not facilitated by increased testosterone as seen in males. Androgens may limit female aggression to a short-term response rather than facilitating it, which may prevent aggression from interfering with reproduction (Elekonich & Wingfield, 2000).

Developmental Effects on Spacing Behavior

Some animals display ***ontogenetic*** shifts in spacing behavior, that is, changes in spacing behavior during maturation or between the time when individuals are juveniles and when they are adults. More typically during early life stages, when individuals are inexperienced and vulnerable, they tend to aggregate, but then aggregation decreases when they are subadults or adults as the costs of competition for critical resources become greater (Krause & Ruxton, 2002). Often, after sexual maturation, aggregation occurs again for reproductive purposes. It is less common to find species in which juveniles are solitary and then aggregate as subadults. Typically juveniles of these species are slow and use ***crypsis*** to avoid predator detection. Thus, aggregation while still a defenseless juvenile would increase detection and, therefore, the risk of predation (Tinbergen et al., 1967; Butler et al., 1999). New Zealand spiny lobsters (*Jasus edwardsii*) are solitary when they are young and small but become more social when they get older and larger, settling with others in dens (Butler et al., 1999). Aggregation occurs as a response to conspecific chemical cues, but the response to chemical cues does not develop until the lobsters are adolescents. Aggregation of subadults and adults appears to be adaptive. Mark Butler and colleagues (1999) found no difference in survival in solitary or aggregated juvenile lobsters, but survival more than doubled when subadults were grouped as compared to when they were solitary. In a few cases such as the spiny lobster, this ontogenetic shift in spacing behavior may be due to developmental constraints. For example, some marine fish do not begin to school until metamorphosis. At this time, the larvae begin to show pigmentation and are more visible to predators (Gallego & Heath, 1994), so it is adaptive to school at this time.

One study of brown trout (*Salmo trutta*) showed that despite rearing environment, aggression occurred in all dyadic encounters, but the level of aggression differed depending on early rearing environment (Sundstrum et al., 2003). Brown trout that had been reared under high-density conditions (in a hatchery) displayed more aggression and initiated aggressive encounters sooner than individuals from low-density populations in the wild, in contrast to what had been predicted. The authors suggested that this may be due to lack of experience in territorial encounters and in assessing fighting ability among hatchery-reared fish compared to wild trout. Similar results were seen in another study where wild and hatchery-reared brown trout actually competed for territories (Deverill et al., 1999). Interestingly, although hatchery-reared fish initiated more aggressive encounters, they were not more successful at obtaining better feeding territories than wild trout (Sundstrom et al., 2003; Deverill et al., 1999). It is not clear what specific factor in the environments may have caused the difference in levels of territorial aggression in these trout, but it appeared that rearing environment did affect subsequent aggression. Somewhat similar results were found in a study of the effects of population density on schooling tendency in guppies (Chapman et al., 2008). Trinadadian guppies raised in low densities showed a greater tendency to school than guppies raised in high densities. The investigators hypothesized that, in the absence of predator cues, higher densities resulted in more aggression and thus increased the costs of schooling relative to benefits during rearing.

In some species such as the desert locust (*Schistocerca gregaria*), which can exist in a solitary or a gregarious phase, crowding or high densities not only increases gregariousness during an individual's lifetime but also in its offspring if crowding occurs during *oviposition* (Islam et al., 1994). Alan McCaffery and colleagues (1998) reported that this maternal effect is due to a chemical in the foam from the egg pods that influences subsequent behavior of offspring.

Consequences of Intraspecific Variation in Spacing Behavior

Intraspecific variation in spacing can have a substantial impact on reproduction, population structure, and demography. An improved knowledge of the ecological and genetic determinants of intraspecific variation in space use will influence our understanding of many aspects of population and behavioral ecology (e.g., mating and social systems, carrying capacity) as well as contribute to conservation of threatened or endangered species.

Effects on Mating Patterns and Reproductive Success

Spacing patterns can influence the social and probably also the genetic mating systems in populations. Determining the *social mating system* (who is sharing space with or living with whom) can be difficult for species like rodents that can be difficult to observe. We used to think that a reliable index of the mating system was the spacing patterns of individuals within a population (Emlen & Oring, 1977; Ostfeld, 1985; Anderson & Titman, 1992), but numerous studies have shown that the *genetic mating patterns* differ significantly from the mating system inferred from trapping or tracking data. We now realize that observations, live trapping, and radio tracking reveal only the social mating systems in animals.

In some species of mammals (Getz et al., 1993; Wolff, 2008; Schradin et al., 2012), birds (for a review see Penteriani et al., 2011), amphibians (Howard, 1978), fish (Gross & Charnov, 1980; Taborsky et al., 2008), and insects (O'Neill, 1983; Peixoto & De Marco, 2009), some individuals, usually males, exhibit territorial behavior and occupy a fixed area of residence while other same-sex adults in the same population do not. Relative to territorial animals, the nonterritorial individuals (also referred to as *wanderers, floaters*, or satellites) typically have much larger home ranges that encompass territories of multiple opposite-sex conspecifics. In the majority of species where populations contain territorial and nonterritorial individuals, individuals may switch roles during their lifetime (Taborsky, 2001; Taborsky et al., 2008; McGuire & Getz, 2010; Penteriani et al., 2011). These nonterritorial individuals have generally been portrayed as individuals of poor quality that were excluded from territories by superior competitors and have been forced to adopt the less preferred nonterritorial tactic as a means of making the best of a bad situation when there is intense competition for suitable territories (Newton, 1992; Taborsky, 1998, 2001). The relative fitness of territorial and nonterritorial individuals has been measured in only a few species, and for some species territorial individuals do have higher reproductive success (greater sac-winged bat, *Saccopteryx bilineata*; Heckel & von Helversen, 2002), while in others such as the prairie vole results differ between studies (Ophir et al., 2008; Solomon et al., unpublished data). However, genetic analyses of paternity showed that wandering and territorial male African stripped mice (*Rhabdomys pumilio*) had similar reproductive success during a year when population density was intermediate (Schradin & Lindholm, 2011), and, in natural populations, wandering male prairie voles sired more offspring than resident males (Solomon et al., unpublished data). Therefore, additional studies using genetic measures of the relative reproductive success of territorial and nonterritorial individuals are critical to understand the factors underlying the evolution and maintenance of these two alternative patterns of space use within populations. In these studies, it is critical to include males that do not sire any offspring (Shuster, 2011) because if some males gain a disproportionately large number of matings, and hence reproductive success, then other males must lose matings or be excluded from mating. An evolutionary genetics approach requires that the relative fitness of all males, "losers" and "winners," be included in the calculations. For example, if most nonterritorial males do not sire any offspring, then excluding these males may create the appearance that there is no difference in the relative fitness of each type of male when, actually, there is a statistically significant difference (Shuster, 2009).

As has been suggested for males, it also has been hypothesized that there should be a relationship between spacing behavior of females and their reproductive success because changes in spacing patterns can result in changes in aggressive behavior (Monaghan & Metcalf, 1985), infanticide (Mappes et al., 1995; Jonsson et al., 2002), or cooperative behavior (Lambin & Krebs, 1991). In bank voles (*Myodes glareolus*), greater overlap of female home ranges negatively affected reproductive success of females, particularly offspring survival (Jonsson et al., 2002). Offspring survival may have decreased due to infanticide by nearby females.

Effects on Population-level Processes

Spacing patterns may be critical to population regulation through their effects on aggression (Watts, 1969) and breeding status (Montgomery et al., 1997). Territoriality can have a strong influence on population dynamics (Davies & Houston, 1984; Sutherland, 1996). If territories are a fixed size, only a certain number of them can fit within a defined area, and if some individuals are excluded from territories, we expect that the number of breeding males and females in the population would be limited (i.e., the population will be regulated by density-dependent processes; Wolff, 1985; Both & Visser, 2003; López-Sepulcre & Kokko, 2005). For example, many species of spiders are territorial, and at least some inhabit territories of fixed sizes (Reichert, 1981). Because territory size is fixed, additional individuals will not be able to establish territories between already existing territories.

When the number of territorial individuals is limited due to constraints of fixed territories, there would be an increase in the number of floaters, which were assumed to be nonbreeders (Smith & Arcese, 1989; Kempenaers et al., 2001). If they do not breed, an increase of floaters, instead of breeders, in the population would lead to stable or decreased population density. A decrease in population density would be expected if floaters cause direct disturbance or conflict for space (López-Sepulcre & Kokko, 2005) such as has been reported in a number of species of birds (Arcese et al., 1992; Komdeur, 1996), ayu fish (*Plecoglossus altivelis*; Iguchi & Hino, 1996), and red squirrels (*Sciurus vulgaris*; Wauters & Lens, 1995). A limit on the number of breeders depends on whether or not territory size is compressible (i.e., whether resources are divisible, Both & Visser, 2003) and how reproductive success is affected by territory size (López-Sepulcre & Kokko, 2005). Now that we are finding that floaters do reproduce in at least some species (Kempenaers et al., 2001; Solomon & Keane, unpublished data), we will need to rethink their influence on population density. The effect of reproductive floaters on population density will depend on their contribution to the population versus the contribution of resident breeders.

In species where the size of the territory is flexible, individuals may establish territories between the boundaries of previously established territories or cause territorial residents to defend smaller territories, which would allow previous floaters or recruits to establish territories (e.g., Eurasian oystercatchers, *Haematopus ostralegus*; Heg et al., 2000). This pattern could result in increasing population density through the settlement of floaters or decreasing density through decreasing birth or survival of offspring from established breeders. If food is the limiting resource for females and the amount of food affects reproduction, we expect there to be density-dependent reproduction that influences population size or density by means of territoriality. At high density, we expect that the optimal territory size will be decreased, and thus reproduction or reproductive success should decrease (Both & Visser, 2003). In areas where great tits (*Parus major*) have large, contiguous territories, there is strong density-dependent reproduction (Both, 1998).

The results from a number of studies on the effects of territoriality on population density are inconsistent, which may be due to the lack of certainty about whether or not the resource being defended was critical or whether the size of territories was flexible (Adams, 2001; Both & Visser, 2003). No effect on population size has been found in some of the studies of rodents. Ylönen and Viitala (1991) housed bank voles in experimental enclosures where they were given either clumped, predictable food or evenly distributed food. Although spacing behavior of females differed depending on the distribution of food resources and females in the clumped food treatment produced their first litter earlier in the spring, there were no differences in population size in late spring. It was assumed that the critical resource that females were competing for was food, but high-quality nest sites could also have been important. In addition, the length of the study may have been too short for an effect to be seen. It is also not known what would have happened to population sizes if the study had continued later into the breeding season. A subsequent study by Alice Rémy (2011) has shown that female bank voles in plots with clumped food were more aggregated compared to females in the dispersed-food treatment. Female aggregation was a strong predictor of population size during the breeding as well as the nonbreeding season. Analyses suggested that voles in all treatments had similar access to food resources. Therefore, it was most likely that differences in social interactions, due to differences in spacing, resulted in differences in population size.

Effects on Conservation

Anthropogenic activities are substantially altering many natural ecosystems. A better understanding of the extent and basis of intraspecific variation in space use is vital for accurately predicting how species will respond to environmental change as well as for the effective conservation and management of a species. Spacing patterns can influence a number of factors that are relevant to conservation including growth rates of a population, susceptibility of a population to disease, responses of a population to exploitation and recovery, and recolonization of habitats (Dobson & Poole, 1998). In a number of species, conspecific aggregation or attraction is responsible for at least some of these effects. In many species, individuals use the presence of cues from conspecifics to determine whether or not a particular habitat is suitable for survival and reproduction (see work by Stamps, 1988, on conspecific attraction).

Another effect that has received some attention in the last decade is the **Allee effect**, which describes a phenomenon where the reproductive rate of a population declines with declining population density (Allee, 1931). If population growth depends on encounter rates between opposite-sex conspecifics or group size, then spacing patterns will have conservation consequences.

If the encounter rates, particularly with members of the opposite sex, or group size fall below a certain threshold, then the population may decline until reproduction ceases and that population becomes extinct (Courchamp et al., 1999). In addition, when the size of a group or population becomes very small, the time it takes for that population to go extinct will decrease (Dobson & Poole, 1998).

Dispersion of individuals within a habitat can also affect the likelihood and timing of parasite or pathogen transmission, which also can lead to severe population decreases or extinction. For example, rabies and canine distemper affected African wild dogs (*Lycaon pictus*) in the early 1990s, which resulted in local population extinction (Alexander & Appel, 1994; Kat et al., 1995). Differences in the sizes of nearest-neighbor distances among individuals living in groups can result in differences in pathogen establishment and transmission rates. For example, African wild dogs and lions (*Panthera leo*) live in relatively small groups, but the rate of spread of pathogens between groups was slow compared to the spread in North Sea harbor seal (*Phoca vitulina*) colonies, where many individuals died because these seals live in large groups (Dobson & Poole, 1998).

Exploitation of elephant populations has resulted in a continent-wide decrease in numbers of greater than 95 percent in the past 200 years (Millner-Gulland et al., 1993). The size of the group in which females live contributes to the problem of finding a suitable mate. Females in widely dispersed, small groups had approximately a 39 percent reproductive rate compared to females living in a single large group. Females in the latter situation had an 87 percent reproductive rate (Dobson & Poole, 1998).

If a population of any particular species were to go extinct in a suitable habitat, the presence of conspecifics or their cues may be necessary when attempting to reestablish a population in that habitat. Without the presence of conspecifics, newly released or translocated individuals may not realize that the habitat is suitable. To solve this problem, biologists have used decoys and playback of species-specific calls to reestablish populations of extirpated birds in suitable habitat patches (Kress, 1983; Ward & Schlossberg, 2004; Parker et al., 2007). Thus, presence or spacing of individuals of the correct sex and age class could have a positive impact on reintroduction programs.

CONCLUSIONS

A considerable body of information describing animal spacing systems across a broad range of taxa has accumulated over the last 50 years. Flexibility in spacing behavior among individuals within a species in response to environmental and individual conditions is commonly reported, but our understanding of factors underlying intraspecific variation in space use still needs improvement. The next challenge for research examining animal spacing is to improve our understanding of the ultimate and proximate bases of variation in spatial systems. Intraspecific variation in spatial structure can be a valuable tool in elucidating the environmental and genetic determinants of spatial dispersion and space use, and we suggest that progress in this area can be achieved using two general research approaches. First, much of the empirical support for ecological determinants of space-use patterns is based on correlations between ecological variables and spacing patterns. Controlled field experiments in which ecological or other variables (e.g., early experience, hormone levels) are manipulated are needed to identify casual rather than correlational relationships between space use and specific factors affecting spacing. Such studies will also be crucial for testing predictions of theoretical models proposed to explain intraspecific variation in spacing (e.g., Maher & Lott, 2000; McLoughlin et al., 2000; Adams, 2001; Both & Visser, 2003). Second, numerous molecular techniques are now readily available that allow investigators to assess genetic variation among large numbers of individuals with relative ease. Genetic data can be used to determine if intra- or interpopulation variation in space use has a genetic component. Ultimately, studies simultaneously examining multiple ecological and individual variables as well as genetic factors should be most enlightening because patterns of space use are likely to influenced to varying degrees by a combination of factors.

ACKNOWLEDGMENTS

We thank former postdoctoral associates and students for assistance with research, which has allowed us to get a glimpse into the spacing patterns of prairie voles and sparked our interest in this topic. We thank numerous colleagues and former students for making us aware of some of the exciting developments in spacing behavior. The National Science Foundation and National Institutes of Health funded our work, which gave us the opportunity to investigate aspects of spacing patterns. Any opinions, findings, and conclusions or recommendations expressed in this material are those of the authors and do not necessarily reflect the views of the National Science Foundation. We thank Ken Yasukawa for the opportunity to learn more about this important topic in the writing of this chapter.

REFERENCES AND SUGGESTED READING

Adams, E. S. (2001). Approaches to the study of territory size and shape. *Annual Review of Ecology and Systematics*, 32, 277–303.

Alexander, K. A. & M. J. G. Appel. (1994). African wild dogs (*Lycaon pictus*) endangered by a canine distemper epizootic among domestic dogs near the Masai Mara National Reserve, Kenya. *Journal of Wildlife Disease*, 30, 481–485.

Allee, W. C. (1931). *Animal Aggregations: A Study in General Sociology*. Chicago: University of Chicago Press.

Anderson, D. C. & P. J. Hodum. (1993). Predator behavior favors clumped nesting in an oceanic seabird. *Ecology*, 74, 2462–2464.

Anderson, M. G. & R. D. Titman. (1992). Spacing patterns. In B. D. J. Batt, A. D. Afton, M. G. Anderson, C. D. Ankney, D. H. Johnson, J. A. Kadlec, & G. L. Krapu (eds.), *Ecology and Management of Breeding Waterfowl* (pp. 251–289). Minneapolis: University of Minnesota Press.

Andersson, M. & F. Götmark. (1980). Social organization and foraging ecology in the arctic skua, *Stercorarius parasiticus*: A test of the food defendability hypothesis. *Oikos*, 35, 63–71.

Arcese, P., J. N. M. Smith, W. M. Hochachka, C. M. Rogers, & D. Ludwig. (1992). Stability, regulation, and the determination of abundance in an insular song sparrow population. *Ecology*, 73, 805–822.

Back, S. R., L. A. Beeler, R. L. Schaefer, & N. G. Solomon. (2002). Testing functional hypotheses for the behavior of resident pine voles, *Microtus pinetorum*, toward non-residents. *Ethology*, 108, 1023–1039.

Baldi, R., C. Campagna, S. Pedraza, J. Burney, & B. J. LeBoeuf. (1996). Social effects of space availability on the breeding behavior of elephant seals in Patagonia. *Animal Behaviour*, 51, 717–724.

Beiswenger, R. E. (1975). Structure and function in aggregations of tadpoles of the American toad, *Bufo americanus*. *Herpetologica*, 31, 222–233.

Boonstra, R., J. E. Lane, S. Boutin, A. Bradley, L. Desantis, A. E. M. Newman, & K. K. Soma. (2008). Plasma DHEA in wild, territorial red squirrels: Seasonal variation and effect of ACTH. *General and Comparative Endocrinology*, 158, 61–67.

Both, C. (1998). Experimental evidence for density dependence of reproduction in great tits. *Journal of Animal Ecology*, 67, 667–674.

Both, C. & M. E. Visser. (2003). Density dependence, territoriality, and divisibility of resources: From optimality models to population processes. *American Naturalist*, 161, 326–336.

Bourke, A. F. G. (1997). Sociality and kin selection in insects. In J. R. Krebs & N. B. Davies (eds.), *Behavioral Ecology: An Evolutionary Approach*, Fourth Edition. (pp. 203–227). Cambridge, MA: Blackwell Science.

Bowen, W. D. (1981). Variation in coyote social organization: The influence of prey size. *Canadian Journal of Zoology*, 59, 639–652.

Bowyer, R. T. (1987). Coyote group size relative to predation on mule deer. *Mammalia*, 51, 515–526.

Bradbury, J. W. (1977). Social organization and communication. In W. A. Wimsatt (ed.), *Biology of Bats*, Vol. 3 (pp. 1–72). New York: Academic Press.

Branchi, I. (2009). The mouse communal nest: Investigating the epigenetic influences of the early social environment on brain and behavior development. *Neuroscience and Biobehavioral Reviews*, 33, 551–559.
Brotherton, P. N. M. & P. E. Komers. (2003). Mate guarding and the evolution of social monogamy in mammals. In U. H. Reichard & C. Boesch (eds.), *Monogamy: Mating Strategies and Partnerships in Birds, Humans and Other Mammals* (pp. 42–58). Cambridge, UK: Cambridge University Press.
Brown, C. R. (1986). Cliff swallow colonies as information centers. *Science*, 234, 83–85.
Brown, J. L. (1964). The evolution of diversity in avian territorial systems. *Wilson Bulletin*, 76, 160–169.
Brown, J. L. (1975). *The Evolution of Behavior*. New York: Norton.
Brown, J. L. & G. H. Orians. (1970). Spacing patterns in mobile animals. *Annual Review of Ecology and Systematics*, 1, 239–262.
Butler, M. J., A. B. MacDiarmid, & J. D. Booth. (1999). The cause and consequence of ontogenetic changes in social aggregation in New Zealand spiny lobster. *Marine Ecology Progress Series*, 188, 179–191.
Caldwell, G. S, S. E. Glickman, & E. R. Smith. (1984). Seasonal aggression independent of seasonal testosterone in wood rats. *Proceedings of the National Academy of Sciences, USA*, 81, 5255–5257.
Carpenter, F. L. (1987). Food abundance and territoriality: To defend or not to defend? *American Zoologist*, 27, 387–399.
Carpenter, F. L. & R. E. MacMillen. (1976). Threshold model of feeding territoriality and test with a Hawaiian honeycreeper. *Science*, 194, 639–642.
Chapman, B. B., A. J. W. Ward, & J. Krause. (2008). Schooling and learning: Early social environment predicts social learning ability in the guppy, *Poecilia reticulata*. *Animal Behaviour*, 76, 923–929.
Chiozzi, G., G. de Marchi, & D. Semere. (2011). Coloniality in the crab plover *Dromas ardeola* does not depend on nest site limitation. *Waterbirds*, 34, 77–81.
Clutton-Brock, T. H. (1991). *The Evolution of Parental Care*. Princeton, NJ: Princeton University Press.
Cochran, G. R. & N. G. Solomon. (2000). Effects of food supplementation on the social organization of prairie voles (*Microtus ochrogaster*). *Journal of Mammalogy*, 81, 746–757.
Cole, K. S. & D. L. G. Noakes. (1980). Development of early social behavior of rainbow trout, *Salmo gairdneri* (Pisces, Salmonidae). *Behavioural Processes*, 5, 97–112.
Corbet, N. U. & R. J. van Aarde. (1996). Social organization and space use in the Cape porcupine in a southern African savanna. *African Journal of Ecology*, 34, 1–14.
Courchamp, F. T., T. Clutton-Brock, & B. Grenfell. (1999). Inverse density dependence and the Allee effect. *Trends in Ecology and Evolution*, 14, 405–410.
Craig, J. L. & M. E. Douglas. (1986). Resource distribution, aggressive asymmetries and variable access to resources in a nectar feeding bellbird. *Behavioral Ecology and Sociobiology*, 18, 231–240.

Cresswell, W. (1994). Flocking is an effective anti-predation strategy in redshanks, *Tringa totanus*. *Animal Behaviour*, 47, 433–442.

Cutrera, A. P., C. D. Antinuchi, M. S. Mora, & A. I. Vassallo. (2006). Home-range and activity patterns of the South American subterranean rodent *Ctenomys talarum*. *Journal of Mammalogy*, 87, 1183–1191.

Danchin, E. & R. H. Wagner. (1997). The evolution of coloniality: The emergence of new perspectives. *Trends in Ecology and Evolution*, 12, 342–347.

Danchin, E., L.-A. Giraldeau, & R. H. Wagner. (2008). Animal aggregations. In E. Danchin, L.-A. Giraldeau, & F. Cézilly (eds.), *Behavioural Ecology: An Evolutionary Perspective on Behaviour* (pp. 503–545). Oxford, UK: Oxford University Press.

Davies, N. B. & A. I. Houston. (1984). Territory economics. In J. R. Krebs & N. B. Davies (eds.), *Behavioural Ecology: An Evolutionary Approach*, Second Edition (pp. 148–169). Sunderland, MA: Sinauer Associates.

Davis, E. S. & C. A. Marler. (2003). The progesterone challenge: Steroid hormone changes following a simulated territorial intrusion in female *Peromyscus californicus*. *Hormones and Behavior*, 44, 185–198.

Davis, R. B., C. F. Herreid II, & H. L. Short. (1962). Mexican free-tailed bats in Texas. *Ecological Monographs*, 32, 311–346.

Deverill, J. I., C. E. Adams, & C. W. Bean. (1999). Prior residence, aggression and territory acquisition in hatchery-reared and wild brown trout. *Journal of Fish Biology*, 55, 868–875.

Dobson, A. & J. Poole. (1998). Conspecific aggregation and conservation biology. In T. Caro (ed.), *Behavioral Ecology and Conservation Biology* (pp. 193–208). Oxford, UK: Oxford University Press.

Eberhard, J. R. & P. W. Ewald. (1994). Food availability, intrusion pressure and territory size: An experimental study of Anna's hummingbirds (*Calypte anna*). *Behavioral Ecology and Sociobiology*, 34, 11–18.

Edwards, D. H. & E. A. Kravitz. (1997). Serotonin, social status and aggression. *Current Opinion in Neurobiology*, 7, 812–819.

Elekonich, M. M. & J. C. Wingfield. (2000). Seasonality and hormonal control of territorial aggression in female song sparrows (Passeriformes: Emberizidae: *Melospiza melodia*). *Ethology*, 106, 493–510.

Elgar, M. A. (1989). Predator vigilance and group size in mammals and birds: A critical review of the empirical evidence. *Biological Review*, 64, 13–33.

Emlen, S. T. & L. W. Oring. (1977). Ecology, sexual selection and the evolution of mating systems. *Science*, 197, 215–223.

Fairbanks, B. & F. S. Dobson. (2007). Mechanisms of the group-size effect on vigilance in Columbian ground squirrels: Dilution versus detection. *Animal Behaviour*, 73, 115–123.

Ferron, J. & J. P. Ouellet. (1989). Temporal and intersexual variation in the use of space with regard to social organization in the woodchuck (*Marmota monax*). *Canadian Journal of Zoology*, 67, 1642–1649.

Fietz, J. (1999). Monogamy as a rule rather than exception in nocturnal lemurs: The case of the fat-tailed dwarf lemur, *Cheirogaleus medius*. *Ethology*, 105, 255–272.
Gallego, A. & M. R. Heath. (1994). The development of schooling behavior in Atlantic herring *Clupea harengus*. *Journal of Fish Biology*, 45, 569–588.
Getz, L. L., J. E. Hofmann, & C. S. Carter. (1987). Mating system and population fluctuations of the prairie vole, *Microtus ochrogaster*. *American Zoologist*, 27, 909–920.
Getz, L. L., B. McGuire, T. Pizzuto, J. E. Hofmann, & B. Frase. (1993). Social organization of the prairie vole, *Microtus ochrogaster*. *Journal of Mammalogy*, 74, 44–58.
Gilchrist, J. S., E. Otali, & F. Mwanguhya. (2004). Why breed communally? Factors affecting fecundity in a communal breeding mammal: The banded mongoose (*Mungos mungo*). *Behavioral Ecology and Sociobiology*, 57, 119–131.
Godin, J.-G. J., L. J. Classon, & M. V. Abrahams. (1988). Group vigilance and shoal size in a small characin fish. *Behaviour*, 104, 29–40.
Goodson, J. L. (1998a). Territorial aggression and dawn song are modulated by septal vasotocin and vasoactive intestinal polypeptide in male field sparrows (*Spizella pusilla*). *Hormones and Behavior*, 34, 67–77.
Goodson, J. L. (1998b). Vasotocin and vasoactive intestinal polypeptide modulate aggression in a territorial songbird, the violet-eared waxbill (Estrilididae: *Uraginthus granatina*). *General and Comparative Endocrinology*, 111, 233–244.
Grant, J. W. A. (1993) Whether or not to defend? The influence of resource distribution. *Marine & Freshwater Behavior and Physiology*, 23, 137–153.
Grant, J. W. A. & R. T. Guha. (1993). Spatial clumping of food increases its monopolization and defense by convict cichlids, *Cichlasoma nigrofasciatum*. *Behavioral Ecology*, 4, 293–296.
Gross, M. R. (1984). Sunfish, salmon, and the evolution of alternative reproductive strategies and tactics in fishes. In R. Wooten & G. Potts (eds.), *Fish Reproduction: Strategies and Tactics* (pp. 55–75). London: Academic Press.
Gross, M. R. & E. L. Charnov. (1980). Alternative male life histories in bluegill sunfish. *Proceedings of the National Academy of Sciences, USA*, 77, 6937–6940.
Gross, M. R. & A. M. MacMillan. (1981). Predation and evolution of colonial nesting in bluegill sunfish (*Lepomis macrochirus*). *Behavioral Ecology and Sociobiology*, 8, 163–174.
Gwinner, E., T. Roedl, & H. Schwabl. (1994). Pair territoriality of wintering stonechats: Behaviour, function and hormones. *Behavioral Ecology and Sociobiology*, 34, 321–327.
Hamilton, W. D. (1971). Geometry for the selfish herd. *Journal of Theoretical Biology*, 31, 295–311.
Hammock, E. A. D., & L. J. Young. (2005). Microsatellite instability generates diversity in brain and sociobehavioral traits. *Science*, 308, 1630–1634.
Hannon, S. J., R. L. Mumme, W. D. Koenig, S. Spon, & F. A. Pitelka. (1987). Poor acorn crop, dominance, and decline in numbers of acorn woodpeckers. *Journal of Animal Ecology*, 56, 197–207.

Hart, D. D. (1985). Causes and consequences of territoriality in a grazing stream insect. *Ecology*, 66, 404–414.

Hau, M., S. T. Stoddard, & K. K. Soma. (2004). Territorial aggression and hormones during the non-breeding season in a tropical bird. *Hormones and Behavior*, 45, 40–49.

Hayes, L. D. (2000). To nest communally or not to nest communally: A review of rodent communal nesting and nursing. *Animal Behaviour*, 59, 677–688.

Heckel, G. & O. von Helversen. (2002). Male tactics and reproductive success in the harem polygynous bat *Saccopteryx bilineata*. *Behavioral Ecology*, 13, 750–756.

Heg, D., B. J. Ens, H. P. van der Jeugd, & L. W. Bruinzeel. (2000). Local dominance and territorial settlement of nonbreeding oystercatchers. *Behaviour*, 137, 473–530.

Herbert-Read, J. E., A. Perna, R. P. Mann, T. M. Schaerf, D. J. T. Sumpter, & A. J. W. Ward. (2011). Inferring the rules of interaction of shoaling fish. *Proceedings of the National Academy of Sciences, USA*, 108, 18726–18731.

Hixon, M.A. (1980). Food production and competitor density as the determinants of feeding territory size. *American Naturalist*, 115, 510–530.

Hixon, M. A., F. L. Carpenter, & D. C. Patton. (1983). Territory area, flower density, and time budgeting in hummingbirds: An experimental and theoretical analysis. *American Naturalist*, 122, 366–391.

Hoogland, J. L. (1981). The evolution of coloniality in white-tailed and black-tailed prairie dogs (Sciuridae: *Cynomys leucurus* and *C. ludovicianus*). *Ecology*, 62, 252–272.

Howard, H. E. (1920). *Territory in Bird Life*. New York: Dutton & Co.

Howard, R. D. (1978). The evolution of mating strategies in bullfrogs, *Rana catesbeiana*. *Evolution*, 32, 850–871.

Iguchi, K. & T. Hino. (1996). Effect of competitor abundance on feeding territoriality in the grazing fish, the ayu *Plecoglossus altivelis*. *Ecological Research*, 11, 165–173.

Ims, R. A. (1987). Responses in spatial organization and behaviour to manipulations of the food resource in the vole *Clethrionomys rufocanus*. *Journal of Animal Ecology*, 56, 585–596.

Insel, T. R., Z. X. Wang, & C. F. Ferris. (1994). Patterns of brain vasopressin receptor distribution associated with social organization in microtine rodents. *Journal of Neuroscience*, 14, 5381–5392.

Islam, M. S., P. Roessingh, S. J. Simpson, & A. L. Mccaffery. (1994). Effects of population density experienced by parents during mating and oviposition on the phase of hatchling desert locusts, *Schistocerca gregaria*. *Proceedings of the Royal Society of London, B*, 257, 93–98.

Jarvis, J. U. M. (1981). Eusociality in a mammal: Cooperative breeding in naked mole-rat colonies. *Science*, 212, 571–573.

Johnson, D. D. P., D. W. Macdonald, & A. J. Dickman. (2000). An analysis and review of models of the sociobiology of the Mustelidae. *Mammal Review*, 30, 171–196.

Jonsson, P., T. Hartikainen, E. Koskela, & T. Mappes. (2002). Determinants of reproductive success in voles: Space use in relation to food and litter size manipulation. *Evolutionary Ecology*, 16, 455–467.

Kat, P. W., K. A. Alexander, J. S. Smith, & L. Munson. (1995). Rabies and African wild dogs in Kenya. *Proceedings of the Royal Society of London, B*, 262, 229–233.

Katz, Y., K. Tunstrøm, C. C. Ioannou, C. Huepe, & I. D. Couzin. (2011). Inferring the structure and dynamics of interactions in schooling fish. *Proceedings of the National Academy of Sciences, USA*, 108, 18720–18725.

Keeley, E. R. & J. W. A. Grant. (1995). Allometric and environmental correlates of territory size in juvenile Atlantic salmon (*Salmo salar*). *Canadian Journal of Fisheries and Aquatic Sciences*, 52, 186–196.

Keeley, E. R. & J. D. McPhail. (1998). Food abundance, intruder pressure, and body size as determinants of territory size in juvenile steelhead trout (*Oncorhynchus mykiss*). *Behaviour*, 135, 65–82.

Kempenaers, B., S. Everding, C. Bishop, P. Boag, & R. J. Robertson. (2001). Extra-pair paternity and the reproductive role of male floaters in the tree swallow (*Tachycineta bicolor*). *Behavioral Ecology and Sociobiology*, 49, 251–259.

Kishimoto, R. (2003). Social monogamy and social polygyny in a solitary ungulates the Japanese serow (*Capricornis crispus*). In U. H. Reichard & C. Boesch (eds.), *Monogamy, Mating Strategies and Partnerships in Birds, Humans, and Other Mammals* (pp. 147–158). Cambridge, UK: Cambridge University Press.

Kleiman, D. G. (1977). Monogamy in mammals. *Quarterly Review of Biology*, 52, 39–69.

Komdeur, J. (1996). Breeding of the Seychelles magpie robin *Copsychus sechellarum* and implications for its conservation. *Ibis*, 138, 485–498.

Krause, J. & G. D. Ruxton. (2002). *Living in Groups*. Oxford, UK: Oxford University Press.

Krekorian, C. O. (1976). Home range size and overlap and their relationship to food abundance in the desert iguana, *Dipsosaurus dorsalis*. *Herpetologica*, 32, 405–412.

Kress, S. W. (1983). The use of decoys, sounds recordings, and gull control for re-establishing a tern colony in Maine. *Colonial Waterbirds*, 6, 185–196.

Lacey, E. A. & P. W. Sherman. (1991). Social organization of naked mole-rat colonies: Evidence for divisions of labor. In P. W. Sherman, J. U. M. Jarvis, & R. D. Alexander (eds.), *The Biology of the Naked Mole-rat* (pp. 275–336). Princeton, NJ: Princeton University Press.

Lambin, X. & C. J. Krebs. (1991). Spatial organization and mating system of *Microtus townsendii*. *Behavioral Ecology and Sociobiology*, 28, 353–363.

Lank, D. B., C. M. Smith, O. Hanotte, T. Burke, & F. Cooke. (1995). Genetic polymorphism for alternative mating behaviour in lekking male ruff *Philomachus pugnax*. *Nature*, 378, 59–62.

Lazarus, J. (1979). Early warning function of flocking in birds: An experimental study with captive quelea. *Animal Behaviour*, 27, 855–865.

Lederer, R. J. (1981). Facultative territoriality in Townsend's solitaire (*Myadestes townsendi*). *Southwestern Naturalist*, 25, 461–467.

Lewis, S. E. & A. E. Pusey. (1997). Factors influencing the occurrence of communal care in plural breeding mammals. In N. G. Solomon & J. A. French (eds.), *Cooperative Breeding in Mammals* (pp. 335–363). New York: Cambridge University Press.

Lima, S. L. (1995). Collective detection of predatory attack by social foragers: Fraught with ambiguity. *Animal Behaviour*, 50, 1097–1108.

Logan, C. A. (1992). Testosterone and reproductive adaptations in the autumnal territoriality of northern mockingbirds *Mimus polyglottos*. *Ornis Scandinavica*, 23, 277–283.

López-Sepulcre A. & H. Kokko. (2005). Territorial defense, territory size, and population regulation. *American Naturalist*, 166, 317–329.

Lott, D. F. (1984). Intraspecific variation in the social systems of wild vertebrates. *Behaviour*, 88, 266–325.

Lott, D. F. (1991). *Intraspecific Variation in the Social Systems of Wild Vertebrates*. Cambridge, UK: Cambridge University Press.

Lucia, K., B. Keane, L. D. Hayes, Y. K. Lin, R. L. Schaefer, & N. G. Solomon. (2008). Philopatry in prairie voles: An evaluation of the habitat saturation hypothesis. *Behavioral Ecology*, 19, 774–783.

Madison, D. M. (1984). Group nesting and its ecological and evolutionary significance in overwintering microtine rodents. In J. F. Merritt (ed.), *Winter Ecology of Small Mammals* (pp. 267–274). Pittsburgh, PA: Carnegie Museum of Natural History.

Madison, D. M., J. P. Hill, & P. E. Gleason. (1984). Seasonality in the nesting behavior of *Peromyscus leucopus*. *American Midland Naturalist*, 112, 201–204.

Magurran, A. E. & A. Higham. (1988). Information transfer across fish shoals under predator threat. *Ethology*, 78, 153–158.

Maher, C. R. (1994). Pronghorn male spatial organization: Population differences in degree of nonterritoriality. *Canadian Journal of Zoology*, 72, 455–464.

Maher, C. R. & D. F. Lott. (1995). Definitions of territoriality used in the study of variation in vertebrate spacing systems. *Animal Behaviour*, 49, 1581–1597.

Maher, C. R. & D. F. Lott. (2000). A review of ecological determinants of territoriality within vertebrate species. *American Midland Naturalist*, 143, 1–29.

Mappes, T, H. Ylönen, & J. Viitala. (1995). Higher reproductive success among kin groups of bank voles (*Clethrionomys glareolus*). *Ecology*, 76, 1276–1282.

Mares, M. A., T. E. Lacher, M. R. Willig, N. A. Bitar, R. Adams, A. Klinger, & D. Tazik. (1982). An experimental analysis of social spacing in *Tamias striatus*. *Ecology*, 63, 267–273.

Martin, J. K., K. A. Handasyde, A. C. Taylor, & G. Coulson. (2007). Long-term pair-bonds without mating fidelity in a mammal. *Behaviour*, 144, 1419–1445.

Matthews, L. M. (2002). Territorial cooperation and social monogamy: Factors affecting intersexual behaviours in pair-living snapping shrimp. *Animal Behaviour*, 63, 767–777.

McCaffery, A. R., S. J. Simpson, M. S. Islam, & P. Roessingh. (1998). A gregarizing factor present in the egg pod foam of the desert locust *Schistocerca gregaria*. *Journal of Experimental Biology*, 201, 347–363.

McGuire, B. & L. L. Getz. (2010). Alternative male reproductive tactics in a natural population of prairie voles *Microtus ochrogaster*. *Acta Theriologica*, 55, 261–270.

McLoughlin, P. D., S. H. Ferguson, & F. Messier. (2000). Intraspecific variation in home range overlap with habitat quality: A comparison among brown bears. *Evolutionary Ecology*, 14, 39–60.

Meadows, P. S. & J. I. Campbell. (1972). Habitat selection by aquatic invertebrates. *Advances in Marine Biology*, 10, 271–382.

Michener, C. D. (1974). *The Social Behavior of the Bees: A Comparative Study*. Cambridge, MA: Belknap Press of Harvard University Press.

Milner-Gulland, E. J. & J. R. Beddington. (1993). The exploitation of elephants for the ivory trade: An historical perspective. *Proceedings of the Royal Society London, B*, 252, 29–37.

Mock, D. W., T. C. Lamey, & D. B. A. Thompson. (1988). Falsifiability and the information center hypothesis. *Ornis Scandinavica*, 19, 231–248.

Møller, A. P. (1987). Advantages and disadvantages of coloniality in the swallow, *Hirundo rustica*. *Animal Behaviour*, 35, 819–832.

Monaghan, P. & N. B. Metcalfe. (1985). Group foraging in wild brown hares: Effects of resource distribution and social status. *Animal Behaviour*, 33, 993–999.

Montgomery, W. I., W. L. Wilson, & R. W. Elwood. (1997). Spatial regulation and population growth in the wood mouse *Apodemus sylvaticus*: Experimental manipulations of males and females in natural populations. *Journal of Animal Ecology*, 66, 755–768.

Moore, M. C., D. K. Hews, & R. Knapp. (1998). Hormonal control and evolution of alternative male phenotypes: Generalizations of models for sexual differentiation. *American Zoologist*, 38, 133–151.

Moore, M. C. & C. A. Marler. (1987). Effects of testosterone manipulations on non-breeding season territorial aggression in free-living male lizards, *Sceloporus jarrovi*. *General and Comparative Endocrinology*, 65, 225–232.

Morton, E. S., L. Forman, & M. Braun. (1990). Extrapair fertilizations and the evolution of colonial breeding in purple martins. *Auk*, 107, 275–283.

Munshi-South, J., L. H. Emmons, & H. Bernard. (2007). Behavioral monogamy and fruit availability in the large treeshrew (*Tupaia tana*) in Sabah, Malaysia. *Journal of Mammalogy*, 88, 1427–1438.

Myers, J. P., P. G. Connors, & F. A. Pitelka. (1979). Territory size in wintering sanderlings: The effects of prey abundance and intruder density. *Auk*, 96, 551–561.

Myrberg, A. A. Jr. (1972). Social dominance and territoriality in the bicolor damselfish, *Eupomacentrus partitus* (Poey) (Pisces: Pomacentridae). *Behaviour*, 41, 207–231.

Nelson, R. J. (2000). *An Introduction to Behavioral Endocrinology*. Second Edition. Sunderland, MA: Sinauer Associates.

Newton, I. (1992). Experiments on the limitation of bird numbers by territorial behaviour. *Biological Reviews*, 67, 129–173.

Nilsen, E. B., I. Herfindal, & J. D. C. Linnell. (2005). Can intra-specific variation in carnivore home-range size be explained using remote-sensing estimates of environmental productivity? *Ecoscience*, 12, 68–75.

Norman, M. D. & G. P. Jones. (1984). Determinants of territory size in the pomacentrid reef fish, *Parma victoriae*. *Oecologia*, 61, 60–69.

Olsthoorn, J. C. M. & J. B. Nelson. (1990). The availability of breeding sites for some British seabirds. *Bird Study*, 37, 145–164.

O'Neill, K. M. (1983). Territoriality, body size, and spacing in males of the beewolf *Philanthus basilaris* (Hymenoptera: Sphecidae). *Behaviour*, 86, 295–321.

Ophir, A. G., J. O. Wolff, & S. M. Phelps. (2008). Variation in neural V1aR predicts sexual fidelity and space use among male prairie voles in semi-natural settings. *Proceedings of the National Academy of Sciences, USA*, 105, 1249–1254.

Ostfeld, R. S. (1985). Limiting resources and territoriality in microtine rodents. *American Naturalist*, 126, 1–15.

Ostfeld, R. S. (1986). Territoriality and mating system of California voles. *Journal of Animal Ecology*, 55, 691–706.

Parker, M. W., S. W. Kress, R. T. Golightly, H. R. Carter, E. B. Parsons, S. E. Schubel, J. A. Boyce, G. J. McChesney, & S. M. Wisely. (2007). Assessment of social attraction techniques to restore a common murre colony in central California. *Waterbirds*, 30, 17–28.

Partridge, B. L. (1982). Structure and function of fish schools. *Scientific American*, 246, 114–123.

Peixoto, P. E. C. & P. De Marco Jr. (2009). No size or density effect on alternative mate-locating tactics in the tropical damselfly *Hetaerina rosea* males (Odonata: Calopterygidae). *Revista de Biologia Tropica*, 57, 361–370.

Penteriani, V., M. Ferrer, & M. M. Delgado. (2011). Floater strategies and dynamics in birds, and their importance in conservation biology: Towards an understanding of nonbreeders in avian populations. *Animal Conservation*, 14, 233–241.

Phelps, S. M. (2010). From endophenotypes to evolution: Social attachment, sexual fidelity and the *avpr1a* locus. *Current Opinion in Neurobiology*, 20, 795–802.

Powell, R. A. (1987). Black bear home range overlap in North Carolina and the concept of home range applied to black bears. *International Conference on Bear Research and Management*, 7, 235–242.

Reichart, S. E. (1981). The consequences of being territorial: Spiders, a case study. *American Naturalist*, 117, 871–892.

Rémy, A. (2011). Linking behaviour with individual traits and environmental conditions, and the consequences for small rodent populations. Ph.D. dissertation, University of Oslo, Oslo, Norway.

Richner, H. & P. Heeb. (1995). Is the information center hypothesis a flop? *Advances in the Study of Behavior*, 24, 1–45.

Richner, H. & P. Heeb. (1996). Communal life: Honest signaling and the recruitment center hypothesis. *Behavioral Ecology*, 7, 115–119.

Riehl, C. (2011). Living with strangers: Direct benefits favour non-kin cooperation in a communally nesting bird. *Proceedings of the Royal Society, London, B,* 278, 1728–1735.
Robertson, D. R., H. P. A. Sweatman, E. A. Fletcher, & M. G. Cleland. (1976). Schooling as a mechanism for circumventing the territoriality of competitors. *Ecology,* 57, 1208–1220.
Rogers, L. L. (1987). The effects of food supply and kinship on social behavior, movements, and population growth of black bears in northeastern Minnesota. *Wildlife Monographs,* 97, 1–72.
Rolland, C., E. Danchin, & M. de Fraipont. (1998). The evolution of coloniality in birds in relation to food, habitat, predation, and life-history traits: A comparative analysis. *American Midland Naturalist,* 151, 514–529.
Rubenstein, D. I. (1981). Population density, resource patterning, and territoriality in the Everglades pygmy sunfish. *Animal Behaviour,* 29, 155–172.
Sandell, M. (1989). The mating tactics and spacing patterns of solitary carnivores. In J. L. Gittleman (ed.), *Carnivore Behavior, Ecology, and Evolution* (pp. 164–182). Ithaca, NY: Comstock Publishers and Associates of Cornell University Press.
Schoener, T. W. (1983). Simple models of optimal feeding-territory size: A reconciliation. *American Naturalist,* 121, 608–629.
Schradin, C. & A. K. Lindholm. (2011). Relative fitness of alternative male reproductive tactics in a mammal varies between years. *Journal of Animal Ecology,* 80, 908–917.
Schradin, C., A. K. Lindholm, J. Johannesen, I. Schoepf, C. H. Yuen, B. König, & N. Pillay. (2012). Social flexibility and social evolution in mammals: A case study of the African striped mouse (*Rhabdomys pumilio*). *Molecular Ecology,* 21, 541–553.
Seghers, B. H. (1974). Schooling behavior in the guppy (*Poecilia reticulata*): An evolutionary response to predation. *Evolution,* 28, 486–489.
Shuster, S. M. (2009). Sexual selection and mating systems. *Proceedings of the National Academy of Sciences, USA,* 106, 10009–10016.
Shuster, S. M. (2011). Differences in relative fitness among alternative mating tactics might be more apparent than real. *Journal of Animal Ecology,* 80, 905–907.
Shuster, S. M. & M. J. Wade (1991). Equal mating success among male reproductive strategies in a marine isopod. *Nature,* 350, 608–610.
Siegel-Causey, D. & S. P. Kharitonov. (1990). The evolution of coloniality. *Current Ornithology,* 7, 285–330.
Simon, C. A. (1975). The influence of food abundance on territory size in the iguanid lizard *Sceloporus jarrovi*. *Ecology,* 56, 993–998.
Sinervo, B., D. B. Miles, W. A. Frankino, M. Klukowski, & D. F. DeNardo. (2000). Testosterone, endurance, and Darwinian fitness: Natural and sexual selection on the physiological bases of alternative male behaviors in side-blotched lizards. *Hormones and Behavior,* 38, 222–233.
Smith, J. N. M. & P. Arcese. (1989). How fit are floaters? Consequences of alternative territorial behaviors in a nonmigratory sparrow. *American Naturalist,* 133, 830–845.

Smith, S. M. & J. W. Cain III. (2009). Foraging efficiency and vigilance behavior of impala: The influence of herd size and neighbor density. *African Journal of Ecology*, 47, 109–118.

Solomon, N. G., A. R. Richmond, P. A. Harding, A. Fries, S. Jacquemin, R. L. Schaefer, K. E. Lucia, & B. Keane. (2009). Polymorphism at the *avpr1a* locus in male prairie voles correlated with genetic but not social monogamy in field populations. *Molecular Ecology*, 18, 4680–4695.

Soma, K. K., K. Sullivan, & J. Wingfield. (1999). Combined aromatase inhibitor and antiandrogen treatment decreases territorial aggression in a wild songbird during the nonbreeding season. *General and Comparative Endocrinology*, 115, 442–453.

Soma, K. K. & J. C. Wingfield. (2001). Dehydroepiandrosterone in songbird plasma: Seasonal regulation and relationship to territorial aggression. *General and Comparative Endocrinology*, 123, 144–155.

Sperry, T. S., C. K. Thompson, & J. C. Wingfield. (2003). Effects of acute treatment with 8-OH-DPAT and fluoxetine on aggressive behavior in male song sparrows (*Melospiza melodia morphna*). *Journal of Neuroendocrinology*, 15, 150–160.

Stamps, J. (1988). Conspecific attraction and aggregation in territorial species. *American Naturalist*, 131, 329–347.

Stamps, J. (1994). Territorial behavior: Testing the assumptions. *Advances in the Study of Behavior*, 23, 173–232.

Steury, T. D. & D. L. Murray. (2003). Causes and consequences of individual variation in territory size in the American red squirrel. *Oikos*, 101, 147–156.

Stimson, J. (1973). The role of the territory in the ecology of the intertidal limpet *Lottia gigantea* (Gray). *Evolution*, 54, 1020–1030.

Streatfeild, C. A., K. E. Mabry, B. Keane, T. O. Crist, & N. G. Solomon. (2011). Intraspecific variability in the social and genetic mating systems of prairie voles, *Microtus ochrogaster*. *Animal Behaviour*, 82, 1387–1398.

Sundström, L. F., M. Löhmus, & J. I. Johnsson. (2003). Investment in territorial defense depends on rearing environment in brown trout (*Salmo trutta*). *Behavioral Ecology and Sociobiology*, 54, 249–255.

Sutherland, W. J. (1996). *From Individual Behavior to Population Ecology*. New York: Oxford University Press.

Taborsky, M. (1998). Sperm competition in fish: "Bourgeois" males and parasitic spawning. *Trends in Ecology and Evolution*, 13, 222–227.

Taborsky, M. (2001). The evolution of bourgeois, parasitic, and cooperative reproductive behaviors in fishes. *Journal of Heredity*, 92, 100–110.

Taborsky, M., R. F. Oliveira, & H. J. Brockmann. (2008). The evolution of alternative reproductive tactics: Concepts and questions. In R. F. Oliveira, M. Taborsky, & H. J. Brockmann (eds.), *Alternative Reproductive Tactics: An Integrated Approach* (pp. 1–21). Cambridge, UK: Cambridge University Press.

Ten Eyck, G. R. (2008). Serotonin modulates vocalizations and territorial behavior in an amphibian. *Behavioural Brain Research*, 193, 144–147.

Tinbergen, N., M. Impekoven, & D. Franck. (1967). An experiment on spacing-out as a defense against predation. *Behaviour*, 28, 307–321.
Treherne, J. E. & W. A. Foster. (1981). Group transmission of predator avoidance behaviour in a marine insect: The Trafalgar effect. *Animal Behaviour*, 29, 911–917.
Tricas, T. C. (1989). Determinants of feeding territory size in the corallivorous butterflyfish, *Chaetodon multicinctus*. *Animal Behaviour*, 37, 830–841.
Turner, G. F. & T. J. Pitcher. (1986). Attack abatement: A model for group protection by combined avoidance and dilution. *American Naturalist*, 128, 228–240.
Uetz, G. W. & K. R. Cangialosi. (1986). Genetic differences in social behavior and spacing in populations of *Metepeira spinipes*, a communal-territorial orb weaver (Araneae, Araneidae). *Journal of Arachnology*, 14, 159–173.
van Schaik, C. P., M. A. van Noordwijk, B. Warsono, & E. Sutriono. (1983). Party size and early detection of predators in Sumatran forest primates. *Primates*, 24, 211–221.
Vanson, A., A. P. Arnold, & B. A. Schlinger. (1996). 3ß-hydroxysteroid dehydrogenase/isomerase and aromatase activity in primary cultures of developing zebra finch telencephalon: Dehydroepiandrosterone as substrate for synthesis of androstendione and estrogens. *General and Comparative Endocrinology*, 102, 342–350.
Veen, J. (1977). Functional and causal aspects of nest distribution in colonies of the Sandwich tern (*Sterna s. sandvicensis* Lath.). *Behaviour Supplement*, 20, 1–193.
Vehrencamp, S. L. & J. S. Quinn. (2004). Joint laying systems. In W. D. Koenig & J. L. Dickenson (eds.), *Ecology and Evolution of Cooperative Breeding in Birds* (pp. 177–196). Cambridge, UK: Cambridge University Press.
Wagner, R. H. (1993). The pursuit of extra-pair copulations by female birds: A new hypothesis of colony formation. *Journal of Theoretical Biology*, 163, 333–346.
Wagner, R. H., M. D. Schug, & E. S. Morton. (1996). Condition-dependent control of paternity by female purple martins: Implications for coloniality. *Behavioral Ecology and Sociobiology*, 38, 379–389.
Wang, Z., L. J. Young, G. J. De Vries, & T. R. Insel. (1998). Voles and vasopressin: A review of molecular, cellular, and behavioral studies of pair bonding and paternal behaviors. *Progress in Brain Research*, 119, 483–499.
Ward, M. P. & S. Schlossberg. (2004). Conspecific attraction and the conservation of territorial songbirds. *Conservation Biology*, 18, 519–525.
Ward, P. & A. Zahavi. (1973). Importance of certain assemblages of birds as "information centres" for food-finding. *Ibis*, 115, 517–534.
Watts, C. H. S. (1969). The regulation of wood mouse (*Apodemus sylvaticus*) numbers in Wytham Woods, Berkshire. *Journal of Animal Ecology*, 38, 285–304.
Wauters, L. & A. A. Dhondt. (1992). Spacing behaviour of red squirrels, *Sciurus vulgaris*: Variation between habitats and the sexes. *Animal Behaviour*, 43, 297–311.
Wauters, L. A & L. Lens. (1995). Effects of food availability and density on red squirrel (*Sciurus vulgaris*) reproduction. *Ecology*, 76, 2460–2469.
West, S. D. (1977). Midwinter aggregations in the northern red-backed vole, *Clethrionomys rutilus*. *Canadian Journal of Zoology*, 55, 1404–1409.

Wiktander, U., O. Olsson, & S. G. Nilsson. (2001). Seasonal variation in home-range size, and habitat area requirement of the lesser spotted woodpecker (*Dendrocopos minor*) in southern Sweden. *Biological Conservation*, 100, 387–395.
Wilson, E. O. (1971). *The Insect Societies*. Cambridge, MA: Belknap Press of Harvard University Press.
Wingfield, J. (1994). Regulation of territorial behavior in the sedentary song sparrow, *Melospiza melodia morphna*. *Hormones and Behavior*, 28, 1–15.
Wingfield, J. C., G. F. Ball, A. M. Dufty Jr., R. E. Hegner, & M. Ramenofsky. (1987). Testosterone and aggression in birds: Tests of the "challenge hypothesis." *American Scientist*, 75, 602–608.
Wingfield, J. C., R. E. Hegner, A. M. Dufty Jr., & G. F. Ball. (1990). The "challenge hypothesis": Theoretical implications for patterns of testosterone secretion, mating systems, and breeding strategies. *American Naturalist*, 136, 829–846.
Wingfield, J. C., S. Lynn, & K. K. Soma. (2001). Avoiding the "costs" of testosterone: Ecological bases of hormone-behavior interactions. *Brain, Behavior and Evolution*, 57, 239–251.
Wittenberger, J. F. (1981). *Animal Social Behavior*. Boston: Duxbury Press.
Wittenberger, J. F. & G. L. Hunt Jr. (1985). The adaptive significance of coloniality in birds. D. S. Farner, J. R. King & K. C. Parkes (eds.), *Avian Biology*, Vol. 8 (pp. 1–78). Orlando, FL: Academic Press.
Wittenberger, J. F. & R. L. Tilson. (1980). The evolution of monogamy: Hypotheses and evidence. *Annual Review of Ecology and Systematics*, 11, 197–232.
Wolff, J. O. (1985). The effects of density, food and interspecific interference on home range size in *Peromyscus leucopus* and *Peromyscus maniculatus*. *Canadian Journal of Zoology*, 63, 2657–2662.
Wolff, J. O. (1993). Why are female small mammals territorial? *Oikos*, 68, 364–370.
Wolff, J. O. (2008). Alternative reproductive tactics in nonprimate male mammals. In R. F. Oliveira, M. Taborsky, & H. J. Brockmann (eds.), *Alternative Reproductive Tactics: An Integrative Approach* (pp. 356–372). Cambridge, UK: Cambridge University Press.
Woodroffe, R. & A. Vincent. (1994). Mothers little helpers: Patterns of male care in mammals. *Trends in Ecology and Evolution*, 9, 294–297.
Wright, D., L. B. Rimmer, V. L. Pritchard, R. K. Butlin, & J. Krause. (2003). Inter and intra-population variation in shoaling and boldness in the zebrafish (*Danio rerio*). *Journal of Fish Biology*, 63, 258–259.
Yahner, R. H. (2012). *Wildlife Behavior and Conservation*. New York: Springer.
Ylönen, H., T. Kojola, & J. Viitala. (1988). Changing female spacing behaviour and demography in an enclosed breeding population of *Clethrionomys glareolus*. *Ecography*, 11, 286–292.
Ylönen, H. & J. Viitala. (1991). Social overwintering and food distribution in the bank vole *Clethrionomys glareolus*. *Holarctic Ecology*, 14, 131–137.
Zahavi, A. (1971). The function of pre-roost gatherings and communal roosts. *Ibis*, 113, 106–109.

9

Let's Get Together:
The Evolution of Social Behavior

Walter D. Koenig and Janis L. Dickinson

INTRODUCTION

Social behavior encompasses a wide variety of interactions among animals, usually of the same species, ranging from simple attraction between individuals, to temporary feeding aggregations and mating swarms, to multigenerational family groups with cooperative brood care. Although some types of social behavior are intuitively obvious, others are not, because social behavior is defined by the kind of interaction involved, not by how organisms are distributed in space. Thus, individuals can be aggregated but not engaged in social behavior; conversely, clumping is not a requirement for individuals to behave socially, although it clearly increases the opportunity for meaningful social interactions to take place. Examples of individuals engaging in social behavior while not appearing particularly social include a lone female tiger moth (family Arctiidae) emitting a pheromone designed to attract potential male mates (Cardé & Millar, 2009) and a male red deer (*Cervus elaphus*) roaring in order to signal dominance (Clutton-Brock & Albon, 1979), attract females (Charlton et al., 2007), and enhance female fertility (McComb, 1987).

One of the most important features of social behavior is that it is not necessarily "social" in the positive sense of involving an amicable interaction or being directed toward some mutually beneficial end. Indeed, social organisms are often fiercely competitive and aggressive—sometimes toward the same individuals with whom they are highly cooperative under different circumstances. As one notable example, female acorn woodpeckers (*Melanerpes formicivorus*), a group-living bird found in the western United States, Mexico, and Central America, nest communally (Koenig et al., 1995b). Two, and rarely three, related females, usually sisters or a mother and daughter, lay eggs in the same nest cavity. Even though they are relatives, females compete with one another to lay their own eggs in the communal nest by removing and subsequently eating their cobreeders' eggs prior to laying their own eggs (Koenig et al., 1995a), a process resulting in each female contributing equally to the final clutch (Haydock & Koenig, 2002). After all birds have commenced laying, however, overt competition ends and is replaced by cooperative sharing of incubation, brooding, and feeding of nestlings.

Social behavior has evolved to enhance an individual's ability to garner resources and form the alliances that help it survive and increase the number of gene copies it places in future generations, whether doing so involves cooperating with others or fighting against them. As a result, animal societies frequently involve a delicate balance between cooperative and competitive behaviors, potentially switching from one to the other depending on the ever-changing costs and benefits of collaborative compared to competitive behaviors.

CATEGORIZING THE DIVERSITY OF SOCIAL BEHAVIOR

The diversity of behaviors that can be considered social renders the classification of ***sociality*** difficult. Among insects, within which there is a wide range of social behavior often within a single taxon, there is a long history of categorizing sociality going back nearly a century based primarily on the involvement of insect parents with their young (Michener, 1969, 2007). Two sequences are recognized, the ***parasocial*** and ***subsocial*** routes, both encompassing a range of societies that culminate in ***eusociality***, a system in which overlapping generations of individuals live in the colony together, brood care is prolonged, and there is a reproductive division of labor where at least some offspring act as workers rather than reproduce.

In the parasocial route, females of the same generation assist one another, in some cases cooperating in nest construction but otherwise rearing their broods separately, whereas in others the broods are attended cooperatively, although each female may still reproduce. In derived eusocial species, brood care is still cooperative among the older generation, but generations overlap and some members of the younger generation are workers that do not reproduce. Some of these have achieved a complex and highly advanced eusocial system with overlapping generations, functionally sterile workers, and distinct morphological and behavioral *castes*.

In the subsocial route, females associate to a varying degree with their offspring, ranging from building a nest and laying an egg on a prey item that is sufficient to allow the larva to develop into an adult, to providing direct care for a prolonged period of time but then departing before the young emerge as adults, to the situation where mothers stay at the nest until offspring emerge and those newly emerged offspring then assist their mother in the rearing of additional broods. As with the parasocial route, this route culminates in cases where the offspring differentiate into a permanently sterile worker caste, a stage mirroring the eusocial outcome achieved by the parasocial sequence. Although useful, much of this terminology has not found wide favor outside of entomological circles and even there is not easily applied to the entire eusociality continuum (Sherman et al., 1995; Michener, 2007).

Apart from the attempt to classify routes to eusociality, however, there are few other general categorizations of social behavior. One of the more comprehensive attempts, encompassing not only reproduction but a wide range of social behavior, is by the American entomologist and science writer Edward O. Wilson, who compiled a set of 10 essential qualities of sociality (Wilson, 1975). These included (1) group size, (2) distributions of different age and sex classes, (3) cohesiveness, (4) amount and pattern of connectedness, (5) *permeability*, or the degree to which societies interact with one another, (6) *compartmentalization*, or the extent to which subgroups operate as discrete units, (7) differentiation of roles among group members, (8) integration of behaviors within groups, (9) communication and information flow, and (10) the fraction of time devoted to social behavior as opposed to individual maintenance. The overlapping nature of these items provides a good indication of the complexities involved with classifying the highly varied social behavior exhibited by animals.

While categorizing social behavior can be a useful exercise, such classifications can also be confusing and misleading. The current tendency is to view sociality as a multifaceted continuum from simple aggregations to the highly organized and complex levels of social organization found in the most derived eusocial species.

PROXIMATE VERSUS ULTIMATE CAUSATION

Social behavior, like other adaptations for survival and reproduction, is best understood by differentiating its *proximate* causes—how the behavior arises—from its *ultimate* causes—its evolutionary history and functional utility. Proximate causes include the mechanisms directly underlying the behavior and include the hereditary, developmental, structural, cognitive, psychological, and physiological aspects associated with the behavior. For example, chacma baboons (*Papio ursinus*) live in relatively large social groups that are subject to high rates of predation by lions, leopards, crocodiles, and hyenas. Adult female baboons experience considerable stress following predator attacks, particularly when attacks result in animals becoming dispersed and separated from the larger troop (Cheney & Seyfarth, 2009). Stress is indicated by significantly high glucocorticoids in their blood—hormones released by the adrenal glands in reponse to corticotropic-releasing hormones coming from the hypothalamus when the brain perceives a physical or psychological stressor (Sapolsky, 1998). High levels of circulating glucocorticoids have the effect of increasing glucose availability in the bloodstream by promoting glucose production and limiting glucose uptake and storage, thus mobilizing energy, sharpening memory, and limiting nonessential but energetically costly functions like digestion, growth, and reproduction (Nelson, 2011). The consequence of these behaviors is to increase both the wariness of the animal and its ability to search for and reunite with its troop, changes that increase its chances of surviving under conditions of relative vulnerability to predation.

What causes these behaviors? At one level of analysis, these fear-based reactions are the result of hormonal changes, specifically stress responses in the hypothalamic-pituitary-adrenal (HPA) axis triggered by the predator attacks. This underlying hormonal mechanism, which is triggered by the predator attacks and separation from the herd, is a proximate cause of these and other fear-based behaviors. At another level of analysis, the ultimate cause of these behaviors is a range of selective processes that have shaped their past and current expression. In this example, the increased survival that arises as a function of reuniting with the troop has selected for individuals that exhibit and respond to high levels of glucocorticoids following predator attacks by fleeing and then quickly seeking to rejoin the group.

The Dutch ethologist Nikolaas Tinbergen was the first to clarify these levels of analysis (or levels of explanation) and named four, which he referred to as "survival value," "causation," "development," and "evolutionary history" (Tinbergen, 1963; see Volume 1, Chapters 1 and 2). Tinbergen further emphasized the importance of addressing questions at the appropriate level of explanation. The critical lesson to keep in mind is that determining the underlying mechanism of a behavior does not, nor can it, address hypotheses regarding the behavior's historical origin or current survival value. Similarly, understanding a behavior's current survival value does not provide insight into the cause of the behavior in terms of the proximate mechanisms that produce it. Several examples of the confusion that can arise when mixing levels of analysis by attempting to address evolutionary questions at an inappropriate explanatory level are described by Sherman (1988), while a discussion of how levels can complement each other is provided by MacDougall-Shackelton (2011).

THE PROXIMATE MECHANISMS OF SOCIAL BEHAVIOR
Genes and Behavior: The Nature-Nurture Debate

Proximate mechanisms trigger the onset of a behavior. For example, rats (*Rattus norvegicus*) whose ovaries and adrenal glands have been removed recover some aspects of their sexual behavior when they are injected with the hormone estradiol (Davidson et al., 1968), and testosterone implants induce **prenuptial molt**—a key signal for mate choice by females—in male superb fairy-wrens (*Malurus cyaneus*) (Peters et al., 2000). As discussed above, proximate mechanisms do not explain the evolutionary basis of a behavior: the fact that testosterone causes male fairy-wrens to molt does not tell us anything about the function of bright breeding plumage in this species. Such information does, however, provide insight into the ways in which organisms are adapted to perform these and other intricate and complex social behaviors.

As an example of genetic effects on an important form of social communication, male zebra finches (*Taeniopygia guttata*) exposed to the song of an unfamiliar male initiate the expression of a transcription factor-encoding gene (*egr1*) in the auditory forebrain devoted to hearing, suggesting that neural processes linked to song pattern are important to song recognition, discrimination, and the formation of auditory associations (Mello et al., 1992). This same gene is also important in achieving the neural plasticity that allows subordinate male *Astatotilapia burtoni* (a fish in the family Cichlidae) to become dominant within minutes of an opportunity to do so, during which they exhibit dramatic changes in body coloration and behavior (Burmeister et al., 2005).

Relatively few genes have been identified thus far that directly influence social organization, but many clearly await discovery. One notable case is found in European honey bees (*Apis mellifera*), where the switch from working in the hive to foraging that takes place when bees are about two to three weeks old is associated with a significant increase in expression of the *foraging* gene, which encodes a protein kinase inducing changes in brain structure and behavior (Ben-Shahar et al., 2002). Another example is the *Gp-9* gene in the fire ant (*Solenopsis invicta*), which encodes a pheromone-binding protein crucial in chemical recognition of conspecifics that determines their ability to recognize queens and regulate their numbers—a social polymorphism that has been key to the explosive range expansion of this species (Krieger & Ross, 2002).

Even more dramatic genetic effects on social behavior have been identified in the social amoeba (*Dictyostelium discoideum*), which grows solitarily while feeding but when starved cooperatively forms a functionally multicellular "slug" consisting of dead stalk cells and viable spores. A genome-wide screen detected mutations in over 100 genes that facultatively produced cheaters—cells that disproportionately attempt to differentiate into spores rather than the stalk during aggregation when in competition with other genotypes—but not when clonal (Santorelli et al., 2008). Such behavioral genetics studies demonstrate the complexities of genetic influences on social behavior and are part of a growing field with significant potential for uncovering new information on the relative importance of nature and nurture to behavioral development and social behavior (Robinson et al., 2008).

A classic debate in behavioral circles is whether genetic predisposition or the environment is more important to the development of social behavior. As we hope is clear from the above discussion, brief as it is, this "nature-nurture" debate, stretching back decades, has been resolved: both are invariably important, and the interesting questions generally focus on how nature and nurture interact (Rutter, 2006).

Vocal Learning

Environmental influences on behavior include factors like nutrition, hormones, an animal's experience of the outside world, and the social context in which it lives. The development of bird song provides key examples of several of these. Songbirds (order Passeriformes), in contrast to all but a handful of other groups that include cetaceans, bats, humans, and two other avian orders—parrots and hummingbirds—learn their vocal signals without having to hear model signalers (Beecher & Brenowitz, 2005). This feat of learning is accomplished by the presence of two main neural pathways in the songbird brain. The first is a motor pathway involved in song production, and the second is a pathway in the anterior forebrain, often referred to as a ***template***, that is involved in song learning and recognition. In some species, learning is restricted to the first year of life. In others, learning is open-ended and continues long after the first year. The experience of hearing and practicing a song provides individuals with the necessary link between the auditory and song systems required to sing properly as adults.

In the white-crowned sparrow (*Zonotrichia leucophyrs*), for example, young males exposed to **conspecific** (same-species) tutor songs during a critical period of 10 to 50 days of age develop normal songs, whereas individuals raised in isolation or exposed to **allospecific** (other-species) songs during the critical period do not. It was originally thought that conspecific tutoring allowed birds to learn by selectively matching their vocal output to their internal, species-specific template (Marler, 1970; Beecher & Brenowitz, 2005). Subsequent studies demonstrated that birds are, in fact, capable of learning allospecific songs, including the radically different song of the unrelated strawberry finch (*Amandava amandava*), over an extended period beyond 50 days of age if a live, rather than a tape-recorded, tutor is used (Baptista & Petrinovich, 1984). More recent work on song sparrows (*Melospiza melodia*) suggests that young birds may be even more likely to learn songs from **eavesdropping** on songs of tutors that birds overhear singing with another young individual than from hearing a live tutor with which they directly interact (Beecher et al., 2007). Meanwhile, studies on other species have led to the conclusion that there is considerable variation in songbird song learning along a variety of dimensions, including when the song is learned, how faithfully birds copy tutors, the importance of early song experience, and the degree to which birds will copy tutor material when it is not produced by or similar to the song of conspecifics (Beecher & Brenowitz, 2005).

The song of the zebra finch provides a good illustration of hormonal influences on song development and singing behavior. There is marked sexual dimorphism in brain development in this species resulting in strikingly greater growth in the vocal-control areas of the brain in males (Nottebohm & Arnold, 1976). Injecting females with estrogen early in life causes them to develop male-like brains and male-typical songs, whereas females given testosterone in adulthood do not. Castrated males, on the other hand, still develop masculinized song nuclei and a song attractive to females (Wade & Arnold, 1996). This indicates that estrogens, rather than androgens (i.e., testosterone), are the primary determinant of sexual differentiation in this species (Adkins-Regan & Ascenzi, 1990). The effects of estrogens appear to be primarily regulated by factors intrinsic to the brain, most likely by the expression of genes that influence the levels of estrogen synthesized in the brain or the responsiveness of brain tissue to circulating levels of estrogen (Wade & Arnold, 2004).

The critical importance of social influences on behavioral development can be seen throughout the period of song learning in the song sparrow, in which each male sings 5 to11 different song types (Nordby et al., 2000). There is a sensitive period in the first summer of life when young birds learn many of their songs, but field studies show that learning also continues through the first year by storing copies of older neighbors' songs in a region of the brain called the forebrain song nuclei. A young song sparrow occupying a territory learns the songs of several of its neighbors, being more likely to learn song types that are shared among two or more neighbors. The end result is that each song sparrow holds roughly half its song types in common with neighbors, particularly older males that have been present longer. The adaptive function of such song-sharing behavior may be that it facilitates the rapid detection of intruders (Beecher et al., 1997). One recent study, for example, found that neighboring song sparrows sharing fewer songs were more aggressive with one another than those sharing more songs (Wilson & Vehrencamp, 2001).

Kin Recognition

Another good example of the importance of proximate mechanisms to social behavior is the study of recognition systems. Mechanisms of recognition are essential for individuals to discriminate members of their own social group, choose a mate of the appropriate sex, avoid incest, locate their parents, care for the right offspring, and engage in nepotism—the preferential treatment of kin. Early work in this area, pioneered by the Austrian zoologist Konrad Lorenz, demonstrated how precocial birds such as geese imprint on their mothers shortly after hatching (Lorenz, 1935; Hess, 1964).

A particularly well-studied area of recognition systems focuses on how parents recognize their offspring. Such offspring recognition probably involves odor in most insects and mammals. Adult birds, however, often use location of the nest as the cue to recognizing their offspring and fail to discriminate against the nestlings of other pairs that are artificially fostered into their nests. Presumably the failure to evolve recognition in this context is due to the fact that nestlings do not move around from one nest to another in the wild.

In species where the mixing of nestlings does occur, however, recognition can be remarkably precise. In Mexican free-tailed bats (*Tadarida brasiliensis*), for example, mothers exhibit spatial memory and return to within a short distance of where they last left their pups. Once in the vicinity, they are capable of using both scent and vocalizations to locate their offspring among thousands huddled in a small area of cave ceiling (Gustin & McCracken, 1987; Balcombe, 1990). In penguin colonies, where there can also be thousands of young huddled in crèches, offspring use vocals cues to recognize their parents. Such cues are based on pitch alone in species that build nests, such as Adélie penguins (*Pygoscelis adeliae*) and gentoo penguins (*P. papua*), but involve more sophisticated acoustic cues including the frequency-modulated shape of calls and the temporal succession of syllables in non-nest-building species such as king penguins (*Aptenodytes patagonicus*) and emperor penguins (*A. forsteri*) (Aubin & Jouventin, 2002; Jouventin & Aubin, 2002).

Kin recognition systems play a particularly important role in the context of nepotism. Three basic mechanisms are recognized that animals can potentially use to recognize kin: environmental cues, prior experience, and ***phenotype matching***—matching the look or smell of an individual to an internal template independent of environmental or other external cues (Sherman et al., 1997). Whichever is used, the mechanism must deal with the possibility of recognition errors, balancing the possibility of being too restrictive and thereby rejecting some individuals that are in fact kin, with that of being too accepting, thereby increasing the probability of treating as kin individuals that are in fact unrelated (Reeve, 1989).

After leaving the nest, some songbirds use vocalizations to recognize individuals, and in a few species, such as the long-tailed tit (*Aegithalos caudatus*), they favor kin when deciding whether and where to help feed nondescendent kin, possibly based on calls learned in the nest that show distinct signatures of kinship (Sharp et al., 2005). In Belding's ground squirrels (*Urocitellus beldingi*), a species in which nepotism is common, kin recognition involves both learning the phenotypes of related individuals during early development and later discriminating these familiar relatives from other unfamiliar individuals that are presumably nonrelatives. This is accomplished by a combination of prior association and the learning of their own phenotype, which they later compare to unknown individuals via phenotype matching (Holmes & Sherman, 1982). Experiments have demonstrated that phenotype matching using odors in this species is very precise, allowing individuals to recognize first-order (mothers related by $r = 0.5$), second-order (grandmothers related by $r = 0.25$), and even third-order (half-aunts related by $r = 0.125$) relatives (Mateo, 2002). Kin recognition is maintained even after long periods of hibernation (Mateo, 2010).

In several systems the genetic basis of kin recognition has been decoded. In cockroaches (*Blattella germanica*), for example, individuals at all developmental stages discriminate siblings from nonsiblings independent of prior association, preferring siblings as social partners and nonsiblings as mating partners. Kin recognition is based on quantitative differences in cuticular hydrocarbons present in their exoskeletons perceived through their anntenae (Lihoreau & Rivault, 2009). In several vertebrates, sophisticated kin recognition systems appear to be based on highly polymorphic genes known as the ***major histocompatibility complex (MHC)***. MHC genes encode cell-surface glycoproteins that bind piptides and present them to lymphocytes, thereby harnessing the mechanism for immunological self-recognition (Wedekind &

Penn, 2000). MHC genes have been implicated in determining odor and mating preferences in house mice (*Mus musculus*) and humans. In both cases, sexual selection appears to be involved, with individuals tending to prefer MHC-dissimilar mates (Potts et al., 1991; Wedekind & Penn, 2000; Jacob et al., 2002).

THE ULTIMATE CAUSES OF SOCIAL BEHAVIOR

While the fitness advantages of behaviors such as mating and caring for offspring are obvious in that they increase the number of an individual's own young, those of social behaviors such as living in groups and helping others are not. Because such behaviors are complex and paradoxical, their ultimate cause remains a key focus of evolutionary biologists.

Social interactions can be characterized as **mutualism** (when both individuals benefit by the behavior), **altruism** (when one individual, the altruist, makes a sacrifice while the other, the recipient, benefits by the behavior), **selfishness** (when the individual performing the behavior benefits at the expense of the recipient), and **spite** (when one individual performs a costly behavior that hurts the recipient, and thus both pay a cost). Mutualisms pose no paradox because both individuals benefit by the interaction. Altruism is more problematical. When individuals behaving altruistically are genetic relatives, as is often the case, **kin selection** is a likely explanation, with altruistic individuals gaining indirect fitness benefits by helping relatives produce additional offspring. As discussed in Chapter 11 of this volume, the conditions for kin altruism to evolve are summarized by Hamilton's rule $rB > C$, where B is the benefit to the recipient, C is the cost to the altruist, and r is the coefficient of relatedness between the two (Hamilton, 1964).

Altruism between unrelated individuals is relatively rare, but it occurs and cannot be explained by kin selection. One solution to the paradox of an individual making sacrifices for a second unrelated individual is **reciprocal altruism** (Trivers, 1971). If individuals interact repeatedly, altruism can be favored as long as the altruist receives a reciprocal benefit outweighing its initial cost.

Although reciprocal altruism is considered a potentially important evolutionary force, a critical caveat is that it can only work if there is a mechanism to punish "cheaters" that accept help without reciprocating. Consequently, unambiguous examples of reciprocal altruism outside of humans are rare. Apparent examples include male coalition formation in baboons (*Papio anubis*) (Packer, 1977), food sharing in vampire bats (*Desmodus rotundus*) (Wilkinson, 1984), and the interactions between cleaner fish (*Labroides dimidiatus*) and the "client" fish they attend, which involves complex ***indirect reciprocity*** (reciprocity by individuals other than those originally helped) by which clients eavesdrop on the cleaning behavior of cleaner fish and subsequently prefer to spend more time with cleaners that engage in cooperative rather than uncooperative cleaning (Bshary & Grutter, 2006). The possibility remains that such forms of complex reciprocity, which are not easy to detect, may be more common than currently recognized.

Considerable recent work has involved investigating the evolution of altruism among unrelated individuals in humans, typically by using "public goods" games in which subjects, usually university students, voluntarily choose a fraction of their "private" goods—money or a money proxy—to add to a public pot that is subsequently multiplied by some factor greater than one but less than the number of players and then divided up among all participants. Although the group's total payoff is maximized when all players contribute all their money to the public pot, the optimal choice for individual players is always to contribute nothing. Thus, contributors are considered altruistic cooperators whereas noncontributors are considered defectors or noncooperators.

The typical result of such games is that individuals initially cooperate and contribute some fraction of their money to the public pot. Because participants are unrelated and often neither know each other nor are likely to interact in the future, neither kin selection nor reciprocity are involved. In the absence of some additional mechanism, however, the extent of cooperation invariably declines rapidly to nearly zero when interactions are repeated and the game is played iteratively.

One mechanism that potentially maintains altruistic behavior is ***altruistic punishment***, in which individuals are given a means to punish noncooperators at a cost to themselves (Fehr & Gächter, 2002). Altruism can also be maintained by the advantages of maintaining a good reputation, a behavior predicated on indirect reciprocity such as was observed in the cleaner fish example discusssed above. Public goods games that incorporate the ability for players to see the history of other players and act on that knowledge—thus allowing for indirect rewarding of cooperators—result in a high level of cooperation compared to games not incorporating such information (Milinski et al., 2002). The evolution of cooperation by indirect reciprocity can be a potent evolutionary force leading to reputation building, morality judgement, and complex social interactions (Nowak & Sigmund, 2005).

Selfish behavior, when one individual benefits at the expense of another, is, perhaps unsurprisingly, not difficult to find. In birds, for example, it is fairly common, although often difficult to observe, for females to lay eggs in nests other than their own, thereby parasitizing the parental care of others. Such

brood parasitism can be interspecific, as in European cuckoos (*Cuculus canorus*) and the North American brown-headed cowbird (*Molothrus ater*) (Davies, 2000) or intraspecific (also known as conspecific brood parasitism), the latter found in birds (Yom-Tov, 1980; Lyon & Eadie, 2008) and insects (Tallamy, 2005). Such parasitic behavior would undoubtedly be even more common were it not for strong counterselection leading to the evolution of a diverse array of defenses designed to guard against or at least reduce the impact of brood parasites (Rothstein, 1990; Davies, 2000).

Spite as a social interaction has traditionally been considered to be at best unlikely and at worst paradoxical (Gadagkar, 1993), but current thinking is that it can evolve by what amounts to the inverse of Hamilton's rule: specifically, when two individuals are negatively related to each other (that is, less related than the average relatedness between two individuals in the population) and the cost to the actor is smaller than the product of the cost to the recipient and its (negative) relatedness to the actor (West & Gardner, 2010). One apparent example is found in the sterile soldier caste of the polyembryonic parasitoid wasp (*Copidosoma floridanum*), in which most larvae develop normally but a few become a sterile soldier morph that seek out and preferentially kill larvae from other eggs, thereby freeing up resources for their clone-mates (Garner et al., 2007). Although such "evolutionary spite" is still thought to be quite rare, "functional spite," in which an individual performs a costly behavior that harms the recipient but that gains the actor some future benefit such as increased social dominance, parasite deterrence, or offspring or sexual partner coercion, is probably much more widespread (Jensen, 2010).

THE RANGE OF SOCIAL BEHAVIOR

Living in Groups: Costs and Benefits

Interacting with other individuals is inherently dangerous and potentially costly. Reasons for this include an increased probability of parasite and disease transmission, increased competition for food and other resources, increased competition for mates, and increased interference among conspecifics. These costs have no automatic counteracting benefits (Alexander, 1974). *Why, then, are animals ever social, and why do they live in groups*? Clearly there must be potential benefits of aggregations that outweigh the inherent costs associated with being in close proximity with conspecifics.

On the positive side, aggregations may provide individuals with increased access to food through information sharing, cooperative defense against predators or non-group members, or simply energy conservation by huddling during cold or inclement weather. In any case, once animals are aggregated, there will inevitably be selection to evolve traits that will better exploit the potential advantages of group living, including mechanisms allowing individuals to better communicate, cooperate, and recognize each other as individuals and as kin in ways that promote their inclusive fitness through increased survivorship and reproductive success. For example, aggregated nymphs of the subsocial treehopper *Umbonia crassicornis* produce vibrational signals in synchronized bursts in response to simulated predator approach. These signals not only initiate a wave of signaling by other individuals within the aggregation but instigate defensive behavior on the part of their mother (Cocroft, 1999). In *Publilia concava*, another species of treehopper, adults live in mutualisms with ants by providing them with honeydew in return for protection against predators. These treehoppers produce vibrational alarm signals following encounters with the predatory ladybird beetle *Harmonia axyridis* that increase the ants' activity, thereby increasing their likelihood of discovering the predator and the effectiveness of their predator protection (Morales et al., 2008). Another classic example of a finely tuned communication system that evolved in the context of group living is that of the honey bee, a highly eusocial species in which workers returning to the hive perform elaborate "dances" that increase foraging efficiency by communicating to hive-mates the direction and distance of high-quality foraging sites (von Frisch, 1967; Seeley, 1995).

Given these complexities, a fruitful way to address the evolution of sociality in any particular case is to determine the costs and benefits of aggregating with others. A good example of a species in which many of these costs and benefits have been studied and identified is the cliff swallow (*Petrochelidon pyrrhonota*), a small, migratory passerine that breeds in colonies of up to several thousand pairs in western North America, primarily underneath overhanging rock ledges, on the sides of cliffs, and (increasingly) on artificial structures such as bridges (Brown & Brown, 1996).

Costs associated with colonial nesting in this species include increased susceptibility to ectoparasites, including both ticks and flies; increased loss of eggs due to disturbance by conspecifics; increased probability of losing paternity due to extrapair copulations; increased incidence of conspecific brood parasitism; increased incidence of food stealing among conspecifics (**kleptoparasitism**); greater attraction of predators; and increased travel distance to foraging areas. Countering these are several benefits, including increased vigilance and detection of predators resulting in greater annual survivorship; greater opportunity for conspecific brood parasitism; increased probability of gaining paternity due to extrapair copulations; and possibly the most significant benefit, that of birds unsuccessful at foraging actively seeking out and subsequently following other birds breeding nearby in the colony whose foraging efforts have been successful, a trait that confers considerable benefits given the highly unpredictable and variable insect food resources on which cliff swallows depend. Note that behaviors are in some cases simultaneously a cost for some individuals but a benefit to others. For example, the incidence of extra-pair matings, which are more frequent in larger colonies, are a benefit for the males that succeed in obtaining them and a cost to the cuckolded males. Similarly, conspecific brood parasitism is beneficial to the females that lay the eggs but a cost to the pair of birds that ends up raising a nestling that is not theirs.

Support for the hypothesis that a colony can act as an ***information center*** (Ward & Zahavi, 1973) has been found not only in cliff swallows but in osprey (*Pandion haliaetus*) (Greene, 1987), evening bats (*Nycticeius humeralis*) (Wilkinson, 1992), and colonial seabirds (Weimerskirch et al., 2010). More generally, the information center hypothesis is an example of animals taking advantage of ***public information***—social or nonsocial information that is accessible to others—for their own benefit. Studies suggest that the use of such public information is potentially widespread, yielding important information used by animals to inform the choices they make in foraging, breeding habitat selection, avoiding predators, mate choice, and the transmission of cultural traits (Danchin et al., 2004). For example, in the collared flycatcher (*Ficedula albicollis*), birds monitor the current reproductive success of conspecifics using such public information to assess local habitat quality and to choose their own subsequent breeding site when they return to an area in a subsequent year (Doligez et al., 2002).

The costs and benefits of group living vary from individual to individual depending on a variety of factors including their age, sex, and status, factors that are often conveniently summarized by the concept of ***dominance***, a key way in which individuals within groups are unequal. As dominant individuals monopolize a larger fraction of a group's resources, group living becomes less beneficial for subordinate individuals in the same group, which then become more likely to leave and try to live on their own. In order for sociality to be maintained in a population, subordinates must gain more from remaining in the group and being social than from leaving the group and trying to reproduce on their own. In the case of the cliff swallows, the delicate balance between these alternatives results in wide variation in group sizes ranging from solitary nesting to nesting in colonies of several thousand pairs.

Passive Aggregations, the Selfish Herd, and the Dilution Effect

A general hypothesis for why individuals might aggregate is predicated on the importance of predation and suggests that animals come together to form a so-called **selfish herd** in which individuals do not directly cooperate with each other but nonetheless passively benefit because each individual's chances of being eaten are reduced (the **dilution effect**), especially for individuals in the interior of the group (Hamilton, 1971). For example, groups of ocean skaters (*Halobates robustus*), a small marine insect that lives on the ocean surface, are depredated by juvenile pilchards (*Sardinops sagax*) at a rate that decreases linearly with increasing group size (Foster & Treherne, 1981). Experiments quantifying shark attacks on rafts of seal decoys also indicate that danger decreases proportionately with group size, thus supporting the importance of the selfish herd effect (De Vos & O'Riain, 2010).

In the simplest example, when a group-living individual encounters a predator that will eat just one prey item, its likelihood of being eaten in a group of size n is reduced from p, the probability when alone, to p/n. This can yield a strong benefit quickly to individuals in aggregations, even if predators are differentially attracted to larger groups, so long as the cost of increased conspicuousness does not overtake the benefit of dilution. Where location within the group matters, social interactions will likely sort out by social status, with some individuals gaining favored positions—typically central rather than peripheral positions—by dominance or nepotism (Krause & Ruxton, 2002).

Social Behavior Based on Protection against Predators and Intruders

Although passive protection against predators can be achieved by any species through the selfish herd and dilution effects, social groups forming to engage in active cooperative defense are more unusual. Examples include muskoxen (*Ovibos moschatus*), which form tight rings and face outwards in defense against wolves (*Canis lupus*) (Tener, 1965; Heard, 1992); mule deer (*Odocoileus hemionus*), which bunch together with conspecifics and aggressively defend against coyotes (*Canis latrans*) (Lingle, 2001); and guanacos (*Lama guanicoe*), which cooperatively defend against the culpeo fox (*Lycalopex culpaeus*) (Novaro et al., 2009). Cooperative defense is also found in invertebrates such as the tent caterpillar (*Malacosoma disstria*), which exhibits collective defense again parasitoids in the form of collective head flicking and biting by groups of instars (McClure & Despland, 2011), and group-living larvae of the Australian sawfly (*Perga affinis*), which store noxious *Eucalyptus* oils in their foregut that they regurgitate when attacked by ants, birds, and mice (Morrow et al., 1976). As another example, resident males of several species of territorial fiddler crabs (*Uca* spp.) will leave their territories and cooperate with neighbors to repel intruders, apparently because the benefits of maintaining neighborhood stability outweigh the costs of repelling intruders (Booksmythe et al., 2012).

Alarm calls and other complex signaling behavior within aggregations can also reduce the likelihood of predation. Such signaling may coordinate a group's escape from danger, confuse a predator, and prompt other nearby individuals to seek protected sites or shelter. Alarm calls may convey information about the type of predator and lead to the appropriate evasive behavior. For example, vervet monkeys (*Chlorocebus pygerythrus*) give short, tonal alarm calls in response to leopards (*Panthera pardus*); low-pitched, staccato grunts in response to martial eagles (*Polemaetus bellicosus*); and high-pitched, chatter-like calls in response to python snakes (*Python sebae*), prompting other monkeys nearby to run into trees for leopards, look up for eagles, and look down for snakes (Seyfarth et al., 1980). Alarm calls might even provide information regarding an individual predator's identity and habits, although this has yet to be demonstrated.

Alarm calling is often considered a good example of an altruistic behavior because it appears to benefit others at some cost to the calling individual. Such calls provide potentially valuable information to others but may endanger the caller by attracting predators, although the evidence for the latter is equivocal, possibly because callers are in some cases simultaneously serving as a ***pursuit-deterrence*** signal, communicating to the predators their ability to evade capture (Blumstein, 2007).

An example of a species whose alarm-calling behavior has been extensively studied is Belding's ground squirrel. Individuals call more frequently when close relatives, including noncollateral kin, are nearby, thus suggesting that alarm calling has evolved at least in part due to kin selection—specifically the indirect fitness benefits of aiding relatives (Sherman, 1977). Similar results have been reported for a variety of other rodents, including several other ground squirrels, the chipmunk (*Eutamias sonomae*), and both black-tailed prairie dogs (*Cynomys ludovicianus*) and Gunnison's prairie dogs (*C. gunnisoni*) (Blumstein, 2007). Alarm calls are given in other contexts besides those likely to entail kin selection, however. For example, those given by birds in mixed-species flocks—common in both birds and monkeys—in some cases appear to convey information regarding the proximity of a predator (Terborgh, 1990; Sharpe et al., 2010), while in others they may be used to selfishly distract flockmates and increase the caller's foraging efficiency (Munn, 1986).

Aggregations have the potential to augment and bolster signaling systems. This is particularly true in ***aposematic*** species that advertise their defenses to potential predators (Orians & Janzen, 1974). There are several potential reasons for this, including proportionately lower detectability, faster learning of the aposematic signal by predators, and increased effectiveness of the aposematic signal (Gamberale & Tullberg, 1998; Riipi et al., 2001). Groups of animals may also confuse predators by looking larger than they actually are or by moving apart in unpredictable ways (Krakauer, 1995), actions that may cause a predator to hesitate just long enough to permit the prey's escape.

Grouping may also serve less direct functions than reducing predation risk. For example, the remarkable shift of the desert locust (*Schistocerca gregaria*) of sub-Saharan Africa from its cryptic solitary phase to its notorious gregarious phase, during which it can form huge swarms of individuals moving cohesively in search of food covering as much as 200 km^2, has been suggested to be a behavioral strategy to reduce predation risk by mobile predators that would otherwise be able to profitably forage in adjacent patches within which locust densities are relatively high (Reynolds et al., 2009).

Social Behavior Based on Increased Foraging Efficiency

Individuals in groups may benefit by cooperating to gain access to food and other resources. In the fruit fly (*Drosophila melanogaster*), for example, males and mated females deposit an aggregation pheromone on rotting fruit that induces aggregated oviposition; larvae subsequently feed on yeasts that develop on the fruit. The apparent benefit of facilitating group oviposition is that at low larval densities, fungi and molds can outcompete the larvae, resulting in high larval mortality (Wertheim et al., 2002). Thus, larvae fare better as initial colonists of fruit when they live in groups. This same strategy can help insect larvae overcome plant defenses, such as in the sawfly (*Neodiprion pratti*), where larvae feed on pine needles and individuals aggregate around the few larvae that succeed in cutting through the tough needle cuticle (Ghent, 1960). In other plant-feeding insects, feeding itself may affect the quality of the food when substances in the insect's saliva that overcome chemical defenses or alter the metabolism of the host plant allow more efficient (for the insect) release of nutrients (Després et al., 2007).

When predators hunt in groups, their prey may become confused, leading to a ***beater effect*** whereby prey flushed out by group activity become easy to capture (Swynnerton, 1915). Where predators hunt cooperatively, such as in the foraging of large carnivores like lions, hyenas, and wolves, they can corner and bring down prey more easily (Packer & Ruttan, 1988). This also occurs in the group-living Harris's hawk (*Parabuteo unicinctus*), where birds form hunting parties of two to six, resulting in the ability to kill prey larger than themselves and flush out rabbits hiding in thick cover, behaviors that result in improved capture success and increased energy intake (Bednarz, 1988).

Group living can select for sophisticated systems of communication and cooperation that enhance overall foraging success. A good example is the highly gregarious eastern tent caterpillar (*Malacosoma americanum*). In this species, hungry individuals that have not yet started to feed follow silk-and-chemical trails laid down by successful foragers returning to their communal tents to attractive feeding sites containing young leaves that are both nutritionally superior and offer caterpillars the ability to better repel predatory ants by regurgitating plant-derived defensive compounds (Peterson et al., 1987; Fitzgerald & Peterson, 1988). An even more sophisticated example is the complex dance of honey bees already mentioned, where successful foragers returning to the hive pass along information on both the direction and distance of food sources to other workers. Such dances, although extraordinary by themselves, are only one component of the complex communication system of this species, which includes chemosensory as well as spatial cues (Grüter & Farina, 2009).

Social Behavior Based on Reproduction

In some species, aggregations are based exclusively on mating. These include various insect aggregations such as are found in some bees and wasps (order Hymenoptera), flies (order Diptera), and butterflies (superfamily Papilionoidea), in which females congregate at conspicuous landmarks (Thornhill & Alcock, 1983), and the aggregations of males at **leks**—display sites used only for mating (Höglund & Alatalo, 1995). The latter is known from nearly 100 species of birds but has been reported in only 13 mammals, nine of which are ungulates, along with the hammer-headed bat (*Hypsignathus monstrosus*) (Bradbury, 1977).

The selective benefits of lek aggregations are a matter of intense debate. One hypothesis is that males congregate in sites where the home ranges of many females overlap (the "hotspot" hypothesis) (Bradbury & Gibson, 1983). Alternatively, female mate choice may be driving male aggregations because leks provide females with a convenient means to quickly assess male quality (Queller, 1987). Yet another hypothesis is that leks form because subdominant and novice males benefit by associating with highly successful, dominant males (the "hotshot" hypothesis) (Beehler & Foster, 1988). One particularly intriguing discovery is that leks in several species are composed of clusters of related kin, raising the possibility that kin selection and inclusive fitness benefits play a role in the evolution of at least some leks (Höglund et al., 1999; Petrie et al., 1999; Shorey et al., 2000). In the case of wild turkeys (*Meleagris gallopavo*), kin selection benefits have been shown to be sufficient to explain the evolution of cooperative courtship whereby several males coordinate displays (Krakauer, 2005). Although only the dominant male in a displaying coalition actually mates in this species, subordinate males gain indirect fitness benefits greater than what they could expect to achieve on their own.

In many cases the aggregation of one sex provides opportunities for the other. For example, in species where females aggregate due to the clumping of food or nest sites, males are likely to aggregate at these sites as well because they are the most efficient places to find females with which to mate. In still others, males and females aggregate in both space and time, such as with the explosive breeding assemblages of many frogs and toads (Wells, 1977).

Social behavior is involved in social dominance and the maintenance of territories, regardless of whether dominance status or territories are held by individuals or by groups. Species defend territories when they are **economically defendable**, meaning that the benefits of the behaviors associated with having exclusive access to the area outweigh the costs of maintaining and defending it from conspecifics (Emlen & Oring, 1977). In the territorial systems of many species, overt defense in the form of direct aggressive behavior against intruders has given way to indirect defense in the form of vocalizations and scent marking.

Social Behavior Based on Access to Clumped or Limited Resources

In most of the cases we have discussed thus far, the benefits of grouping are apparently derived from the social interactions that take place within aggregations; that is, aggregations have formed because of some direct benefit of sociality such as group defense or information sharing that outweighs the inevitable costs. In some cases, however, the benefits of sociality do not appear to be sufficient to explain aggregations; rather, groups apparently form in order that individuals are able to gain access to localization of some critical and limited resource (Alexander, 1974). Classic examples include safe sleeping sites for hamadryas baboons (*Papio hamadryas*) (Kummer, 1968) and suitable breeding sites for colonial marine birds and mammals. Access to clumped and limited resources rather than intrinsic social benefits appears to be an important driver of group living in many other systems as well, however.

Examples include cases of *female* (or *harem*) *defense polygyny*, in which females aggregate, often because of clumped food or nest sites, providing dominant males the opportunity to defend them and thereby gain multiple mating opportunities (Emlen & Oring, 1977). Female-defense polygyny is found in many ungulates (Clutton-Brock, 1989) and more rarely in other taxa, incuding fishes (Seki et al., 2009); at least one species of bird, the Montezuma oropendola (*Psarocolius montezuma*) (Webster, 1994); various insects (Heinze & Hölldobler, 1993); and some amphipods (Just, 1988).

Cooperative breeding occurs when more than two individuals contribute to the care of young within a single brood. The most well studied of these are helper-at-the-nest systems in which offspring delay dispersal and remain in their natal groups as nonbreeding helpers, often for several years, during which time they assist in raising younger siblings rather than breeding on their own (Brown, 1987). Although in a few of these cooperative breeders young may gain sufficient intrinsic or social benefits by remaining at home to compensate for the reproduction they forgo, in the majority of cases this is not the case. Rather, they forgo breeding not because it is an inferior fitness option but because they are unable to acquire a suitable territory or mate—circumstances referred to as the "ecological constraints" hypothesis (Koenig, 1981; Emlen, 1991).

A good example is the Florida scrub-jay (*Aphelocoma coerulescens*), where long-term demographic studies have shown that helpers gain an estimated 0.14 offspring equivalents of fitness by helping compared to 0.62 offspring equivalents gained by first-time breeders in the population (Woolfenden & Fitzpatrick, 1984). Similar conclusions have been reached by experimental studies in red-cockaded woodpeckers (*Picoides borealis*) and Seychelles warblers (*Acrocephalus sechellensis*). In the former, helpers immediately give up their helper status and occupy artificial nest cavities when they are provided (Walters et al., 1992), while in the latter birds transplanted to an uninhabited island forgo group living and breed as pairs until the population saturates the available high-quality territories (Komdeur, 1992). In all these cases, the evidence indicates that ecological constraints (or "resource access" benefits; S.-F. Shen and S. Emlen, personal communication), rather than social benefits derived from group living, prompt young to remain at home. Once home, young serve as helpers in order to "make the best of a bad job," garnering what inclusive benefits they can while waiting for a territory or breeding vacancy that will allow them to reproduce (Koenig et al., 1992).

A difficulty in assessing the importance of access to a limited resource is that once aggregations form, there will inevitably be selection to take advantage of group living. Thus, various advantages to sociality may emerge and eventually come to outweigh the disadvantages of grouping even though the original impetus for aggregations was primarily or exclusively extrinsic constraints rather than social benefits (Koenig et al., 1992). Further, formation of cooperative territorial groups can lead to supersaturation of habitat with more birds than just one pair per territory; this can impose a new selective pressure favoring group living by increasing the constraints on outside options (Dickinson & Hatchwell, 2004).

Cooperative Breeding and Reproductive Skew

Cooperative breeding is found in approximately 9 percent of birds (Cockburn, 2006), a smaller proportion of mammals, a few fishes, and some insects and arachnids. It is particularly common among birds in Australia, possibly due to the common phylogenetic history of a large proportion of its avifauna (Russell, 1989). In general, however, cooperative breeding is rare because it requires parental care, which outside birds and mammals is itself a relatively uncommon behavior.

Cooperative breeding is generally associated with species in which dispersal is restricted and thus there are opportunities for prolonged contact between close relatives such as occurs in permanent residents inhabiting mild climates (Hatchwell, 2009). In birds, cooperative breeding is generally believed to be a result of a shortage of high-quality territories or mates, and helpers will typically become breeders if given the opportunity to do so, as discussed above. Although such ecological constraints are clearly important in many species that have been well studied, others inhabit highly variable and unpredictable regions where breeding is difficult, at least in some years, and in such cases helpers may frequently be necessary for successful reproduction (Emlen, 1982; Jetz & Rubenstein, 2011). This "hard-life" hypothesis for the evolution of cooperative breeding has been proposed to explain cooperative breeding in African starlings (family Sturnidae) (Rubenstein & Lovette, 2007) and humans, where raising young is a long, difficult process wrought with danger (Hrdy, 2009). It may also be important in some species of cooperatively breeding birds in which helping was previously thought to be driven by ecological constraints as suggested by the finding that helpers are more important when circumstances are unfavorable for breeding because of poor environmental conditions or breeder inexperience (Magrath, 2001). In contrast, in the acorn woodpecker, helpers benefit groups significantly more following a good acorn crop when conditions for breeding are favorable than following poor crops when conditions are not (Koenig et al., 2011).

In vertebrates, most cases of cooperative breeding involve helping at the nest, as already described. Less common are species exhibiting **cooperative polygamy** or mate-sharing, in which multiple cobreeders of one or both sexes share mates of the opposite sex. In species such as the Harris's hawk and Galápagos hawk (*Buteo galapagoensis*), multiple males mate with a female in a system of cooperative polyandry (Faaborg & Patterson, 1981; Faaborg et al., 1995). In others, such as the European dunnock (*Prunella modularis*) and the acorn woodpecker, multiple males may share and mate with more than one female in a system of **cooperative polygynandry**. These two species differ, however, in that dunnock groups are not family based: cobreeding male dunnocks are unrelated and females breed on separate territories (Davies, 1992). In contrast, acorn woodpecker groups, like most cooperative breeders, are family based, and cobreeders within a sex are typically first-order relatives, either siblings or parents and their offspring (Koenig et al., 1995b).

The outcome of mate-sharing—the degree to which reproduction is monopolized by one or a few individuals—is referred to as the degree of **reproductive skew** and is highly variable. In egalitarian or low-skew societies, cobreeders share parentage relatively equally, as is true for joint-nesting female acorn woodpeckers because of the phenomenon of egg destruction mentioned earlier. In contrast, in high-skew societies, reproduction is highly biased toward a single individual, as is the case in most of the eusocial species we discuss next, including ant colonies, honey bees, and naked mole-rats (*Heterocephalus glaber*). The factors determining the degree of reproductive skew is a subject of considerable interest (Johnstone, 2000; Hager & Jones, 2009), but most current models start with the importance of four factors: (1) the expected success of a subordinate that reproduces on its own (corresponding to the degree of ecological constraints), (2) the expected success of the group if the subordinate remains in the group, (3) the genetic relatedness of group members, and (4) the probability that a subordinate would win a fight with a dominant without being severely injured (Keller & Reeve, 1994). These factors attempt to synthesize the relative costs and benefits of group living to a subordinate, taking into account the potential for inclusive fitness benefits of not breeding and the costs associated with competition for reproductive opportunities within the group.

Cooperative breeders include species exhibiting a remarkably diverse range of social organization and behavior (Cockburn, 1998, 2004) and include some of the most intensively studied species in the world (Stacey & Koenig, 1990; Koenig & Dickinson, 2004). Before moving on to a discussion of eusociality, we briefly discuss three, two birds and one mammal. In conjunction with the species already mentioned, they provide an introduction to the bizarre and unusual social behavior exhibited by cooperatively breeding species.

Australian fairy-wrens in the genus *Malurus* encompass a series of spectacularly plumaged species, all of which are cooperative breeders, usually in the form of young males remaining as helpers in their natal groups. Experimental work has shown that ecological constraints are key to this system, with helpers being limited by a shortage of potential mates (Pruett-Jones & Lewis, 1990). More surprising was the discovery that a high proportion of dominant, breeding pairs appeared to consist of close relatives and that incest was apparently common (Rowley et al., 1986). Subsequent molecular work demonstrated that this was not the case; rather, a majority of offspring turn out to be the product of extra-group matings, leading to the unexpected finding that there is virtually no correlation between group composition—the social mating system—and who actually mates with whom—the genetic mating system (Mulder et al., 1994). The driver of this extraordinary situation appears to be sexual selection, with females preferring to mate with older males in the population regardless of their social mate (Double & Cockburn, 2000).

We have already mentioned acorn woodpeckers, which are a cooperatively polygynandrous species in which cooperation and competition—the latter most dramatically in the form of egg destruction by joint-nesting females—are both regular features of their social behavior. Offspring in this species typically remain as nonbreeding helpers assisting their parents in raising subsequent young. In contrast to fairy-wrens there is no extra-group mating, nor do unrelated individuals join established groups. Groups thus consist entirely of closely related males and equally closely related females.

Of particular interest is the question of what determines reproductive roles within groups? The two main possibilities are within-sex *reproductive competition*—that young birds are reproductively suppressed by their older, dominant parents—and *incest avoidance*—that young birds do not breed because they are closely related to the breeders of the opposite sex. Long-term demographic and genetic studies have confirmed that when reproductive vacancies arise, they are filled by unrelated birds from elsewhere rather than by subordinates of the same sex, after which subordinates of the opposite sex are able to inherit and cobreed along with their older, presumably dominant parents (Koenig et al., 1998; Haydock et al., 2001). Thus incest avoidance, rather than reproductive competition, is the primary determinant of reproductive roles in this species. Incest avoidance is so strong that the population loses an estimated 9 to 12 percent in reproductive potential due to this factor alone (Koenig et al., 1999). Incest avoidance is similarly an important factor influencing reproductive roles in the vast majority of, if not all, highly social vertebrates (Koenig & Haydock, 2004).

Kalahari meerkats (*Suricata suricatta*) provide a good example of a cooperatively breeding mammal. Breeders of both sexes live in cooperative groups with dominants monopolizing most of the reproduction (Hodge et al., 2008; Spong et al., 2008). Group augmentation, a positive group-size effect on reproduction, arises because helpers enhance pup growth and survival by babysitting, which is only done by subordinates and sometimes involves remaining in the burrow without food for extended periods of time (Clutton-Brock et al., 2001). There is a measurable cost of helping in terms of weight loss, but helpers of both sexes benefit from living in the group with fitness gains through both direct reproduction and the raising of nondescendant kin (Russell et al., 2007). Although reproductive skew is relatively great, female subordinates sometimes succeed in becoming pregnant, after which they compete for reproductive success within the group through infanticide (Kutsukake & Clutton-Brock, 2008). Meanwhile, subordinate males foray to other groups where they compete to sire extra-group young (Young et al.,

2007). While these strategies are not equivalent to breeding as a dominant, they provide young animals with options that are relatively profitable given the ecological constraints imposed by limited food, high susceptibility to predation, and availability of breeding vacancies.

Eusociality

Eusociality, a social organization in which young are cared for cooperatively, generations overlap, and the society is segregated into distinct **castes** that provide different services to the colony, is often considered the pinnacle of social evolution. Eusociality is found in several orders of insects, including bees and wasps (order Hymenoptera), termites (order Isoptera), gall-making aphids (order Hemiptera), and thrips (order Thysanoptera); *Synalpheus regalis*, a tropical reef-dwelling marine shrimp dependent on sponges for survival (Duffy, 1996); and two species of vertebrates, both African mole-rats in the family Bathyergidae, the naked mole-rat (*Heterocephalus glaber*) and the Damaraland mole-rat (*Cryptomys damarensis*) (Jarvis & Bennett, 1993; Bennett & Faulkes, 2000). Reproductive skew in eusocial species is typically very high—often only a single individual (the "queen") reproduces out of a colony of thousands or tens of thousands. They also exhibit extreme task specialization, which makes colonies potentially very efficient at gathering resources. Workers may never reproduce during their entire lives but nonetheless gain inclusive fitness benefits by aiding the reproduction of the queen, who is typically their mother. With a high proportion of the population forgoing reproduction, often permanently, the evolution of altruism in eusocial species is a topic of particular interest and played a key role in the development of kin-selection theory (Hamilton, 1964, 1972).

In diploid species, including most vertebrates, sisters share half their genes in common, and individuals who help parents produce a benefit tantamount to the fitness they would achieve by producing an offspring of their own. Eusocial Hymenoptera, however, are **haplodiploid**, a system in which unfertilized (**haploid**) eggs become males and fertilized (**diploid**) eggs become females. As a result, a group of sisters with a single mother and father, such as may make up the workers of an ant colony or a beehive, share one-quarter of their genes through their mothers and one-half their genes through their father, as all sperm from a single haploid male are genetically identical. The result is that such sisters share 75 percent of their genes through common ancestry, whereas mothers share only 50 percent of their genes with their own daughters. Female workers potentially transmit more copies of their genes by helping their mother produce sisters than by producing their own daughters and sons, thus providing a potential genetic basis for the evolution of eusociality.

As appealing as this hypothesis may be, subsequent study has indicated that haplodiploidy is likely to be a predisposing, rather than a causal, factor in the evolution of eusociality. Hymenoptera queens often mate with multiple males, and thus sperm is provided by more than one source, diluting the haplodiploidy effect on sister relatedness (Hughes et al., 2008b). In addition, multiple queens may found wasp colonies, further diluting the average genetic relatedness among workers. It is also relevant that many eusocial species are not haplodiploid, including not only the two mole-rats but all termites (order Isoptera).

As a result of these and other considerations, workers in eusocial colonies are thought to have originally given up reproduction due to ecological constraints on independent breeding, the latter being due to high predation rates, a shortage of nest sites, and a short breeding season (Brockmann, 1997). As in the case of cooperatively breeding birds, opportunities to survive and reproduce away from the colony are limited, favoring individuals that stay home. If females remain in their natal groups, within-colony relatedness will be high and kin selection will be a potentially important evolutionary force that favors cooperation.

There remains, however, considerable controversy regarding the evolution of eusociality, particularly in terms of the role of kin selection (Nowak et al., 2010). Future phylogenetic work aimed at determining the evolutionary origins of eusociality are likely to be particularly important in resolving this issue. A recent study of female mating frequencies in eusocial hymenoptera, for example, indicated that monogamy is ancestral in all independent eusocial lineages, supporting the hypothesis that kin selection and inclusive fitness has been key to the evolution of eusociality in this group (Hughes et al., 2008a).

THE STUDY OF SOCIAL BEHAVIOR

As should by now be clear, social behavior is a large field encompassing a variety of issues that are studied at multiple levels using a diverse array of high- and low-tech methods. At all levels, however, the goal is generally to deduce causes using strong inference based on a set of critical predictions (Platt, 1964). If tests of these predictions indicate that the predictions are not met, then the hypothesis is falsified; if the predictions are met, the hypothesis is supported. This does not mean that the hypothesis is true, however, because there are always likely to be alternatives that remain to be tested (see Volume 1, Chapter 1).

Consider the variability in behavior of male step parents, which may (1) kill the offspring of their new mate, (2) tolerate the offspring, or (3) invest in the offspring and treat them as a normal parent (Rohwer, 1986; Rohwer et al., 1999). The first option, infanticide, was long considered aberrant behavior but is now generally considered a form of sexual competition by which a male step parent gains a reproductive advantage through earlier conception by his new female (Hrdy, 1974). More difficult to explain is why male step parents sometimes adopt and feed the offspring of widowed females as if they were the normal genetic parent.

One possible explanation (the current benefits hypothesis) is that they in fact have mated with the female and thus may have genetic offspring in the female's nest (Meek & Robertson, 1991). An alternative (the future benefits hypothesis) is that the adoptive male gains future reproductive benefits by virtue of his foster behavior, increasing the likelihood that his female will mate with him during her next breeding attempt. The current benefits hypothesis predicts that some of a remated female's nestlings have been sired by the adoptive father, whereas the future benefits hypothesis predicts that the adoptive male will mate sooner, usually with the widowed female, and ultimately produce more offspring than an unpaired male that fails to adopt.

Experimental work on western bluebirds (*Sialia mexicana*), where about half of male step parents feed their adoptive young while the other half tolerate them, but are never infanticidal, found that male provisioning had no effect on female condition or survivorship and male step parents that fed young were not more likely to breed with the female on her subsequent attempt than were males that did not feed (Dickinson & Weathers, 1999). This finding rejects the future benefits hypothesis in this species but does not prove that males gain current benefits, which would require either observational data demonstrating that adoptive males had previously achieved extra-pair matings with the widowed female or that one or more of her offspring had been sired by the adoptive male (Meek & Robertson, 1991).

Ideally, hypotheses are mutually exclusive such that only one can be true. In many cases, however, behaviors have more than one current function and more than one hypothesis may be supported. In the case of feeding by step parents, for example, it is possible that males might have both achieved extra-pair matings with the female and be more likely to breed with her in the future, thus supporting both the current and future benefits hypotheses.

Strong inference relies on clear tests and critical predictions to distinguish between alternative hypotheses designed to explain some phenomenon at a particular level of analysis. Predictions can be tested either with data collected from field observations or with experiments, although the latter are often considered preferable because confounding factors are more easily controlled. Unfortunately, experimental manipulations may alter other factors beyond those intended, especially where social behavior is concerned. In order to minimize such problems, researchers generally try to test their hypotheses using multiple lines of evidence.

Social behavior is a complex combination of the costs and benefits of living in groups, dominance interactions, conflict between the sexes, nepotism, competition, and cooperation. The diversity of social behavior continues to provide significant material for evolutionary biologists to understand natural selection and the process of evolution and for mechanistic biologists to understand how the underlying processes governing behavior work. Through behavioral ecology, these questions are framed within an ecological context to peel back the layers and reveal the complex underpinnings of the natural world.

ACKNOWLEDGMENTS

We wish to thank the National Science Foundation for support of our research on avian social behavior. Portions of the chapter have been adapted from Walter D. Koenig and Janis L. Dickinson, Encyclopædia Britannica Online, s. v. "social behaviour, animal," accessed February 28, 2013, http://www.britannica.com/EBchecked/topic/550897/social-behaviour-animal.

REFERENCES AND SUGGESTED READING

Adkins-Regan, E. & M. Ascenzi. (1990). Sexual differentiation of behavior in the zebra finch: Effect of early gonadectomy or androgen treatment. *Hormones and Behavior*, 24, 114–127.

Alexander, R. D. (1974). The evolution of social behavior. *Annual Review of Ecology and Systematics*, 5, 325–383.

Aubin, T. & P. Jouventin. (2002). How to vocally identify kin in a crowd: The penguin model. *Advances in the Study of Behavior*, 31, 243–277.

Balcombe, J. P. (1990). Vocal recognition of pups by mother Mexican free-tailed bats, *Tadarida brasiliensis mexicana*. *Animal Behaviour*, 39, 960–966.

Baptista, L. F. & L. Petrinovich. (1984). Social interaction, sensitive phases and the song template hypothesis in the white-crowned sparrow. *Animal Behaviour*, 32, 172–181.

Bednarz, J. C. (1988). Cooperative hunting Harris' hawks (*Parabuteo unicinctus*). *Science*, 239, 1525–1527.

Beecher, M. D. & E. A. Brenowitz. (2005). Functional aspects of song learning in songbirds. *Trends in Ecology and Evolution*, 20, 143–149.

Beecher, M. D., J. M. Burt, A. L. O'Loghlen, C. N. Templeton, & S. E. Campbell. (2007). Bird song learning in an eavesdropping context. *Animal Behaviour*, 73, 929–935.

Beecher, M. D., J. C. Nordby, S. E. Campbell, J. M. Burt, C. E. Hill, & A. L. O'Loghlen. (1997). What is the function of song learning in songbirds? *Perspectives in Ethology*, 12, 77–97.

Beehler, B. M. & M. S. Foster. (1988). Hotshots, hotspots, and female preference in the organization of lek mating systems. *American Naturalist*, 131, 203–219.

Bennett, N. C. & C. G. Faulkes. (2000). *African Mole-rats: Ecology and Eusociality*. Cambridge, UK: Cambridge University Press.

Ben-Shahar, Y., A. Robichon, M. B. Sokolowski, & G. E. Robinson. (2002). Influence of gene action across different time scales on behavior. *Science*, 296, 741–744.

Blumstein, D. T. (2007). The evolution of alarm communication in rodents: Structure, function, and the puzzle of apparently altruistic calling. In J. Wolff & P. W. Sherman (eds.), *Rodent Societies: An Ecological and Evolutionary Perspective* (pp. 317–327). Chicago: University of Chicago Press.

Booksmythe, I., C. Hayes, M. D. Jennions, & P. R. Y. Backwell. (2012). The effects of neighbor familiarity and size on cooperative defense of fiddler crab territories. *Behavioral Ecology*, 23, 285–289.

Bradbury, J. W. (1977). Lek mating behavor in the hammer-headed bat. *Zeitschrift für Tierpsychologie*, 45, 225–255.

Bradbury, J. W. & R. M. Gibson. (1983). Leks and mate choice. In P. Bateson (ed.), *Mate Choice* (pp. 109–138). Cambridge, UK: Cambridge University Press.

Brockmann, H. J. (1997). Cooperative breeding in wasps and vertebrates: The role of ecological constraints. In J. C. Choe & B. J. Crespi (eds.), *The Evolution of Social Behavior in Insects and Arachnids* (pp. 347–371). Cambridge, UK: Cambridge University Press.

Brown, C. R. & M. B. Brown. (1996). *Coloniality in the Cliff Swallow: The Effect of Group Size on Social Behavior*. Chicago: University of Chicago Press.

Brown, J. L. (1987). *Helping and Communal Breeding in Birds*. Princeton, NJ: Princeton University Press.

Bshary, R. & A. S. Grutter. (2006). Image scoring and cooperation in a cleaner fish mutualism. *Nature*, 441, 975–978.

Burmeister, S. S., E. D. Jarvis, & R. D. Fernald. (2005). Rapid behavioral and genomic responses to social opportunity. *PLoS Biology*, 3, e363. doi:10.1371/journal.pbio.0030363.

Cardé, R. T. & J. G. Millar. (2009). The scent of a female: Sex pheromones of female tiger moths. In W. E. Conner (ed.), *Tiger Moths and Woolly Bears: Behavior, Ecology and Evolution of the Arctiidae* (pp. 127–144). Oxford, UK: Oxford University Press.

Charlton, B. D., D. Reby, & K. McComb. (2007). Female red deer prefer the roars of larger males. *Biology Letters*, 3, 382–385.

Cheney, D. L. & R. M. Seyfarth. (2009). Stress and coping mechanisms in female primates. *Advances in the Study of Behavior*, 39, 1–44.

Clutton-Brock, T. H. (1989). Mammalian mating systems. *Proceedings of the Royal Society of London, B*, 236, 339–372.

Clutton-Brock, T. H. & S. D. Albon. (1979). The roaring of red deer and the evolution of honest advertisement. *Behaviour*, 69, 145–170.

Clutton-Brock, T. H., A. F. Russell, L. L. Sharpe, P. N. M. Brotherton, G. M. McIlrath, S.White, & E. Z. Cameron. (2001). Effects of helpers on juvenile development and survival in meerkats. *Science*, 293, 2446–2449.
Cockburn, A. (1998). Evolution of helping behavior in cooperatively breeding birds. *Annual Review of Ecology and Systematics*, 29, 141–177.
Cockburn, A. (2004). Mating systems and sexual conflict. In W. D. Koenig & J. L. Dickinson (eds.), *Ecology and Evolution of Cooperative Breeding in Birds* (pp. 81–101). Cambridge, UK: Cambridge University Press.
Cockburn, A. (2006). Prevalence of different modes of parental care in birds. *Proceedings of the Royal Society of London, B*, 273, 1375–1383.
Cocroft, R. B. (1999). Offspring-parent communication in a subsocial treehopper (Hemiptera: Membracidae: *Umbonia crassicornis*). *Behaviour*, 136, 1–21.
Danchin, É., L.-A. Giraldeau, T. J.Valone, & R. H. Wagner. (2004). Public information: From nosy neighbors to cultural evolution. *Science*, 305, 487–491.
Davidson, J. M., C. H. Rodgers, E. R. Smith, & G. J. B. Bloch. (1968). Stimulation of female sex behavior in adrenalectomized rats with estrogen alone. *Endocrinology*, 82, 193–195.
Davies, N. B. (1992). *Dunnock Behaviour and Social Evolution*. New York: Oxford University Press.
Davies, N. B. (2000). *Cuckoos, Cowbirds and Other Cheats*. London: T. and A. D. Poyser.
Després, L., J.-P. David, & C. Gallet. (2007). The evolutionary ecology of insect resistance to plant chemicals. *Trends in Ecology and Evolution*, 22, 298–307.
De Vos, A. & M. J. O'Riain. (2010). Sharks shape the geometry of a selfish seal herd: Experimental evidence from seal decoys. *Biology Letters*, 6, 48–50.
Dickinson, J. L. & B. J. Hatchwell. (2004). Fitness consequences of helping, In W. D. Koenig & J. L. Dickinson (eds.), *Ecology and Evolution of Cooperative Breeding in Birds* (pp. 48–66). Cambridge, UK: Cambridge University Press.
Dickinson, J. L. & W. W. Weathers. (1999). Replacement males in the western bluebird: Opportunity for paternity, chick-feeding rules, and fitness consequences of male parental care. *Behavioral Ecology and Sociobiology*, 45, 201–209.
Doligez, B., É. Danchin, & J. Clobert. (2002). Public information and breeding habitat selection in a wild bird population. *Science*, 297, 1168–1170.
Double, M. & A. Cockburn. (2000). Pre-dawn infidelity: Females control extra-pair mating in superb fairy-wrens. *Proceedings of the Royal Society of London, B*, 267, 465–470.
Duffy, J. E. (1996). Eusociality in a coral-reef shrimp. *Nature*, 381, 512–514.
Emlen, S. T. (1982). The evolution of helping. I. An ecological constraints model. *American Naturalist*, 119, 29–39.
Emlen, S. T. (1991). Evolution of cooperative breeding in birds and mammals. In J. R. Krebs & N. B. Davies (eds.), *Behavioural Ecology: An Evolutionary Approach*, Third Edition (pp. 301–337). Oxford, UK: Blackwell Scientific Publications.
Emlen, S. T. & L. W. Oring. (1977). Ecology, sexual selection, and the evolution of mating systems. *Science*, 197, 215–223.

Faaborg, J., P. G. Parker, L. Delay, Tj. de Vries, J. C. Bednarz, S. M. Paz, J. Naranjo, & T. A. Waite. (1995). Confirmation of cooperative polyandry in the Galapagos hawk (*Buteo galapagoensis*). *Behavioral Ecology and Sociobiology*, 36, 83–90.

Faaborg, J. & C. B. Patterson. (1981). The characteristics and occurrence of cooperative polyandry. *Ibis*, 123, 477–484.

Fehr, E. & S. Gächter. (2002). Altruistic punishment in humans. *Nature*, 415, 137–140.

Fitzgerald, T. D. & S. C. Peterson. (1988). Cooperative foraging and communication in caterpillars. *BioScience*, 38, 20–25.

Foster, W. A. & J. E. Treherne. (1981). Evidence for the dilution effect in the selfish herd from fish predation on a marine insect. *Nature*, 293, 466–467.

Gadagkar, R. (1993). Can animals be spiteful? *Trends in Ecology and Evolution*, 8, 232–234.

Gamberale, G. & B. S. Tullberg. (1998). Aposematism and gregariousness: The combined effect of group size and coloration on signal repellence. *Proceedings of the Royal Society of London, B*, 265, 889–894.

Garner, A., I. C. W. Hardy, P. D. Taylor, & S. A. West. (2007). Spiteful soldiers and sex ratio conflict in polyembryonic parasitoid wasps. *American Naturalist*, 169, 519–533.

Ghent, A. W. (1960). A study of the group-feeding behaviour of larvae of the jack pine sawfly, *Neodiprion pratti banksianae* Roh. *Behaviour*, 16, 110–148.

Greene, E. (1987). Individuals in an osprey colony discriminate between high and low quality information. *Nature*, 329, 239–241.

Grüter, C. & W. M. Farina. (2009). The honeybee waggle dance: Can we follow the steps? *Trends in Ecology and Evolution*, 24, 242–247.

Gustin, M. K. & G. F. McCracken. (1987). Scent recognition between females and pups in the bat *Tadarida brasiliensis mexicana*. *Animal Behaviour*, 35, 13–19.

Hager, R. & C. B. Jones (eds.). (2009). *Reproductive skew in vertebrates: Proximate and ultimate causes*. Cambridge, UK: Cambridge University Press.

Hamilton, W. D. (1964). The genetical evolution of social behaviour. I and II. *Journal of Theoretical Biology*, 7, 1–52.

Hamilton, W. D. (1971). Geometry for the selfish herd. *Journal of Theoretical Biology*, 31, 295–311.

Hamilton, W. D. (1972). Altruism and related phenomena, mainly in social insects. *Annual Review of Ecology and Systematics*, 3, 193–232.

Hatchwell, B. J. (2009). The evolution of cooperative breeding in birds: Kinship, dispersal and life history. *Philosophical Transactions of the Royal Society of London, B*, 364, 3217–3227.

Haydock, J. & W. D. Koenig. (2002). Reproductive skew in the polygynandrous acorn woodpecker. *Proceedings of the National Academy of Sciences, USA*, 99, 7178–7183.

Haydock, J., W. D. Koenig, & M. T. Stanback. (2001). Shared parentage and incest avoidance in the cooperatively breeding acorn woodpecker. *Molecular Ecology*, 10, 1515–1525.

Heard, D. C. (1992). The effect of wolf predation and snow cover on musk-ox group size. *American Naturalist*, 139, 190–204.

Heinze, J. & B. Hölldobler. (1993). Fighting for a harem of queens: Physiology of reproduction in *Cardiocondyla* male ants. *Proceedings of the National Academy of Sciences, USA*, 90, 8412–8414.

Hess, E. H. (1964). Imprinting in birds. *Science*, 146, 1128–1139.

Hodge, S. J., A. Manica, T. P. Flower, & T. H. Clutton-Brock. (2008). Determinants of reproductive success in dominant female meerkats. *Journal of Animal Ecology*, 77, 92–102.

Höglund, J. & R. Alatalo. (1995). *Leks*. Princeton, NJ: Princeton University Press.

Höglund, J., R. V. Alatalo, A. Lundberg, P. T. Rintamäki, & J. Lindell. (1999). Microsatellite markers reveal the potential for kin selection on black grouse leks. *Proceedings of the Royal Society of London, B*, 266, 813–816.

Holmes, W. G. & P. W. Sherman. (1982). The ontogeny of kin recognition in two species of ground squirrels. *American Zoologist*, 22, 491–517.

Hrdy, S. B. (1974). Male-male competition and infanticide among the langurs (*Presbytis entellus*) of Abu, Rajasthan. *Folia Primatologica*, 22, 19–58.

Hrdy, S. B. (2009). *Mothers and Others*. Cambridge, MA: Belknap Press.

Hughes, W. O. H., B. P. Oldroyd, M. Beekman, & F. L. W. Ratnieks. (2008a). Ancestral monogamy shows kin selection is key to the evolution of eusociality. *Science*, 320, 1213–1216.

Hughes, W. O. H., F. L. W. Ratnieks, & B. P. Oldroyd. (2008b). Multiple paternity or multiple queens: Two routes to greater intracolonial genetic diversity in the eusocial Hymenoptera. *Journal of Evolutionary Biology*, 21, 1090–1095.

Jacob, S., M. K. McClintock, B. Zelano, & C. Ober. (2002). Paternally inherited HLA alleles are associated with women's choice of male odor. *Nature Genetics*, 30, 175–179.

Jarvis, J. U. M. & N. C. Bennett. (1993). Eusociality has evolved independently in two genera of bathyergid mole-rats—but occurs in no other subterranean mammal. *Behavioral Ecology and Sociobiology*, 33, 253–260.

Jensen, K. (2010). Punishment and spite, the dark side of cooperation. *Proceedings of the Royal Society of London, B*, 365, 2635–2650.

Jetz, W. & D. R. Rubenstein. (2011). Environmental uncertainty and the global biogeography of cooperative breeding in birds. *Current Biology*, 21, 72–78.

Johnstone, R. A. (2000). Models of reproductive skew: A review and synthesis. *Ethology*, 106, 5–26.

Jouventin, P. & T. Aubin. (2002). Acoustic systems are adapted to breeding ecologies: Individual recognition in nesting penguins. *Animal Behaviour*, 64, 747–757.

Just, J. (1988). Siphonoecetinae (Corophiidae) 6: A survey of phylogeny, distribution, and biology. *Crustaceana*, 13, 193–208.

Keller, L. & H. K. Reeve. (1994). Partitioning of reproduction in animal societies. *Trends in Ecology and Evolution*, 9, 98–102.

Koenig, W. D. (1981). Reproductive success, group size, and the evolution of cooperative breeding in the acorn woodpecker. *American Naturalist*, 117, 421–443.
Koenig, W. D. & J. L. Dickinson (eds.). (2004). *Ecology and Evolution of Cooperative Breeding in Birds*. Cambridge, UK: Cambridge University Press.
Koenig, W. D. & J. Haydock. (2004). Incest avoidance. In W. D. Koenig & J. L. Dickinson (eds.), *Ecology and Evolution of Cooperative Breeding in Birds* (pp. 142–156). Cambridge, UK: Cambridge University Press.
Koenig, W. D., J. Haydock, & M. T. Stanback. (1998). Reproductive roles in the cooperatively breeding acorn woodpecker: Incest avoidance versus reproductive competition. *American Naturalist*, 151, 243–255.
Koenig, W. D., R. L. Mumme, M. T. Stanback, & F. A. Pitelka. (1995a). Patterns and consequences of egg destruction among joint-nesting acorn woodpeckers. *Animal Behaviour*, 50, 607–621.
Koenig, W. D., F. A. Pitelka, W. J. Carmen, R. L. Mumme, & M. T. Stanback. (1992). The evolution of delayed dipersal in cooperative breeders. *Quarterly Review of Biology*, 67, 111–150.
Koenig, W. D., P. B. Stacey, M. T. Stanback, & R. L. Mumme. (1995b). Acorn woodpecker (*Melanerpes formicivorus*). In A. Poole & F. Gill (eds.), *The Birds of North America*, No. 194. Philadelphia: Academy of Natural Sciences; Washington, DC: American Ornithologists' Union.
Koenig, W. D., M. T. Stanback, & J. Haydock. (1999). Demographic consequences of incest avoidance in the cooperatively breeding acorn woodpecker. *Animal Behaviour*, 57, 1287–1293.
Koenig, W. D., E. L. Walters, & J. Haydock. (2011). Variable helper effects, ecological conditions, and the evolution of cooperative breeding in the acorn woodpecker. *American Naturalist*, 178, 145–158.
Komdeur, J. (1992). Importance of habitat saturation and territory quality for evolution of cooperative breeding in the Seychelles warbler. *Nature*, 358, 493–495.
Krakauer, A. H. (2005). Kin selection and cooperative courtship in wild turkeys. *Nature*, 434, 69–72.
Krakauer, D. C. (1995). Groups confuse predators by exploiting perceptual bottlenecks: A connectionist model of the confusion effect. *Behavioral Ecology and Sociobiology*, 36, 421–429.
Krause, J. & G. D. Ruxton. (2002). *Living in Groups*. Oxford, UK: Oxford University Press.
Krieger, M. J. B. & K. G. Ross. (2002). Identification of a major gene regulating complex social behavior. *Science*, 295, 328–332.
Kummer, H. (1968). Two variations in the social organization of baboons. In P. Jay (ed.), *Primates: Studies in Adaptation and Variability* (pp. 293–312). New York: Holt, Rinehart, & Winston.
Kutsukake, N. & T. H. Clutton-Brock. (2008). The number of subordinates moderates intrasexual competition among males in cooperatively breeding meerkats. *Proceedings of the Royal Society of London, B*, 275, 209–216.

Lihoreau, M. & C. Rivault. (2009). Kin recognition via cuticular hydrocarbons shapes cockroach social life. *Behavioral Ecology*, 20, 46–53.

Lingle, S. (2001). Anti-predator strategies and grouping patterns in white-tailed deer and mule deer. *Ethology*, 107, 295–314.

Lorenz, K. Z. (1935). Der kumpan in der umwelt des vogels. *Journal für Ornithologie*, 83, 137–213, 287–413.

Lyon, B. E. & J. M. Eadie. (2008). Conspecific brood parasitism in birds: A life-history perspective. *Annual Review of Ecology, Evolution, and Systematics*, 39, 343–363.

MacDougall-Shackleton, S. A. (2011). The levels of analysis revisited. *Philosophical Transactions of the Royal Society of London, B*, 366, 2076–2085.

Magrath, R. D. (2001). Group breeding dramatically increases reproductive success of yearling but not older female scrubwrens: A model for cooperatively breeding birds? *Journal of Animal Ecology*, 70, 370–385.

Marler, P. (1970). A comparative approach to vocal learning: Song development in white-crowned sparrows. *Journal of Comparative and Physiological Psychology*, 71, 1–25.

Mateo, J. M. (2002). Kin-recognition abilities and nepotism as a function of sociality. *Proceedings of the Royal Society of London, B*, 269, 721–727.

Mateo, J. M. (2010). Self-referent phenotype matching and long-term maintenance of kin recognition. *Animal Behaviour*, 80, 929–935.

McClure, M. & E. Despland. (2011). Defensive responses by a social caterpillar are tailored to different predators and change with larval instar and group size. *Naturwissenschaften*, 98, 425–434.

McComb, K. (1987). Roaring by red deer stags advances the date of oestrus in hinds. *Nature*, 330, 648–649.

Meek, S. B. & R. J. Robertson. (1991). Adoption of young by replacement male birds: An experimental study of eastern bluebirds and a review. *Animal Behaviour*, 42, 813–820.

Mello, C. V., D. S. Vicario, & D. F. Clayton. (1992). Song presentation induces gene expression in the songbird forebrain. *Proceedings of the National Academy of Sciences, USA*, 89, 6818–6822.

Michener, C. D. (1969). Comparative social behavior of bees. *Annual Review of Entomology*, 14, 299–342.

Michener, C. D. (2007). *The Bees of the World*. Baltimore, MD: Johns Hopkins University Press.

Milinski, M., D. Semmann, & H. J. Krambeck. (2002). Reputation helps solve the "tragedy of the commons." *Nature*, 415, 424–426.

Morales, M. A., J. L. Barone, & C. S. Henry. (2008). Acoustic alarm signalling facilitates predator pretection of treehoppers by mutualist ant bodyguards. *Proceedings of the Royal Society of London, B*, 275, 1935–1941.

Morrow, P. A., T. E. Bellas, & T. Eisner. (1976). *Eucalyptus* oils in the defensive oral discharge of Australian sawfly larvae (Hymenoptera: Pergidae). *Oecologia*, 24, 193–206.

Mulder, R. A., P. O. Dunn, A. Cockburn, K. A. Lazenby-Cohen, & M. J. Howell. (1994). Helpers liberate female fairy-wrens from constraints on extra-pair mate choice. *Proceedings of the Royal Society of London, B*, 255, 223–229.

Munn, C. A. (1986). Birds that "cry wolf." *Nature*, 391, 143–145.

Nelson, R. J. (2011). *An Introduction to Behavioral Endocrinology*, Fourth Edition. Sunderland, MA: Sinauer Associates.

Nordby, J. C., S. E. Campbell, J. M. Burt, & M. D. Beecher. (2000). Social influences during song development in the song sparrow: A laboratory experiment simulating field conditions. *Animal Behaviour*, 59, 1187–1197.

Nottebohm, F. & A. P. Arnold. (1976). Sexual dimorphism in vocal control areas of the songbird brain. *Science*, 194, 211–213.

Novaro, A. J., C. A. Moraga, C. Briceño, M. C. Funes, & A. Marino. (2009). First records of culpeo (*Lycalopex culpaeus*) attacks and cooperative defense by guanacos (*Lama guanicoe*). *Mammalia*, 73, 148–150.

Nowak, M. A. & K. Sigmund. (2005). Evolution of indirect reciprocity. *Nature*, 437, 1291–1298.

Nowak, M. A., C. E. Tarnita, & E. O. Wilson. (2010). The evolution of eusociality. *Nature*, 466, 1057–1062.

Orians, G. H. & D. Janzen. (1974). Why are embryos so tasty? *American Naturalist*, 108, 581–592.

Packer, C. (1977). Reciprocal altruism in *Papio anubis*. *Nature*, 265, 441–443.

Packer, C. & L. Ruttan. (1988). The evolution of cooperative hunting. *American Naturalist*, 132, 159–198.

Peters, A., L. B. Astheimer, C. R. J. Boland, & A. Cockburn. (2000). Testosterone is involved in acquisition and maintenance of sexually selected male plumage in superb fairy-wrens, *Malurus cyaneus*. *Behavioral Ecology and Sociobiology*, 47, 438–445.

Peterson, S. C., N. D. Johnson, & J. L. LeGuyader. (1987). Defensive regurgitation of allelochemicals derived from host cyanogenesis by eastern tent caterpillars. *Ecology*, 68, 1268–1272.

Petrie, M., A. Krupa, & T. Burke. (1999). Peacocks lek with relatives even in the absence of social and environmental cues. *Nature*, 401, 155–157.

Platt, J. R. (1964). Strong inference. *Science*, 146, 347–353.

Potts, W. K., C. J. Manning, & E. K. Wakeland. (1991). Mating patterns in seminatural populations of mice influenced by MHC genotype. *Nature*, 352, 619–621.

Pruett-Jones, S. G. & M. J. Lewis. (1990). Sex ratio and habitat limitation promote delayed dispersal in superb fairy-wrens. *Nature*, 348, 541–542.

Queller, D. C. (1987). The evoluiton of leks through female choice. *Animal Behaviour*, 35, 1424–1432.

Reeve, H., K. (1989). The evolution of conspecific acceptance thresholds. *American Naturalist*, 133, 407–435.

Reynolds, A. M., G. A. Sword, S. J. Simpson, & D. R. Reyolds. (2009). Predator percolation, insect outbreaks and phase polyphenism. *Current Biology*, 19, 20–24.

Riipi, M., R. V. Alatalo, L. Lindström, & J. Mappes. (2001). Multiple benefits of gregariousness cover detectability costs in aposematic aggregations. *Nature*, 413, 512–514.
Robinson, G. E., R. D. Fernald, & D. F. Clayton. (2008). Genes and social behavior. *Science*, 322, 896–900.
Rohwer, S. (1986). Selection for adoption versus infanticide by replacement "mates" in birds. *Current Ornithology*, 3, 353–395.
Rohwer, S., J. C. Herron, & M. Daly. (1999). Stepparental behavior as mating effort in birds and other animals. *Evolution and Human Behavior*, 20, 367–390.
Rothstein, S. I. (1990). A model system for coevolution: Avian brood parasitism. *Annual Review of Ecology and Systematics*, 21, 481–508.
Rowley, I., E. M. Russell, & M. G. Brooker. (1986). Inbreeding: Benefits may outweigh costs. *Animal Behaviour*, 34, 939–941.
Rubenstein, D. R. & I. J. Lovette. (2007). Temporal environmental variability drives the evolution of cooperative breeding in birds. *Current Biology*,17, 1414–1419.
Russell, A. F., A. J. Young, G. Spong, N. R. Jordan, & T. H. Clutton-Brock. (2007). Helpers increase the reproductive potential of offspring in cooperative meerkats. *Proceedings of the Royal Society of London, B*, 274, 513–520.
Russell, E. M. (1989). Co-operative breeding: A Gondwanan perspectic. *Emu*, 89, 61–62.
Rutter, M. (2006). *Genes and Behavior: Nature-Nurture Interplay Explained*. Oxford, UK: Blackwell Publishing.
Santorelli, L. A., C. R. L. Thompson, E. Villegas, J. Svetz, C. Dinh, A. Parikh, et al. (2008). Facultative cheater mutants reveal the genetic complexity of co-operation in social amoebae. *Nature*, 451, 1107–1110.
Sapolsky, R. M. (1998). *Why Zebras Don't Get Ulcers*. New York: W. H. Freeman and Co.
Seeley, T. D. (1995). *The Wisdom of the Hive: The Social Physiology of Honey Bee Colonies*. Cambridge, MA: Harvard University Press,
Seki, S., M. Kohda, G. Takamoto, K. Karino, Y. Nakashima, & T. Kuwamura. (2009). Female defense polygyny in the territorial triggerfish *Sufflamen chrysopterum*. *Journal of Ethology*, 27. 215–220.
Seyfarth, R. M., D. L. Cheney, & P. Marler. (1980). Monkey responses to three different alarm calls: Evidence of predator classification and semantic communication. *Science*, 210, 801–803.
Sharp, S. P., A. McGowan, M. J. Wood, & B. J. Hatchwell. (2005). Learned kin recognition cues in a social bird. *Nature*, 434, 1127–1130.
Sharpe, L. L., A. S. Joustra, & M. I. Cherry. (2010). The presence of an avian co-forager reduces vigilance in a cooperative mammal. *Biology Letters*, 6, 475–477.
Sherman, P. W. (1977). Nepotism and the evolution of alarm calls. *Science*, 197, 1246–1253.
Sherman, P. W. (1988). The levels of analysis. *Animal Behaviour*, 36, 616–619.

Sherman, P. W., E. A. Lacey, H. K. Reeve, & L. Keller. (1995). The eusociality continuum. *Behavioral Ecology*, 6, 102–108.
Sherman, P. W., H. K. Reeve, & D. W. Pfennig. (1997). Recognition systems. In J. R. Krebs & N. B. Davies (eds.), *Behavioural Ecology* (pp. 69–96). Oxford, UK: Blackwell.
Shorey, L., S. Piertney, J. Stone, & J. Höglund. (2000). Fine-scale genetic structuring on *Manacus manacus* leks. *Nature*, 408, 352–353.
Spong, G. F., S. J. Hodge, A. J. Young, & T. H. Clutton-Brock. (2008). Factors affecting the reproductive success of dominant male meerkats. *Molecular Ecology*, 17, 2287–2299.
Stacey, P. B. & W. D. Koenig. (1990). *Cooperative Breeding in Birds: Long-term Studies of Ecology and Behavior*. Cambridge, UK: Cambridge University Press.
Swynnerton, C. F. M. (1915). Mixed bird parties. *Ibis*, 3, 346–354.
Tallamy, D. W. (2005). Egg dumping in insects. *Annual Review of Entomology*, 50, 347–370.
Tener, J. S. (1965). *Muskoxen*. Ottawa, Canada: Queen's Printer.
Terborgh, J. (1990). Mixed flocks and polyspecific associations: Costs and benefits of mixed groups to birds and monkeys. *American Journal of Primatology*, 21, 87–100.
Thornhill, T. & J. Alcock. (1983). *The Evolution of Insect Mating Systems*. Cambridge, MA: Harvard University Press.
Tinbergen, N. (1963). On aims and methods of ethology. *Zietschrift für Tierpsychologie*, 20, 410–433.
Trivers, R. L. (1971). The evolution of reciprocal altruism. *Quarterly Review of Biology*, 46, 35–57.
von Frisch, K. (1967). *The Dance Language and Orientation of Bees*. Cambridge, MA: Harvard University Press.
Wade, J. & A. P. Arnold. (1996). Functional testicular tissue does not masculinize development of the zebra finch song system. *Proceedings of the National Academy of Sciences, USA*, 93, 5264–5268.
Wade, J. & A. P. Arnold. (2004). Sexual differentiation of the zebra finch song system. *Annals of the New York Academy of Sciences*, 1016, 540–559.
Walters, J. R., C. K. Copeyon, & J. H. Carter III. (1992). Test of the ecological basis of cooperative breeding in red-cockaded woodpeckers. *Auk*, 109, 90–97.
Ward, P. & A. Zahavi. (1973). The importance of certain assemblages of birds as "information-centres" for food-finding. *Ibis*, 115, 517–534.
Webster, M. S. (1994). Female-defence polygyny in a neotropical bird, the Montezuma oropendola. *Animal Behaviour*, 48, 779–794.
Wedekind, C. & D. Penn. (2000). MHC genes, body odours, and odour preferences. *Nephrology Dialysis Transplantation*, 15, 1269–1271.
Weimerskirch, H., S. Bertrand, J. Silva, J. C. Marques, & E. Goya. (2010). Use of social information in seabirds: Compass rafts indicate the heading of food patches. *PLoS One*, 5, e9928. doi:10.1371/journal.pone.0009928.

Wells, K. D. (1977). The social behaviour of anuran amphibians. *Animal Behaviour*, 25, 666–693.

Wertheim, B., J. Marchais, L. E. M. Vet, & M. Dicke. (2002). Allee effect in larval resource exploitation in *Drosophila*: An interaction among density of adults, larvae, and micro-organisms. *Ecological Entomology*, 27, 608–617.

West, S. A. & A. Gardner. (2010). Altruism, spite, and greenbeards. *Science*, 327, 1341–1344.

Wilkinson, G. S. (1984). Reciprocal food sharing in vampire bats. *Nature*, 308, 181–184.

Wilkinson, G. S. (1992). Information transfer at evening bat colonies. *Animal Behaviour*, 44, 501–518.

Wilson, E. O. (1975). *Sociobiology: The New Synthesis*. Cambridge, MA: Belknap Press.

Wilson, P. L. & S. L. Vehrencamp. (2001). A test of the deceptive mimicry hypothesis in song-sharing song sparrows. *Animal Behaviour*, 62, 1197–1205.

Woolfenden, G. E. & J. W. Fitzpatrick. (1984). *The Florida Scrub Jay: Demography of a Cooperative-breeding Bird*. Princeton, NJ: Princeton University Press.

Yom-Tov, Y. (1980). Intraspecific nest parasitism in birds. *Biological Reviews*, 55, 93–108.

Young, A. J., G. Spong, & T. Clutton-Brock. (2007). Subordinate male meerkats prospect for extra-group paternity: Alternative reproductive tactics in a cooperative mammal. *Proceedings of the Royal Society of London, B*, 274, 1603–1609.

10

Ecological and Evolutionary Feedbacks in the Evolution of Aggression

Renée A. Duckworth

INTRODUCTION

Acts of aggression are ubiquitous among animals and play a key role in survival and reproduction. Animals often use aggressive behaviors to defend or usurp a resource (Stamps & Krishnan, 1997; Garcia & Arroyo, 2002), to compete for mates (Bartoš, 1986; Hagelin, 2002), to fend off predators (Andersson et al., 1980; Redondo & Carranza, 1989), and during foraging (Riechert, 1993). However, even though aggression is important in a number of distinct contexts, it is also a costly behavior. These costs range from direct costs in terms of energy expenditure and injury during aggressive conflicts to more subtle costs of disrupting social bonds in communally living species. A balance between costs and benefits has traditionally been the main explanation for variation in aggression observed within and among individuals as well as between populations and species; however, recent work on animal personalities has introduced the possibility that constraints may also be important to understanding this variation.

Aggression, by definition, is a social (or antisocial) behavior in that it only occurs in the context of two or more interacting individuals. At the most basic level, aggression is used to either subdue another individual (as during predation or aggressive courtship) or to repel another individual (as during territorial defense or when fending off predators). By influencing territorial spacing, predator-prey dynamics, and social-group membership, aggressive interactions can influence how individuals arrange themselves in space and time and can have large-scale ecological consequences. The importance of evolutionary feedback effects resulting from ecological consequences of aggression is a new and exciting area of research that may prove critical to our understanding of why populations and species vary.

Partly because of the importance of aggression to fitness and partly because aggressive behaviors are relatively easy to observe and characterize in natural populations, aggression is well studied in diverse contexts and species, and thus evolutionary studies of aggression provide unique insight into the evolution of behavior more generally. In this chapter, I will first summarize recent work on personality differences in aggression to explore the novel insights this work poses for understanding the evolution of aggression, and I will then discuss the role of ecological and evolutionary feedbacks in maintaining population- and species-level differences in aggression.

WHAT IS AGGRESSIVE PERSONALITY VARIATION?

Personality variation refers to consistent differences in behavior among individuals. The study of nonhuman animal personalities is a recent addition to the field of behavioral ecology and is set against a background of several decades of research on the evolution of animal behavior from an optimality perspective (Sih et al., 2004; Bell, 2007; McNamara et al., 2009), where it is assumed that animals strategically adjust their behavior to maximize fitness given existing trade-offs (Roff, 1994). Behavioral ecologists have used this framework—which assumes that constraints to behavioral flexibility are weak or nonexistent and that behavior of individuals within a population will converge on a single optimal expression in a particular context—to predict when and how individuals should reversibly adjust their behavior in different contexts (Krebs & Davies, 1991). Yet recent studies demonstrating the ubiquity of animal personalities challenge these basic assumptions because they show that individuals are often limited in the flexibility of their behavior, that there are often pronounced differences in behavior among individuals in the same context, and that distinct behaviors are often closely correlated in expression (Dingemanse & Réale, 2005; Bergmüller & Taborsky, 2010).

A particularly puzzling component of personality variation is the remarkable consistency in behavior across contexts even when changing behavior would confer higher fitness (Sih et al., 2004). For example, in fishing spiders

(*Dolomedes triton*), females that are most aggressive in a foraging context are also most aggressive in a mating context, and aggressive females are likely to cannibalize prospective mates before copulating with them to the extent that some females attack every single male and thus fail to mate and produce offspring (Arnqvist & Henriksson, 1997; Johnson & Sih, 2005). In another example, in many species of birds, the most aggressive males invest the least in parental care, and this often results in lower reproductive success (e.g., Tuttle, 2003; Duckworth, 2006a; see Ketterson & Nolan, 1994, for review). These costs to inflexible expression of aggression raise the question of why individuals that behave aggressively in one context cannot modify their overall behavioral strategy to behave less aggressively in a different context.

Variation in aggressiveness has been one of the most commonly described axes of personality variation. Consistent differences in aggression have been documented in insects, spiders, fish, birds, and mammals. Such stability in behavior is usually detected as significant repeatability either over time or in distinct functional contexts (e.g., aggression toward a rival male versus aggression towards a predator). A ***meta-analysis*** of studies that compared repeatability of 13 different classes of behavior showed that aggressive behavior was one of most highly repeatable (Bell et al., 2009). Yet there is a rich history of work on aggression showing that it is often context dependent and strongly influenced by environmental conditions. For example, increases in group size and experimental decreases of food resources both independently increase the number of aggressive interactions in red deer (*Cervus elaphus*) stags (Appleby, 1980; Bartoš, 1986). Moreover, many studies of territorial animals show that individuals are more aggressive toward strangers than toward neighbors (Jaeger, 1981; Temeles, 1994), demonstrating that individuals often modulate aggressiveness depending on social context. Even abiotic conditions have been shown to influence aggressiveness—in coral reef fish (*Pomacentrus* spp.) aggressiveness increased with increases in water temperature (Biro et al., 2010). *How can we reconcile such seeming flexibility of aggression with the growing body of work that shows aggression to be highly repeatable within individuals?* The key to resolving this puzzle is to recognize that most studies investigating the effects of environmental variation on aggression often ignore individual variation and focus instead on mean population-level changes. Moreover, high repeatability does not preclude the possibility that individuals modify their behavior in different situations but simply means that the rank order of aggressiveness of a group of individuals is consistent over time or across different contexts such that, even if all individuals decrease their aggressive response in a particular context, the most aggressive individuals in one context are still the most aggressive in the other (see Figure 10.1 for an example).

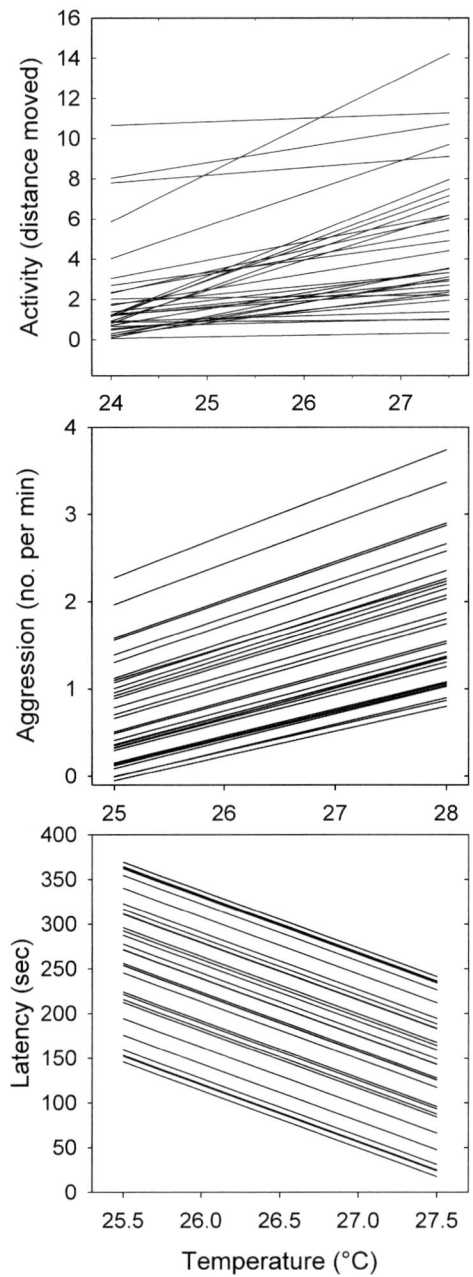

Figure 10.1. Examples of consistent individual differences in a highly plastic behavior. Coral reef fish show plasticity in activity (top graph), aggression (middle graph), and latency (bottom graph) in relation to temperature. Rank order of individuals is constant across temperatures for aggression and latency but not for activity levels. (Figure from Biro et al., 2010. Used by permission of the Royal Society.)

Such consistent differences in aggressive behavior among individuals raise several important questions. *Why do individuals show consistency in the expression of aggression even when placed in different situations where the costs and benefits of displaying aggressive behavior vary? To what extent is aggression correlated with other traits? Why do such correlations evolve, and what are their evolutionary implications? Are there any general patterns across species regarding which traits are associated with aggression? Finally, how do individual-, population-, and species-level variation in aggression impact ecological and evolutionary processes?*

WHY ARE INDIVIDUALS CONSISTENT IN EXPRESSION OF AGGRESSION?

There are two main answers to these questions: **natural selection** favors constancy in expression (hereafter "adaptive hypothesis") or **developmental constraints** limit flexibility of aggression (hereafter "constraint hypothesis"). The adaptive hypothesis assumes there is unlimited potential for flexibility of aggression within individuals but that stability is adaptive. In other words, if extreme flexibility of aggression were adaptive, then it would easily evolve given enough time. Stability of aggression could be adaptive if natural selection favors either its correlation with other, less flexible traits (see below for examples) or its predictability (McElreath & Strimling, 2006; Wolf et al., 2007; McNamara et al., 2009). In contrast, under the constraint hypothesis, there are intrinsic limits to flexibility of aggression, and thus selection is assumed to play a minor or no role in the evolution of consistency (Duckworth, 2010). From this perspective, the physical structures that underlie variation in aggression, including variation in brain anatomy, neuronal connectivity, neurotransmitter synthesis and degradation, hormone secretion patterns, hormone receptor distribution, and endocrine gland function, may be limited in their flexibility, and in turn this limits flexibility of aggression. In other words, under this view, no matter the intensity of natural selection, there is a limit to how much flexibility of aggression can evolve.

The relative importance of selection and constraint in the evolution of stable differences in aggression between individuals is not clear. Many of the adaptive hypotheses predict strong correlations between aggression and other traits, and while such correlations are often found (see below for examples), it is unclear whether these correlations are a cause or consequence of limited flexibility of aggression (Duckworth, 2010). Adaptive hypotheses also predict variability among species in whether they express consistent differences in aggression, whereas the constraint hypothesis predicts that limits to flexibility should be widespread across a diversity of taxa, especially if they arise from limitations on organismal design due to physical laws (*sensu* Maynard Smith et al., 1985; Brakefield, 2006). Evidence for high repeatability of aggression across studies supports the constraint hypothesis; however, more rigorous comparison of flexibility of aggression across a diversity of taxa as well as a comparison of developmental mechanisms underlying individuals differences in behavior are needed before any conclusions on this topic can be drawn.

Unfortunately, very little is known about the developmental basis of individual differences in aggression as most studies of behavioral development focus on larger-scale differences between the sexes, between normal and mutant phenotypes, or between species. Thus, there is currently very little data linking naturally occurring individual variation in behavior to neural and endocrine differences among individuals. However, there is some preliminary support for the idea that constraints are important. The constraint hypothesis predicts that similar developmental pathways would underlie individual variation in aggression across disparate taxa (Duckworth, 2010). Recent reviews suggest that variation in aggression is linked to variation in the serotonin signaling system (Figure 10.2) across a wide variety of species, from

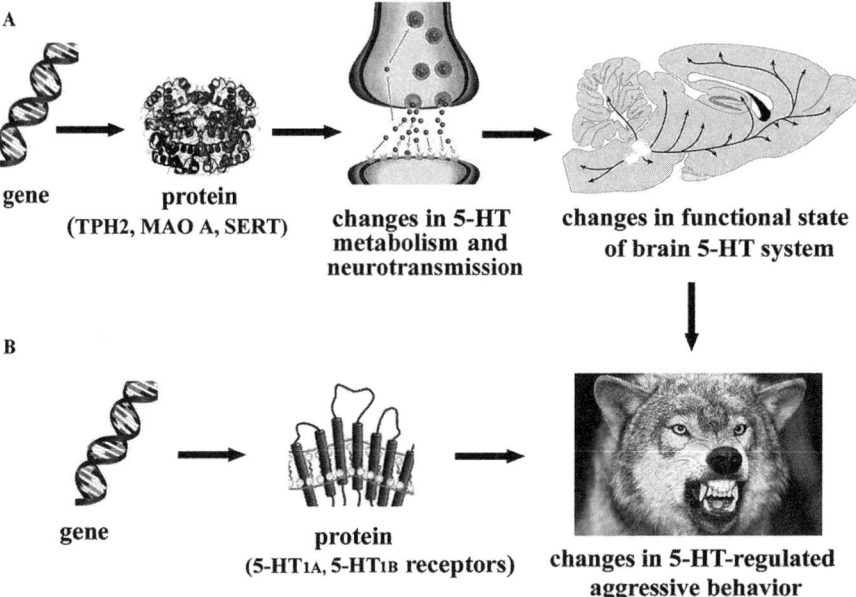

Figure 10.2. Serotonin (5-HT) pathways that underlie variation in aggression are remarkably conserved across disparate taxa. (a) One pathway from genes to aggression involves key enzymes involved in synthesis (TPH2), degradation (MAO A), and transport (SERT) of serotonin. (b) A shorter pathway involves changes in 5-HT receptors. (Figure from Popova, 2006. Used by permission of John Wiley and Sons.)

crayfish to foxes to humans (Popova, 2006), providing some preliminary evidence for this prediction. Yet, at the same time, other physiological systems that are known to influence aggression, such as hormone variation, show extensive flexibility over evolutionary time in their association with aggression (Hau, 2007; Wingfield et al., 2007). For example, aggression is closely linked to testosterone levels in males of many temperate songbird species but is very low and not responsive to territorial challenges in tropical species (Wingfield et al., 2007). Moreover, many bird species that are territorial in the nonbreeding season when gonads are regressed (reduced in size and nonfunctional) also show dissociation between testosterone and aggression (e.g., Schwabl & Kriner, 1991; Soma et al., 2000). Finally, in a recent study, there was no correlation between testosterone levels and aggressive personality differences in western bluebirds (*Sialia mexicana*) (Duckworth & Sockman, 2012). These studies suggest that the association between testosterone and aggression is flexible over an evolutionarily timescale and that endocrine responsiveness in adulthood is not likely to be an important constraint on the evolution of behavioral flexibility. However, hormones may still act during early development to influence aggressive personality as recent research on both humans and model lab organisms has demonstrated a link between personality variation and hormone exposure early in ontogeny (Carere & Balthazart, 2007; Hines, 2008). In sum, more work is necessary to understand how natural selection shapes aggression and how individual differences in aggression develop in order to understand why consistent differences in aggression are so common across a wide variety of animal species.

EVOLUTION OF CORRELATIONS BETWEEN AGGRESSION AND OTHER TRAITS

Individuals vary in morphological, behavioral, and life history traits that influence the costs and benefits of aggressive behavior either through functional links or through trade-offs. Thus, associations between other traits and aggression can evolve if selection favors their coexpression. Once formed, correlations between traits can constrain their independent evolution and can significantly affect each trait's future response to selection (Riska, 1989; Roff, 1997; Lynch & Walsh, 1998). In this section, I will review evidence for associations between aggression and other traits such as body size, dominance, parental care, and other personality traits as these are the traits most commonly predicted to be linked to aggression. I will also discuss the implications of these correlations for understanding the function and evolution of aggression and the extent to which these correlations might constrain adaptive evolution of aggression.

Body Size and Aggression

Body size and aggression are often functionally linked because the costs of initiating an aggressive encounter are less for larger compared to smaller animals. Moreover, differences in aggression can cause differences in body size. For example, in the desert spider (*Agelenopsis aperta*), more aggressive individuals are better foragers and thus reach a larger body size because of their ability to acquire more food (Riechert & Johns, 2003). These clear functional links between aggression and body size have led to the general prediction of a positive correlation between aggression and body size; however, even though larger individuals are more aggressive in some species (e.g., Zack, 1975; Brace & Pavey, 1978; Dowds & Elwood, 1985; Herrel et al., 2009), in many other species aggression and body size are either unlinked or smaller individuals are more aggressive (Just & Morris, 2003; Morrell et al., 2005). For example, in two species of swordtail fishes (*Xiphophorus nigrensis* and *X. multilineatus*), when the difference in size between fish was very large, contests were settled without fights and the smaller animal retreated; however, when individuals were more closely matched in size, 78 percent of observed fights were initiated by the smaller individual, and in 70 percent of the fights, the fish that attacked first lost the contest because body size was a better predictor of winning than aggression (Morris et al., 1995). In a recent study of zebra finches (*Taeniopygia guttata*), aggression was only weakly correlated with body size, and this relationship differed among the sexes such that the two traits were positively correlated in males and negatively correlated in females (Bolund et al., 2007). In more than 50 percent of staged interactions between velvet swimming crabs (*Necora puber*), aggression was initiated by the smaller crab, even though it was less likely to win (Smith et al., 1994). Finally, in western bluebirds, aggression and body size are not correlated, and moreover, western bluebirds, even though smaller than their sister species, mountain bluebirds (*S. currucoides*), are more aggressive and are competitively superior to them in territorial disputes (Duckworth & Badyaev, 2007). Thus, the main theme that emerges from a review of studies measuring the relationship between body size and aggression is that there is no consistent pattern across species in the association of these traits.

The lack of a reliable relationship between aggression and body size suggests that there is no consistent strategy that large and small animals pursue in conflict situations. In fact, the only consistent pattern that emerges is that fights usually do not occur when individuals are extremely different in body size as visual cues apparently provide adequate information about competitive differences such that individuals do not need to fight to determine dominance.

Dominance and Aggression

Dominance refers to a status of power of one individual over another individual, whereas aggression refers to a specific set of behaviors such as attacks, chases, or displacements that are elicited with an intent to inflict harm on or instill fear into another individual. Dominance relationships are often assessed by observing agonistic interactions between individuals and determining which one relents. Dominance and aggression are so intricately linked that many studies use the terms interchangeably (Drews, 1993). While aggression and dominance are frequently correlated (e.g., Anestis, 2005; Colléter & Brown, 2011; Riebli et al., 2011), there are many exceptions. For example, in pronghorn antelope (*Antilocapra americana*), bank voles (*Myodes glareolus*), and domestic pigs (*Sus scrofa*), aggressive personality and dominance status are unrelated (Fairbanks, 1994; Bolhuis et al., 2005; Korpela et al., 2011). Moreover, Richard Francis (1983) showed that in paradise fish (*Macropodus opercularis*), males that were generally dominant across multiple encounters did not differ in their aggressiveness from males that were generally submissive. In fact, after subjecting paradise fish to five generations of bidirectional selection for dominance, he showed that even though the lines diverged significantly in their dominance by the end of the experiment, they did not differ in aggression (Figure 10.3). Wolves (*Canus lupus*) provide another illustrative example. Despite the prevailing view that a wolf pack is a group of individuals aggressively vying for dominance, in naturally occurring wolf packs aggression is rarely used to establish dominance (Mech, 1999). In fact, in one free-living pack observed over a 13-year period, no aggressive interactions were ever observed (Mech, 1999). This example appears to characterize the majority of wild pack behavior as the typical wolf pack is a family and the breeding pair is able to maintain its status without aggression. Submissive behavior, rather than being a response to aggression, is performed by the offspring toward the breeding pair or occasionally by the breeding female to the breeding male. In the wild, only larger packs including nonkin show aggressive behavior (Bradshaw et al., 2009).

To explain the lack of a general relationship between dominance and aggression, Francis (1988) put forward an intriguing possibility—that dominance status is not related to individual differences in aggression despite the fact that dominance relationships are often established through use of aggressive behavior. To understand this perspective it is important to distinguish between measurements of an individual's overall level of aggressiveness outside the context of a specific dominance interaction and what individuals actually do during a dominance interaction. In other words, even though dominant

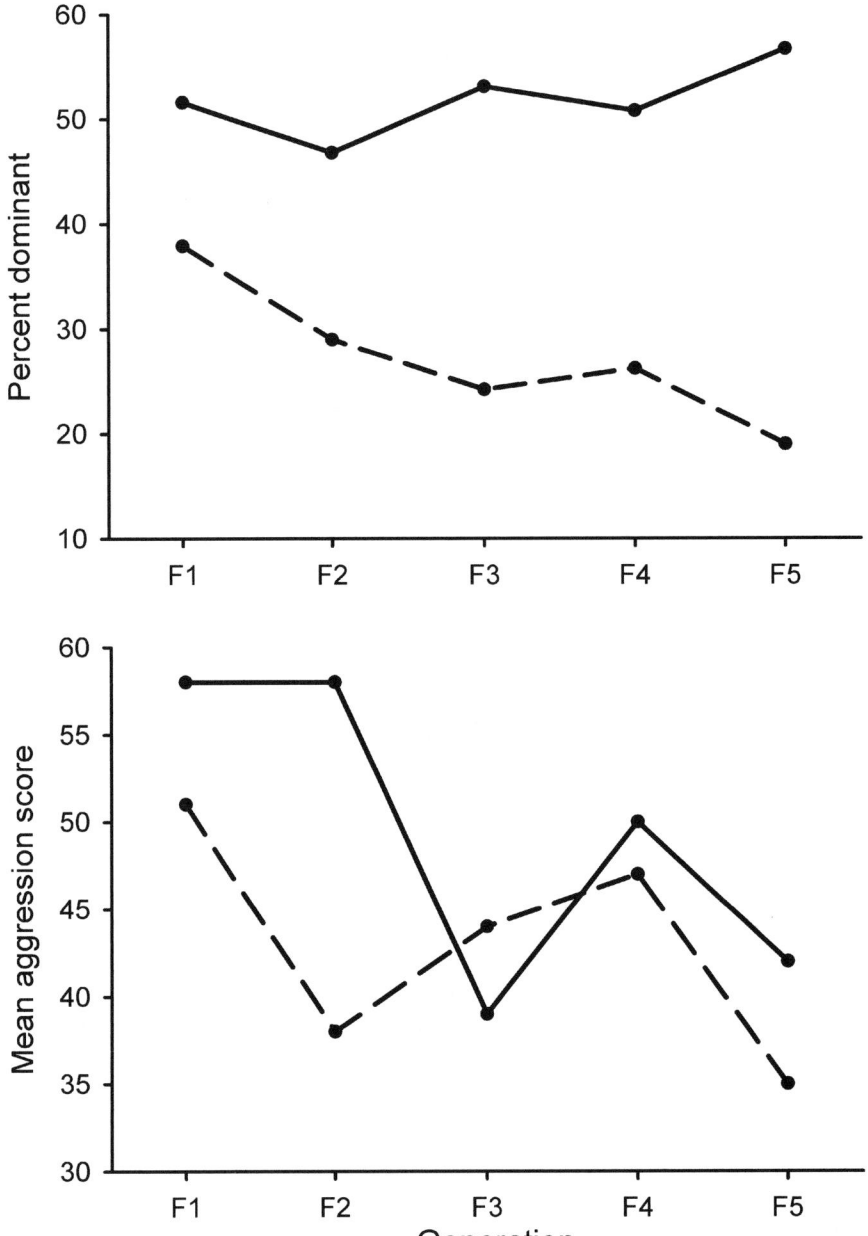

Figure 10.3. Divergent selection lines. Artificial selection experiments are a tool frequently used by evolutionary biologists to gain insight into the genetics of trait variation. These experiments start with a group of individuals that vary in the trait of interest. "High" and "low" selection lines are created from this initial group by selecting only individuals that display either the highest or lowest expression of the focal trait to reproduce (e.g., only individuals in the top or bottom 25 percent of trait values are selected for breeding). These experiments can provide insight into many aspects of the genetics of a trait, including the symmetry of response to upward and downward selection (Pitnick & Miller, 2000) and whether there are correlated responses in other traits to selection on the focal trait (Schwarzkopf et al., 1999). Correlated responses are commonly caused by pleiotropy, in which one gene affects more than one trait. Thus, artificial selection experiments can shed light on how traits are linked. In a study by Francis (1984), an artificial selection experiment was carried out in paradise fish (*Macropodus opercularis*) for five generations. Both high and low lines for dominance were created where only individuals ranking in the top and bottom ~20 percent in dominance status were selected to breed. The results showed a significant response to selection in the downward but not the upward selected lines (top graph). However, when the same selection lines were tested for aggression by exposing them to an intruder fish (either another male separated by a glass partition or a mirror), they showed no difference in aggression from one another (bottom graph). This experiment showed that divergence in the dominance scores of the two lines was not accompanied by any changes in aggressiveness, showing that these traits are not linked in this species. (Figures adapted from Francis, 1984)

individuals may occasionally use aggression to establish or maintain dominance, this does not mean that they are generally more aggressive than other individuals.

If aggression plays only a limited role, then *what factors are most important in influencing dominance*? Dominance is an emergent property influenced by multiple interacting factors including both intrinsic characteristics of the individual as well as social context (Weiß et al., 2011). Intrinsic characteristics include differences in sex, body size, motivation, prior experience, age, and sexual traits (e.g., Watt, 1986; Lemel & Wallin, 1993; Elwood et al., 1998; Nosil, 2002; Duckworth et al., 2004). In general, males are dominant to females, larger individuals are dominant to smaller ones; older, more experienced individuals are dominant to younger, less experienced ones; and motivation can overturn any of these generalities (Cristol, 1992; Lemel & Wallin, 1993), especially if there is a large difference between individuals in the benefits of accessing a particular resource (Enquist & Leimar, 1987). For example, in house crickets (*Acheta domesticus*), body size usually determines the outcome of dominance interactions; however, when motivation is maximized by food deprivation, this overrides the effects of male body size (Nosil, 2002).

Furthermore, in natural populations of birds, the effects of motivation have been shown to be an important determinant of dominance even in species where, all else being equal, the largest or most ornamented males win contests (Lemel & Wallin, 1993).

Recently, a more nuanced view of aggression's relation to dominance is emerging where it is less important in establishing long-term dominance relationships and instead is more important for short-term or initial interactions between individuals. Support for this idea comes from studies of aggression in primate species. Multiple experiments that attempted to elicit aggressive interactions from stable primate groups in captivity (e.g., by making food or space more scarce) failed (Bernstein & Gordon, 1974). The only predictable trigger of aggression occurred when a foreign individual or individuals were introduced into a stable group. Typically, the intruding animal was severely attacked, but initial high levels of aggression declined rapidly, especially when the intruding animal behaved submissively, suggesting that aggression was motivated primarily by the need to maintain social order rather than as a means for mediating competition for resources (Bernstein & Gordon, 1974; de Waal, 1986). In domestic cats (*Felis catus*), dominance sustained without the use of aggression was more stable than dominance formed on the basis of aggressive display (Fonberg, 1988). Many studies of birds have shown that once individuals get to know each other well, they do not use aggression to maintain dominance (Temeles, 1994). Finally, in *Chasmognathus* crabs, Silvia Pedetta and colleagues found that aggression is the main determinant of dominance between size-matched individuals (Pedetta et al., 2010). In fact, aggression even overrides prior experience in this species—in lab trials, even when pitting the same individuals against one another multiple times, the crabs fight anew, and there seems to be no memory of previous encounters. They suggest this makes sense in the context of this species's natural history as interactions with conspecifics occur mainly in the context of burrow disputes between resident and wandering crabs, and it is rare that fights would occur between the same opponents multiple times. Thus, there is no benefit for ***individual recognition*** mechanisms to evolve in this species. These studies all suggest that aggression may be more important for short-term or one-time interactions and less of a factor for maintaining long-term dominance hierarchies. The implications are that it may be important to understand the expected duration of conflict situations in order to understand when and why individual variation in aggression evolves.

Parental Care and Aggression

Aggression is costly in terms of time, energy, and risk, and thus is expected to trade off with investment in other costly behaviors and traits (Bennett

& Houck, 1983; Robertson, 1986; Johnstone & Norris, 1993; Duckworth, 2006a; Rosvall, 2011). In fact, one of the most consistent relationships between aggression and other traits across a wide variety of species is a negative correlation between aggression and *parental care*. For example, in many passerine birds, males that show high levels of aggression often invest the least in parental care (Ketterson & Nolan, 1994). Because, in these species, males and females form strong pair bonds and both sexes contribute to offspring care, this means that highly aggressive males often have lower reproductive success (e.g., Duckworth, 2006a). In some of these species, variation in aggression is related to alternative male mating strategies within a population, where aggressive males compensate for fitness costs of low parental care by actively pursuing mating opportunities outside the pair bond, whereas other males are nonaggressive and good fathers (Ketterson & Nolan, 1994). Why aggression, mating behavior, and parental care are frequently linked is not clear. It is known that all of these behaviors are influenced by circulating testosterone levels—testosterone is necessary for sperm production and hence mating, frequently correlates with aggression, and is antagonistic to the expression of parental behavior. These links have led to the idea that negative correlations between aggression and parental care are primarily due to the joint effects of testosterone. However, whether aggression and parental behavior are linked directly or whether they are instead only correlated through testosterone is still an open question. If the latter, then in species where testosterone and aggression are uncoupled, there should not be a trade-off between aggression and parental behavior. Direct links between aggression and parental care could occur if there is a fundamental neural trade-off where individuals "wired" to be aggressive cannot also be parental, and vice versa. Finally, it is also possible that there is a basic time or energy trade-off that links these two behaviors such that aggressive individuals spend so much time fighting they do not have time or energy left for offspring care.

Recent studies suggest that the relationship between aggression and parental behavior may be more complex than originally thought and that the antagonistic effects of testosterone on parental behavior may not always be present (see Lynn, 2008, for review). For example, in western bluebirds, aggressive personality differences are unrelated to natural variation in testosterone even though there is a negative relationship between aggression and male parental care in this species (Duckworth, 2006a; Duckworth & Sockman 2012). Moreover, in California mice (*Peromyscus californicus*), testosterone is actually required to maintain high levels of paternal behavior, paternal behavior and aggression are positively correlated, and testosterone and aggression are not related in a simple way (Trainor & Marler, 2001). Both castration and experimentally increased testosterone did not influence aggressive response in this species; instead only the control males increased their aggression in response to a social challenge. The authors suggest that this counterintuitive result may be because this was the only group whose level of aggression was allowed to fluctuate naturally, suggesting that testosterone responsiveness to social challenge, rather than mean level of testosterone, may be more important in modulating aggressiveness in this species. Finally, even in most bird species, recent studies are showing that some species are "behaviorally insensitive" to testosterone such that experimental increases in testosterone do not increase aggressive behavior and fail to dampen parental behaviors (Lynn, 2008). Thus, taken together, these studies show that testosterone does not universally mediate the trade-off between aggression and parental care across species. Moreover, this relationship was mainly developed in songbirds, and, as shown by the California mice example, more evidence is needed from a broader array of taxa to determine whether the trade-off between aggression and parental care is universal.

Temperament Traits and Aggression

In a recent review, Denis Réale and colleagues characterized aggressiveness as one of five temperament (or personality) categories that also include shyness-boldness, exploration-avoidance, activity, and sociability (Réale et al. 2007). Correlations among these distinct behavioral axes are widespread and are referred to as ***behavioral syndromes*** (Sih et al., 2004). Many studies have documented an aggression-boldness syndrome—where animals that are more aggressive are also bolder and more explorative in novel environments (first described by Huntingford, 1976; see Norton et al., 2011, and citations therein for examples). Although correlations between these personality axes are common, they are not ubiquitous and appear to be maintained by natural selection. For example, in threespine sticklebacks (*Gasterosteus aculeatus*) the presence of this behavioral syndrome is correlated with predation pressure across populations such that in populations with high predation the correlation between boldness and aggression is strong, but it breaks down in populations with less intense predation (Bell & Stamps, 2004; Dingemanse et al., 2007). As an adaptive explanation for such a pattern, Niels Dingemanse and colleagues suggested that in ponds with predators, spatial variation in predation risk might favor evolution of alternative solitary or shoaling strategies where solitary individuals monopolize a habitat patch that is poor in food but safe and shoaling individuals roam patches of habitat that are relatively dangerous but rich in food (Dingemanse et al., 2007). Competition for safe patches would be intense, and so solitary individuals should be aggressive and also more exploratory because they have to find patches on their own, whereas shoaling individuals need to be more tolerant of neighbors and can also rely on them to acquire foraging information, so they do not need to be as explorative.

Such adaptive explanations for correlations between aggression and exploratory behavior still need to be tested, but trade-offs originating from differences in social strategy might be the key to understanding correlations between personality traits as these correlations are often found in species where there is wide variation in social strategy between individuals (Cote & Clobert, 2007; Bergmüller & Taborsky, 2010). Examples include freshwater fish, where solitary individuals are both more explorative and aggressive than social individuals (Ward et al., 2004); *Myrmica* ants, where there are strong correlations between aggression, boldness, activity, and sociability at the individual, caste, and colony levels (Chapman et al., 2011); the cooperatively breeding cichlid (*Neolamprologus pulcher*), where an aggression-boldness-explorativeness syndrome was linked to female helping behavior (Schürch & Heg, 2010); and the socially polymorphic comb-footed spider (*Anelosimus studiosus*), where social individuals were less aggressive, less active, and less responsive to prey (Pruitt et al., 2008). Social conflict can select for stable coexistence of different behavioral types, and this has led to the idea that personality differences may evolve primarily in response to social environment (Bergmüller & Taborsky, 2010). While this is an intriguing idea, the evidence so far is only correlative, and the direction of causality between different personality axes is not clear. After all, differences in aggression between individuals might be as likely to lead to differences in sociability as the reverse.

Correlations between Aggression and Other Traits: Ever-present, but Not Consistent

Two main themes emerge from a survey of correlations between aggression and other traits: (1) such correlations are widespread across a diverse array of taxa and (2) there are no consistent patterns across taxa in the specific traits that are correlated with aggression. Aggression is closely linked to body size and dominance in some species but not others. It frequently covaries with aspects of life history investment, such as parental care, but not consistently across species. Many species show strong correlations between aggression and other components of personality variation, but the strength and presence of these correlations vary across species and even across populations within a species.

Such diversity in the strength and direction of correlations has important implications for understanding the evolution of distinct aggressive phenotypes. First, it reinforces the importance of aggression in a wide variety of contexts and life histories. Second, it suggests that the evolution of aggression is not constrained by correlations with other traits, as these correlations can break up and be reformed in a relatively short time span (~10,000 years in the case of stickleback populations that differ in the expression of behavioral correlations; Dingemanse et al., 2007). Finally, it supports the notion that aggression, rather than being an emergent property of other components of the phenotype, is a trait in its own right. After all, if aggression was always consistently correlated with other traits such as boldness or activity levels, this could indicate that these are not really separate traits at all but simply the distinct responses reflecting a common underlying temperament or coping style. In the next two sections, I discuss how individual, population, and species differences in aggression can influence ecological and evolutionary processes and how ecological and evolutionary feedbacks on aggression might provide the key to understanding both the diversity and ubiquity of correlations between aggression and other traits.

ECOLOGICAL CONSEQUENCES OF AGGRESSION

There is a long history of research that investigates the role of aggressive interactions in population and community ecology (Walls, 1990; Amarasekare, 2002; Peiman & Robinson, 2010). By directly affecting competitive interactions between individuals, variation in aggression can influence individual spacing patterns, population dynamics, and population cycles. Moreover, recent studies have found links between aggression and dispersal behavior (e.g., Rusu & Krackow, 2005; Duckworth & Badyaev, 2007; Raihani et al., 2008). Thus, aggression may also influence population connectivity and the dynamics of colonization. In fact, recent studies have indicated that distinct aggressive phenotypes may play a role in species range expansions as well as the success of invasive species. In this section, I will review the evidence that variation in aggressive behavior has strong impacts on ecological dynamics.

Aggression, Individual Spacing, and Population Cycles

In territorial species, aggression can influence individual spacing, which in turn can have large effects on population density and potentially even population cycles (Adams, 2001). One of the earliest proponents of linking aggression to population cycles was Dennis Chitty (1952), who suggested that at high density selection favors large, aggressive animals with low reproductive rates, and at low densities it favors smaller, less aggressive animals with high reproductive rates (Chitty, 1967). Subsequent researchers expanded on these ideas to take into account the potential role of kin interactions (Charnov & Finerty, 1980). However, experimental tests of these ideas have not held up in cycling populations of microtine voles (Boonstra & Hogg, 1988; Boonstra et al., 1994), leading to the conclusion that intrinsic changes in behavior alone could not account for populations cycles (Stenseth & Łomnicki, 1990; Stenseth et al., 1996).

Recent work in birds has shown that, in conjunction with extrinsic factors, aggression can play a key role in population cycles. In red grouse (*Lagopus lagopus scoticus*), there has long been a debate about the relative importance of intrinsic changes in aggression versus extrinsic fluctuations in parasite loads in driving population cycles. While experimental studies have shown that changes in parasitism can produce cycles through their effects on breeding productivity (Hudson, 1986), there is also strong evidence that population cycles are caused by annual variation in the aggressiveness of males through its effects on population density and recruitment of new individuals into the population (Mougeot et al., 2003). Researchers were able to mimic population cycles in a Scottish population by experimentally increasing aggressiveness with testosterone implants in four separate populations. As aggression increased, males expanded their territories and recruitment of new males into the population declined, resulting in a breeding density that was reduced by 50 percent, changing the populations' trajectories from increasing to declining. However, while the effects of aggression on population cycles were clear, it was not clear what ultimately drove changes in aggression. Subsequent studies showed that high levels of testosterone increased parasite infection, which in turn decreased male aggressive behavior (Fox & Hudson, 2001; Seivwright et al., 2005). Thus, the most recent consensus suggests that a combination of extrinsic fluctuations in parasite loads and intrinsic fluctuations in aggression are necessary to explain the observed population cycles (New et al., 2009).

Aggression is also linked to population density and population cycles in bluebirds (*Sialia* spp.). Western bluebirds depend on tree cavities to breed—a limiting resource that historically was patchily distributed and ephemeral. Nest cavities occur at high densities following forest fires, which create suitable habitats for bluebirds by opening up understory vegetation and creating dead snags where nest holes are abundant. Eventually, as the forest regrows, bluebirds are no longer able to breed in these habitat patches because snag density decreases and regrowth of the forest eliminates the open meadows bluebirds depend on to forage for insect prey (Power & Lombardo, 1996; Guinan et al., 2000). Western bluebirds' sister species, mountain bluebirds, are frequently among the earliest colonizers following forest fires (Hutto, 1995), whereas, western bluebirds often show delayed patterns of colonization (Saab et al., 2004; Kotliar et al., 2007). Competition for nest cavities among these and other secondary-cavity-nesting species is intense and often involves aggressive displacement (Gowaty, 1984; Newton, 1994; Merilä & Wiggins, 1995). Western bluebirds, while less dispersive and slower to find new habitat, are on average more aggressive than mountain bluebirds and rapidly displace them when they colonize new areas (Duckworth & Badyaev, 2007). The maintenance of these cycles of species replacement depends at least partly on the evolution of two distinct dispersal strategies in western bluebirds in which dispersal and aggression are closely linked. Highly aggressive males tend to leave their natal populations and disperse to new areas to breed—these newly colonized areas initially have a very low population density, and this enables aggressive males to obtain large territories. On the other hand, nonaggressive males tend to remain near where they were born, which usually is a much older population with a higher density of western bluebirds than newly colonized areas (Duckworth & Badyaev, 2007; Duckworth, 2008). These patterns of biased dispersal with respect to aggression produce a strong correlation between population age, density, and aggressive behavior. Newly colonized population are less dense but highly aggressive, whereas older, well-established populations are less aggressive and have higher densities.

These avian examples show that aggression can be an important determinant of population cycles, however, not in the ways originally envisioned by Chitty (1967)—body size and aggression are not linked in these species, and in both examples, aggression was associated with lower population density because more aggressive individuals are more likely to space themselves farther apart and social tolerance among related individuals enables them to breed at higher density (Lambin & Krebs, 1991). It remains to be seen whether cycles of aggression might play an important role in other classic systems that show populations cycles, such as snowshoe hares (*Lepus americanus*) and the Norway lemming (*Lemmus lemmus*).

Aggression, Invasion, and Range Limits

Recent studies have linked differences in aggression between populations and species to the dynamics of invasion and range expansion. The Argentine ant (*Linepithema humile*) is an invasive species whose success has been attributed at least in part to its aggressive displacement of other species. Introduced populations have undergone **genetic bottlenecks** that have led to very low genetic diversity and the growth of "super-colonies" with nests that essentially function as a single colony spread over many kilometers (Suarez et al., 2008). Nests within these super-colonies are not aggressive toward one another due to their genetic similarity but are very aggressive to other species. In contrast, in the native range, Argentine ants do not form such super-colonies, display much higher levels of intraspecific aggression, and coexist with a diverse community of ant species. These patterns suggest that changes in colony structure and aggression in introduced populations have facilitated invasion success by decreasing intraspecific competition and thus enabling them to outcompete other species (Holway, 1999; Tsutsui et al., 2003). Finally, the invasive red-eared slider (*Trachemys scripta elegans*) and endangered native Spanish terrapin (*Mauremys leprosa*) in the Iberian Peninsula provide another example linking aggression and invasiveness. Red-eared sliders are more aggressive than Spanish terrapins and outcompete them during foraging, significantly restricting their access to food resources (Polo-Cavia et al., 2011).

Aggressive differences between species are not just linked to successful invasions but have also been shown to be important in the dynamics of natural range expansions and in determining species' range limits (Peiman & Robinson, 2010). Scott Pearson and Sievert Rowher (2000) found that competitive superiority of Townsend's warblers (*Dendroica townsendi*) over hermit warblers (*D. occidentalis*) is causing a hybrid zone to move in these species, thereby expanding Townsend's warblers' range at the expense of hermit warblers. They showed that this competitive difference between the species was largely attributable to differences in aggression—Townsend's warblers are more aggressive than hermit warblers. Similarly, the recent range expansion of the barred owl (*Strix varia*) at the expense of the threatened northern spotted owl is at least partly due to its higher aggression (Van Lanen et al., 2011). In bluebirds, competitive superiority of western bluebirds over mountain bluebirds is largely due to the highly aggressive nature of western bluebirds that colonize new populations (Duckworth, 2008). This competitive difference was most obvious during western bluebirds' recent range expansion, in which nest-box programs enabled them to rapidly recolonize areas in the northwestern United States where they had previously gone extinct due to the loss of natural nest cavities. The expansion of their range back to their historical range limits was accompanied by the rapid displacement of lower-elevation mountain bluebird populations (Duckworth & Badyaev, 2007). The range expansion was a natural experiment that provided insight into the competitive dynamics of these species—it showed that mountain bluebirds are limited at the lower edge of their range by competition with western bluebirds rather than by abiotic or other ecological factors.

Such competitive exclusion through direct aggressive interactions is a common theme, especially in the context of species range limits across elevational gradients. In two tropical bird genera—*Catharus* thrushes and *Henicorhina* wrens—asymmetries in aggressiveness explained nonoverlapping ranges across an elevational gradient in tropical forests (Jankowski et al., 2010).

Differences in aggression among four *Eutamias* chipmunks occurring across an elevational gradient in the Sierra Nevada Mountains is thought to at least partially explain their nonoverlapping distributions (Heller, 1971). In *Plethodon* salamanders, evidence suggests that high levels of interspecific aggression have evolved in populations in the Great Smoky Mountains, causing elevational range segregation between *P. glutinosus* and *P. jordani*, whereas in the Balsam Mountains interspecific aggression was largely absent and the two species' ranges overlapped extensively (Hairston et al., 1987). One pattern that emerges across these studies is that the more aggressive species usually lives in the more mild ecological conditions. This pattern is not confined to elevational gradients—red foxes (*Vulpes vulpes*; Figure 10.4) are more aggressive than and competitively dominant to arctic foxes (*Alopex lagopus*) (Frafjord et al., 1989). Evidence suggests that the arctic fox is limited at its southern range edge by the more aggressive red fox and the red fox is limited at its northern range edge by its inability to cope with the extreme climactic conditions of the arctic (Hersteinsson & MacDonald, 1992; Tannerfeldt et al., 2002). Thus, high levels of aggression may be necessary to exclude

Figure 10.4. Red foxes (*Vulpes vulpes*) fighting. Red foxes are more aggressive than arctic foxes (*Vulpes lagopus*), and their competitive superiority may be at least partly responsible for declining arctic fox populations in Scandinavia. (Alex Badyaev, www.tenbestphotos.com)

competitors from areas of abundant resources but may come at the cost of surviving and breeding in more extreme ecological conditions.

EVOLUTIONARY CONSEQUENCES OF AGGRESSION

Strong ecological consequences of aggressive interactions can affect future evolutionary trajectory of populations. Aggressive behavior has great potential to affect selection pressures because it is often used to obtain a breeding territory (Stamps & Krishnan, 1997) and therefore can affect individual fitness by determining the quality of environment in which offspring develop. Western bluebirds provide one of the clearest examples of territorial aggression influencing natural selection. As secondary cavity nesters, their nest sites are extremely limited, and as a consequence, males prefer to acquire territories with multiple nest cavities (Meek & Robertson, 1991; Plissner & Gowaty, 1995); however, only the most aggressive males are able to compete successfully for these territories (Duckworth, 2006b). Using this knowledge, I tested the idea that aggressive interactions over nest cavities could cause males to sort into distinct habitats, which in turn could influence evolution of morphology (Duckworth, 2006b). By placing a high density of nest boxes in open habitat with very low tree cover and a low density of nest boxes in closed habitat with high tree cover, I experimentally caused aggressive interactions to sort males into these different habitats. Aggressive males acquired territories with multiple nest boxes in the open habitat, and nonaggressive males were pushed into the closed habitat where they acquired territories with only a single nest box. Most importantly, males experienced differential selection on morphology across these habitat types. Specifically, males with longer tails and legs were favored in open habitats where high agility is required to forage efficiently, whereas in forested habitats, where agility is less important, selection on morphology was weak. These results showed that aggression can affect selection on a local scale by determining individual settlement patterns. Moreover, because such sorting caused a correlation between aggression and selection on body size, this study has important implications for our understanding of how correlations between aggression and other traits might originate—through a nonrandom link between aggression and habitat type.

Aggressive interactions between species can also have evolutionary consequences through *agonistic character displacement* (Grether et al., 2009). Classic *character displacement* occurs when competition between species causes them to diverge in traits in populations where their ranges overlap compared to populations where their ranges do not overlap (Grant, 1994). Such divergence is thought to lessen interspecific competition and enable coexistence in areas of overlap. Agonistic character displacement is a specialized case where divergence in areas of overlap occurs specifically in traits that affect the rate, intensity, or outcome of competitive interactions. One of the clearest examples of such agonistic character displacement is the case of brook and ninespine sticklebacks (*Culaea inconstans* and *Pungitius pungitius*, respectively). Kathryn Peiman and Beren Robinson (2007) showed that brook sticklebacks from populations that overlap with ninespine sticklebacks are more aggressive than those from populations that do not overlap, suggesting that there has been selection for enhanced aggressiveness where these species come into contact. In another example, Dean Adams (2004) found that robustness of head shape—a trait that is associated with enhanced fighting ability—showed increases in areas of overlap for two *Plethodon* salamanders, *P. jordani* and *P. teyahalee*. Moreover, he found a positive correlation between head shape and aggression. He suggested that these differences in head shape stemmed from aggressive interference competition—salamanders that were more aggressive benefited from having a morphology that increased their fighting ability.

The outcome of aggressive interactions will depend not only on an individual's own aggressive behavior but also on the aggressive phenotype of other individuals in the population. This social context of aggression can produce novel evolutionary feedback dynamics for aggression and the traits associated with aggression, particularly because the environment that elicits aggression (other competitors) can evolve. Such influences of genotypes of other individuals in the population on a focal individual's aggression are termed ***indirect genetic effects*** (Wolf et al., 1998). The importance of indirect genetic effects for evolution have only recently been recognized. In one of the few empirical papers showing indirect genetic effects on aggression, Alastair Wilson and colleagues (2009) found a strong positive ***genetic covariance*** between a focal individual's aggression and the aggression of its opponent. Such covariance between aggression and the social environment can lead to positive evolutionary feedbacks and result in rapid evolution of aggression in the presence of strong natural selection (Wolf et al., 1998).

CONCLUSIONS

Animals display aggression in a wide range of circumstances from competition over mates, food, or other resources, to territory defense and offspring protection, to the establishment of dominance hierarchies within social groups. The ubiquity and importance of aggression has made it the focus of an immense amount of research, making studies of the evolution of aggression a rich resource for understanding the evolution of behavior more generally.

While early studies of aggression focused on trying to understand the optimal expression of aggression within a population (Maynard Smith & Price,

1973; Parker, 1974), recent studies showing consistent differences among individuals have shifted the focus to trying to understand the relative importance of selection and constraint in shaping aggressive phenotypes. These studies have shown that aggression is among the most repeatable of behavioral traits, often varies extensively among individuals within populations, and is frequently correlated with other behaviors. Selection for integration of aggression with other traits is often cited as a key component of adaptive hypotheses for the evolution of consistent individual differences, and while the ubiquity of correlations between aggression and other traits supports this idea, it is not clear whether consistent differences in aggression are a cause or consequence of these correlations. Certainly, the diversity of correlations between aggression and other traits suggests that, if selection for integration is the main cause, then there is a diversity of ways for selection to produce consistent individual differences in aggression. At the same time, the idea that developmental constraints play an important role in the evolution of constancy in the expression of aggression needs empirical testing. Under the constraint hypothesis, correlations between aggression and other traits might be a consequence of consistent differences rather than a cause. More work on the patterns of trait correlations as well as the developmental basis for differences in aggression across a diversity of species is needed to test these alternative hypotheses.

Finally, individual, population-, and species-level variations in aggression can have large-scale ecological consequences by influencing population density and species coexistence. Recent studies show that distinct aggressive phenotypes play a role in population cycles, range expansions, and the success of invasive species as well as competitive exclusion at range edges. In turn, these strong ecological consequences of aggressive interactions can influence evolutionary dynamics of populations, ultimately producing feedbacks that further influence the evolution of aggression.

ACKNOWLEDGMENTS

I thank Alex Badyaev for comments, which improved this manuscript. This work was supported by funding from the National Science Foundation (DEB 918095).

REFERENCES AND SUGGESTED READING

Adams, D. C. (2004). Character displacement via aggressive interference in Appalachian salamanders. *Ecology*, 85, 2664–2670.

Adams, E. S. (2001). Approaches to the study of territory size and shape. *Annual Review of Ecology and Systematics*, 32, 277–303.

Amarasekare, P. (2002). Interference competition and species coexistence. *Proceedings of the Royal Society of London, B*, 269, 2541–2550.

Andersson, M., C. G. Wiklund, & H. Rundgren. (1980). Parental defense of offspring: A model and an example. *Animal Behaviour*, 28, 536–542.

Anestis, S. F. (2005). Behavioral style, dominance rank, and urinary cortisol in young chimpanzees (*Pan troglodytes*). *Behaviour*, 142, 1245–1268.

Appleby, M. C. (1980). Social rank and food access in red deer stags. *Behaviour*, 74, 294–309.

Arnqvist, G. & S. Henriksson. (1997). Sexual cannibalism in the fishing spider and a model for the evolution of sexual cannibalism based on genetic constraints. *Evolutionary Ecology*, 11, 255–273.

Bartoš, L. (1986). Dominance and aggression in various sized groups of red deer stags. *Aggressive Behavior*, 12, 175–182.

Bell, A. M. (2007). Future directions in behavioural syndromes research. *Proceedings of the Royal Society of London, B*, 274, 755–761.

Bell, A. M., S. J. Hankison, & K. L. Laskowski. (2009). The repeatability of behaviour: A meta-analysis. *Animal Behaviour*, 77, 771–783.

Bell, A. M. & J. A. Stamps. (2004). Development of behavioural difference between individuals and population of sticklebacks, *Gasterosteus aculeatus*. *Animal Behaviour*, 68, 1339–1348.

Bennett, A. F. & L. D. Houck. (1983). The energetic cost of courtship and aggression in a plethodontid salamader. *Ecology*, 64, 979–983.

Bergmüller, R. & M. Taborsky. (2010). Animal personality due to social niche specialisation. *Trends in Ecology and Evolution*, 25, 504–511.

Bernstein I. & T. Gordon. (1974). The function of aggression in primate societies: Uncontrolled aggression may threaten human survival, but aggression may be vital to the establishment and regulation of primate societies and sociality. *American Scientist*, 62, 304–311.

Biro, P. A., C. Beckmann, & J. A. Stamps. (2010). Small within-day increases in temperature affects boldness and alters personality in coral reef fish. *Proceedings of the Royal Society of London, B*, 277, 71–77.

Bolhuis, J. E., W. G. P. Schouten, J. W. Schrama, & V. M. Weigant. (2005). Individual coping characteristics, aggressiveness and fighting strategies in pigs. *Animal Behaviour*, 69, 1085–1091.

Bolund, E., H. Schielzeth, & W. Forstmeier. (2007). Intrasexual competition in zebra finches: The role of beak colour and body size. *Animal Behaviour*, 74, 715–724.

Boonstra, R., W. M. Hochachka, & L. Pavone. (1994). Heterozygosity, aggression, and population flucuations in meadow voles (*Microtus pennsylvanicus*). *Evolution*, 48, 1350–1363.

Boonstra, R. & I. Hogg. (1988). Friends and strangers: A test of the Charnov-Finerty hypothesis. *Oecologia*, 77, 95–100.

Brace, R. & J. Pavey. (1978). Size-dependent dominance hierarchy in the anemone *Actinia equine*. *Nature*, 273, 752–753.

Bradshaw, J. W. S., E. J. Blackwell, & R. A. Casey. (2009). Dominance in domestic dogs: Useful construct or bad habit? *Journal of Veterinary Behavior*, 4, 135–144.

Brakefield, P. M. (2006). Evo-devo and constraints on selection. *Trends in Ecology and Evolution*, 21, 362–368.

Carere, C. & J. Balthazart. (2007). Sexual versus individual differentiation: The controversial role of avian maternal hormones. *Trends in Endocrinology and Metabolism*, 18, 73–80.

Chapman, B. B., H. Thain, J. Coughlin, & W. O. H. Hughes. (2011). Behavioural syndromes at multiple scales in *Myrmica* ants. *Animal Behaviour*, 82, 391–397.

Charnov, E. L. & J. P. Finerty. (1980). Vole population cycles: A case for kin-selection? *Oecologia*, 45, 1–2.

Chitty, D. (1952). Mortality among voles (*Microtus agrestis*) at Lake Vyrnwy, Montgomeryshire in 1936–9. *Philosophical Transactions of the Royal Society of London, B*, 236, 505–552.

Chitty, D. (1967). The natural selection of self-regulatory behavior in animal populations. *Proceedings of the Ecological Society of Australia*, 2, 51–78.

Colléter, M. & C. Brown. (2011). Personality traits predict hierarchy rank in male rainbowfish social groups. *Animal Behaviour*, 81, 1231–1237.

Cote, J. & J. Clobert. (2007). Social personalities influence natal dispersal in a lizard. *Proceedings of the Royal Society of London, B*, 274, 383–390.

Cristol, D. A. (1992). Food deprivation influences dominance status in dark-eyed juncos, *Junco hyemalis*. *Animal Behaviour*, 43, 117–124.

de Waal, F. (1986). The integration of dominance and social bonding in primates. *Quarterly Review of Biology*, 61, 459–479.

Dingemanse, N. J. & D. Réale. (2005). Natural selection and animal personality. *Behaviour*, 142, 1159–1184.

Dingemanse, N. J., J. Wright, A. J. N. Kazem, D. K. Thomas, R. Hickling, & N. Dawnay. (2007). Behavioural syndromes differ predictably between 12 populations of three-spined stickleback. *Journal of Animal Ecology*, 76, 1128–1138.

Dowds, B. & R. Elwood. (1985). Shell wars 2: The influence of relative size on decisions made during crab shell fights. *Animal Behaviour*, 33, 649–656.

Drews, C. (1993). The concept and definition of dominance in animal behaviour. *Behaviour*, 125, 283–313.

Duckworth, R. A. (2006a). Behavioral correlations across reproductive contexts provide a mechanism for a cost of aggression. *Behavioral Ecology*, 17, 1011–1019.

Duckworth, R. A. (2006b). Aggressive behavior affects selection on morphology by determining the environment of breeding in a passerine bird. *Proceedings of the Royal Society of London, B*, 273, 1789–1795.

Duckworth, R. A. (2008). Adaptive dispersal strategies and the dynamics of a range expansion. *American Naturalist*, 172, S4–S17.

Duckworth, R. A. (2010). Evolution of personality: Developmental constraints on behavioral flexibility. *Auk*, 127, 752–758.

Duckworth, R. A. & A. V. Badyaev. (2007). Coupling of aggression and dispersal facilitates the rapid range expansion of a passerine bird. *Proceedings of the National Academy of Sciences, USA*, 104, 15017–15022.

Duckworth, R. A., M. T. Mendonça, & G. E. Hill. (2004). Condition dependent sexual traits and social dominance in the house finch. *Behavioral Ecology*, 15, 779–784.

Duckworth, R. A. & K. W. Sockman. (2012). Proximate mechanisms of behavioural inflexibility: Implications for the evolution of personality traits. *Functional Ecology*, 26, 559–566.

Elwood, R. W., K. E. Elwood, W. B. Gallagher, & J. T. A. Dick. (1998). Probing motivational state during antagonistic encounters in animals. *Nature*, 393, 66–68.

Enquist, M. & O. Leimar. (1987). Evolution of fighting behavior: The effect of variation in resource value. *Journal of Theoretical Biology*, 107, 187–205.

Fairbanks, W. (1994). Dominance, age and aggression among female pronghorn, *Antilocapra americana* (Family: Antilocapridae). *Ethology*, 97, 278–293.

Fonberg, E. (1988). Dominance and aggression. *International Journal of Neuroscience*, 41, 201–213.

Fox, A. & P. J. Hudson. (2001). Parasites reduce territorial behaviour in red grouse. *Ecology Letters*, 4, 139–143.

Frafjord, K., D. Becker, & A. Angerbjörn. (1989). Interactions between arctic and red foxes in Scandinavia: predation and aggression. *Arctic*, 42, 354–356.

Francis, R. C. (1983). Experiential effects on agonistic behavior in the paradise fish, *Macropodus opercularis*. *Behaviour*, 85, 292–313.

Francis, R. C. (1984). The effects of bidirectional selection for social dominance on agonistic behavior and sex ratios in the paradise fish (*Macropodus opercularis*). *Behaviour*, 90, 25–45.

Francis, R. C. (1988). On the relationship between aggression and social dominance. *Ethology*, 78, 223–237.

Garcia, J. T. & B. E. Arroyo. (2002). Intra- and interspecific agonistic behaviour in sympatric harriers during the breeding season. *Animal Behaviour*, 64, 77–84.

Gowaty, P. A. (1984). House sparrows kill eastern bluebirds. *Journal of Field Ornithology*, 55, 378–380.

Grant, P. R. (1994). Ecological character displacement. *Science*, 266, 746–747.

Grether, G. F., N. Losin, C. N. Anderson, K. Okamoto. (2009). The role of interspecific interference competition in character displacement and the evolution of competitor recognition. *Biological Reviews*, 84, 617–635.

Guinan, J. A., P. A. Gowaty, & E. K. Eltzroth. (2000) Western bluebird (*Sialia mexicana*). In A. Poole & F. Gill (eds.), *The Birds of North America* (No. 510). Philadelphia: The Academy of Natural Sciences; Washington, DC: The American Ornithologists' Union.

Hagelin, J. C. (2002). The kinds of traits involved in male-male competition: A comparison of plumage, behavior, and body size in quail. *Behavioral Ecology*, 13, 32–41.

Hairston, N. G., K. C. Nishikawa, & S. L. Stenhouse. (1987). The evolution of competing species of terrestrial salamanders: Niche partitioning or interference? *Evolutionary Ecology*, 1, 247–262.
Hau, M. (2007). Regulation of male traits by testosterone: Implications for the evolution of vertebrate life histories. *BioEssays*, 29, 133–144.
Heller, H. C. (1971). Altitudinal zonation of chipmunks (Eutamias): –nterspecific aggression. *Ecology*, 52, 312–319.
Herrel, A., D. V. Andrade, J. E. de Carvalho, A. Brito, A. Abe, & C. Navas. (2009). Aggressive behavior and performance in the Tegu lizard *Tupinambis merianae*. *Physiological and Biochemical Zoology*, 82, 680–685.
Hersteinsson, P. & D. W. MacDonald. (1992). Interspecific competition and the geographical distribution of red and artic foxes *Vulpes vulpes* and *Alopex lagopus*. *Oikos*, 64, 505–515.
Hines, M. (2008). Early androgen influences on human neural and behavioral development. *Early Human Development*, 84, 805–807.
Holway, D. A. (1999). Competitive mechanisms underlying the displacement of native ants by the invasive argentine ant. *Ecology*, 80, 238–251.
Hudson, P. J. (1986). The effect of a parasitic nematode on the breeding production of red grouse. *Journal of Animal Ecology*, 55, 85–92.
Huntingford, F. A. (1976). The relationship between anti-predator behaviour and aggression among conspecifics in the three-spined stickleback, *Gasterosteus aculeatus*. *Animal Behaviour*, 24, 245–260.
Hutto, R. L. (1995). Composition of bird communities following stand-replacement fires in northern Rocky Mountain (U.S.A.) conifer forests. *Conservation Biology*, 9, 1041–1058.
Jaeger, R. (1981). Dear enemy recognition and the costs of aggression between salamanders. *American Naturalist*, 117, 962–974.
Jankowski, J. E., S. K. Robinson, & D. J. Levey. (2010). Squeezed at the top: nterspecific aggression may constrain elevational ranges in tropical birds. *Ecology*, 91, 1877–1884.
Johnson, J. & A. Sih. (2005). Precopulatory sexual cannibalism in fishing spiders (*Dolomedes triton*): A role for behavioral syndromes. *Behavioral Ecology and Sociobiology*, 58, 390–396.
Johnstone, R. A. & K. Norris. (1993). Badges of status and the cost of aggression. *Behavioral Ecology and Sociobiology*, 32, 127–134.
Just, W. & M. R. Morris. (2003). The Napoleon complex: Why smaller males pick fights. *Evolutionary Ecology*, 17, 509–522.
Ketterson, E. D. & V. Nolan Jr. (1994). Male parental behavior in birds. *Annual Review of Ecology and Systematics*, 25, 601–628.
Korpela, K., J. Sundell, & H. Ylönen. (2011). Does personality in small rodents vary depending on population density? *Oecologia*, 165, 67–77.
Kotliar, N. B., P. L. Kennedy, & K. Ferree. (2007). Avifaunal responses to fire in southwestern montane forests along a burn severity gradient. *Ecological Applications*, 17, 491–507.

Krebs, J. R. & N. B. Davies (eds.). (1991). *Behavioral Ecology: An Evolutionary Approach*. Oxford, UK: Blackwell Scientific Publications.

Lambin, X. & C. J. Krebs. (1991). Can changes in female relatedness influence microtine population dynamics? *Oikos*, 61, 126–132.

Lemel, J. & K. Wallin. (1993). Status signalling, motivational condition and dominance: An experimental study in the great tit, *Parus major* L. *Animal Behaviour*, 45, 549–558.

Lynch, M. & B. Walsh. (1998). *Genetics and analysis of quantitative traits*. Sunderland, MA: Sinauer Associates.

Lynn, S. E. (2008). Behavioral insensitivity to testosterone: Why and how does testosterone alter paternal and aggressive behavior in some avian species but not others? *General and Comparative Endocrinology*, 157, 233–240.

Maynard Smith, J., R. Burian, S. Kauffman, P. Alberch, J. Campbell, B. Goodwin, R. Lande, D. Raup, & L. Wolpert. (1985). Developmental constraints and evolution: A perspective from the Mountain Lake Conference on development and evolution. *Quarterly Review of Biology*, 60, 265–287.

Maynard Smith, J. & G. R. Price. (1973). The logic of animal conflict. *Nature*, 246, 15–18.

McElreath, R. & P. Strimling. (2006). How noisy information and individual asymmetries can make "personality" an adaptation: A simple model. *Animal Behaviour*, 72, 1135–1139.

McNamara, J. M., P. Stephens, S. Dall, & A. I. Houston. (2009). Evolution of trust and trustworthiness: Social awareness favours personality differences. *Proceedings of the Royal Society of London, B*, 276, 605–613.

Mech, L. D. (1999). Alpha status, dominance, and division of labor in wolf packs. *Canadian Journal of Zoolgy*, 77, 1196–1203.

Meek, S. B. & R. J. Robertson. (1991). How do floater male eastern bluebirds benefit by filling vacancies on territories where females already have young? *Behavioral Ecology*, 3, 95–100.

Merilä, J. & D. A. Wiggins. (1995). Interspecific competition for nest holes causes adult mortality in the collared flycatcher. *Condor*, 97, 445–450.

Morrell, L. J., J. Lindström, & G. D. Ruxton. (2005). Why are small males aggressive? *Proceedings of the Royal Society of London, B*, 272, 1235–1241.

Morris, M. R., L. Gass, & M. J. Ryan. (1995). Assessment and individual recognition of opponents in the swordtails *Xiphophorus nigrensis* and *X. multilineatus*. *Behavioral Ecology and Sociobiology*, 37, 303–310.

Mougeot, F., S. M. Redpath, F. Leckie, & P. J. Hudson. (2003). The effect of aggressiveness on the population dynamics of a territorial bird. *Nature*, 421, 737–739.

New, L. F., J. Matthiopoulos, S. M. Redpath, & S. T. Buckland. (2009). Fitting models of multiple hypotheses to partial population data: Investigating the causes of cycles in red grouse. *American Naturalist*, 174, 399–412.

Newton, I. (1994). The role of nest sites in limiting the numbers of hole-nesting birds: a review. *Biological Conservation*, 70, 265–276.

Norton, W., K. Stumpenhorst, T. Faus-Kessler, A. Folchert, N. Rohner, M. P. Harris, et al.. (2011). Modulation of Fgfr1a signaling in zebrafish reveals a genetic basis for the aggression-boldness syndrome. *Journal of Neuroscience*, 31, 13796–13807.

Nosil, P. (2002). Food fights in house crickets, *Acheta domesticus*, and the effects of body size and hunger level. *Canadian Journal of Zoology*, 80, 409–417.

Parker, G. A. (1974). Assessment strategy and the evolution of fighting behaviour. *Journal of Theoretical Biology*, 47, 223–243.

Pearson, S. F. & S. Rohwer. (2000). Asymmetries in male aggression across an avian hybrid zone. *Behavioral Ecology*, 11, 93–101.

Pedetta, S., L. Kaczer, & H. Maldonado. (2010). Individual aggressiveness in the crab *Chasmagnathus*: Influence in fight outcome and modulation by serotonin and octopamine. *Physiology and Behavior*, 101, 438–445.

Peiman, K. & B. W. Robinson. (2007). Heterospecific aggression and adaptive divergence in brook stickleback (*Culaea Inconstans*). *Evolution*, 61, 1327–1338.

Peiman, K. & B. W. Robinson. (2010). Ecology and evolution of resource-related heterospecific aggression. *Quarterly Review of Biology*, 85, 133–158.

Pitnick, S. & G. T. Miller. (2000). Correlated response in reproductive and life history traits to selection on testis length in *Drosophila hydei*. *Heredity*, 84, 416–426.

Plissner, J. H. & P. A. Gowaty. (1995). Eastern bluebirds are attracted to two-box nest sites. *Wilson Bulletin*, 107, 289–295.

Polo-Cavia, N., P. López, & J. Martín. (2011). Aggressive interactions during feeding between native and invasive freshwater turtles. *Biological Invasions*, 13, 1387–1396.

Popova, N. K. (2006). From genes to aggressive behavior: The role of serotonergic system. *BioEssays*, 28, 495–503.

Power, H. W. & M. P. Lombardo. (1996). Mountain bluebird. In A. Poole & F. Gill (eds.), *The Birds of North America* (No. 222). Philadelphia: The Academy of Natural Sciences; Washington, DC: The American Ornithologists' Union.

Pruitt, J. N., S. E. Riechert, & T. C. Jones. (2008). Behavioural syndromes and their fitness consequences in a socially polymorphic spider, *Anelosimus studiosus*. *Animal Behaviour*, 76, 871–879.

Raihani, N. J., A. R. Ridley, L. E. Browning, M. J. Nelson-Flower, & S. Knowles. (2008). Juvenile female aggression in cooperatively breeding pied babblers: Causes and contexts. *Ethology*, 114, 452–458.

Réale, D., S. M. Reader, D. Sol, P. T. McDougall, & N. J. Dingemanse. (2007). Integrating animal temperament within ecology and evolution. *Biological Reviews of the Cambridge Philosophical Society*, 82, 291–318.

Redondo, T. & J. Carranza. (1989). Offspring reproductive value and nest defense in the magpie (*Pica pica*). *Behavioral Ecology and Sociobiology*, 25, 369–378.

Riebli, T., B. Avgan, A.-M. Bottini, C. Duc, M Taborsky., & D. Heg. (2011). Behavioural type affects dominance and growth in staged encounters of cooperatively breeding cichlids. *Animal Behaviour*, 81, 313–323.

Riechert, S. E. (1993). Investigation of potential gene flow limitation of behavioral adaptation in an aridlands spider. *Behavioral Ecology and Sociobiology*, 32, 355–363.

Riechert, S. E. & P. M. Johns. (2003). Do female spiders select heavier males for the genes for behavioral aggressiveness they offer their offspring?*Evolution*, 57, 1367–1373.

Riska, B. (1989). Composite traits, selection response and evolution. *Evolution*, 43, 1172–1191.

Robertson, R. J. (1986). Spitefulness, altruism, and the cost of aggression: Evidence against superterritoriality in tree swallows. *Condor*, 88, 104–105.

Roff, D. A. (1994). Optimality modeling and quantitative genetics: A comparison of the two approaches. In C. Boake (ed.), *Quantitative Genetic Studies of Behavioral Evolution* (pp. 49–66). Chicago: University of Chicago Press.

Roff, D. A. (1997). *Evolutionary Quantitative Genetics*. New York: Longman.

Rosvall, K. (2011). Cost of female intrasexual aggression in terms of offspring quality: A cross-fostering study. *Ethology*, 117, 332–344.

Rusu, A. & S. Krackow. (2005). Agonistic onset marks emotional changes and dispersal propensity in wild house mouse males (*Mus domesticus*). *Journal of Comparative Physiology*, 119, 58–66.

Saab, V. A., J. Dudley, & W. L. Thompson. (2004). Factors influencing occupancy of nest cavities in recently burned forests. *Condor*, 106, 20–36.

Schürch, R. & D. Heg. (2010). Life history and behavioral type in the highly social cichlid *Neolamprologus pulcher*. *Behavioral Ecology*, 21, 588–598.

Schwabl, H. & E. Kriner. (1991). Territorial aggression and song of male Europian robins (*Erithacus rubecula*) in autumn and spring: Effects of antiandrogen treatment. *Hormones and Behavior*, 25, 180–194.

Schwarzkopf, L., M. W. Blows, & M. J. Caley. (1999). Life history consequences of divergent selection on egg size in *Drosophila melanogaster*. *American Naturalist*, 54, 333–340.

Seivwright, L. J., S. M. Redpathl, F. Mougeot, F. Leckie, & P. J. Hudson. (2005). Interactions between intrinsic and extrinsic mechanisms in a cyclic species: Testosterone increases parasite infection in red grouse. *Proceedings of the Royal Society of London, B*, 272, 2299–2304.

Sih, A., A. Bell A., & J. C. Johnson. (2004). Behavioral syndromes: An ecological and evolutionary overview. *Trends in Ecology and Evolution*, 19, 372–378.

Smith, I. P., F. A. Huntingford, R. J. A.Atkinson, & A. C. Taylor. (1994). Strategic decisions during agonistic behaviour in the velvet swimming crab, *Necora puber* (L.). *Animal Behaviour*, 47, 885–894.

Soma, K. K., A. D. Tramontin, & J. C. Wingfield. (2000). Oestrogen regulates male aggression in the non-breeding season. *Proceedings of the Royal Society of London, B*, 267, 1089–1096.

Stamps, J. A. & V. V. Krishnan. (1997). Functions of fights in territory establishment. *American Naturalist*, 150, 393–405.

Stenseth, N. C., O. N. Bjornstad, & W. Falck. (1996). Is spacing behaviour coupled with predation causing the microtine density cycle? A synthesis of current process-oriented and pattern-oriented studies. *Proceedings of the Royal Society of London, B*, 263, 1423–1435.

Stenseth, N. C. & A. Łomnicki. (1990). On the Charnov-Finerty hypothesis: The unproblematic transition from docile to aggressive and the problematic transition from aggressive to docile. *Oikos*, 58, 234–238.

Suarez, A. V., D. A. Holway, & N. D. Tsutsui. (2008). Genetics and behavior of a colonizing species: The invasive Argentine ant. *American Naturalist*, 172, S72–S84.

Tannerfeldt, M., B. Elmhagen, & A. Angerbjörn. (2002). Exclusion by interference competition? The relationship between red and arctic foxes. *Oecologia*, 132, 213–220.

Temeles, E. (1994). The role of neighbours in territorial systems: When are they "dear enemies"? *Animal Behaviour*, 47, 339–350.

Trainor, B. C. & C. A. Marler. (2001). Testosterone, paternal behavior, and aggression in the monogamous california mouse (*Peromyscus californicus*). *Hormones and Behavior*, 40, 32–42.

Tsutsui, N. D., A. V. Suarez, & R. K. Grosberg. (2003). Genetic diversity, asymmetrical aggression, and recognition in a widespread invasive species. *Proceedings of the National Academy of Sciences, USA*, 100, 1078–1083.

Tuttle, E. M. (2003). Alternative reproductive strategies in the white-crowned sparrow: Behavioral and genetic evidence. *Behavioral Ecoogy.* 14, 425–432.

Van Lanen, N. J., A. B. Franklin, K. P. Huyvaert, R. F. Reiser, & P. C. Carlson. (2011). Who hits and hoots at whom? Potential for interference competition between barred and northern spotted owls. *Biological Conservation*, 144, 2194–2201.

Walls, S. C. (1990). Interference competition in postmetamorphic salamanders: Interspecific differences in aggression by coexisting species. *Ecology*, 71, 307–314.

Ward, A. J. W., P. Thomas, & P. J. B. Hart. (2004). Correlates of boldness in the three-spined sticklebacks (*Gasterosteus aculeatus*). *Behavioral Ecology and Sociobiology*, 55, 561–568.

Watt, D. J. (1986). Relationship of plumage variability, size and sex to social dominance in Harris' sparrows. *Animal Behaviour*, 34, 16–27.

Weiß, B., K. Kortschal, & K. Foerster. (2011). A longitudinal study of dominance and aggression in greylag geese (*Anser anser*). *Behavioral Ecology*, 22, 616–624.

Wilson, A. J., U. Gelin, M. C. Perron, & D. Réale. (2009). Indirect genetic effects and the evolution of aggression in a vertebrate system. *Proceedings of the Royal Society of London, B*, 276, 533–541.

Wingfield, J. C., S. L. Meddle, I. Moore, S. Busch, D. Wacker, S. Lynn, et al. (2007). Endocrine responsiveness to social challenges in northern and southern hemisphere populations of *Zonotrichia*. *Journal of Ornithology*, 148, S435–S441.

Wolf, J. B., E. D. Brodie, J. M. Cheverud, A. J. Moore, & M. J. Wade. (1998). Evolutionary consequences of indirect genetic effects. *Trends in Ecology and Evolution*, 13, 64–69.

Wolf, M., G. S. van Doorn, O. Leimar, & F. J. Weissing. (2007). Life-history trade-offs favour the evolution of animal personalities. *Nature*, 447, 581–585.

Zack, S. (1975). A description and analysis of agonistic behaviour patterns in an opisthobranch mollusc, *Hermissenda crassicornis*. *Behaviour*, 5, 238–267.

11
Altruism and Kinship

Lee Alan Dugatkin

INTRODUCTION

Animal behaviorists have a long-standing interest in understanding the evolution of **altruism**. It all started when Charles Darwin obsessed over the "altruism question" in his classic book *On the Origin of Species*. In particular, Darwin was very worried about the behavior of honey bee workers. The problem was the self-sacrificial—indeed, suicidal—behavior that worker bees display when their nest is threatened. *How could **natural selection** ever favor this sort of behavior?* Darwin's biographer, Janet Browne (2002), describes him as "specially exercised over honey bees" (p. 203), and Darwin paints himself as "half mad" over honey bee behavior. If Darwin (1859) was correct that "natural selection acts only by the preservation and accumulation of small inherited modifications, each profitable to the preserved being" (p. 95), then *what was going with these suicidal honeybees?* And *how could he explain other prosocial behaviors such as guard duty and food sharing?*

Darwin's solution to the altruism problem revolved around kinship. But, before we can understand what Darwin proposed, we need to be clear about some terminology. Webster's dictionary defines "kin" as "one's relatives." But what animal behaviorists mean when they talk about **kinship** is something much more specific, namely **genetic relatedness**. Spouses and in-laws are kin under the everyday usage of the word but not the more stringent, evolution-based definition adopted by animal behaviorists. There is a similar distinction to be made about the everyday definition of *altruism* and the definition used in the animal behavior literature. While the Webster's definition of altruism is "unselfish regard for or devotion to the welfare of others," the evolutionary definition is more operational—altruism is an act that is costly to the donor and beneficial to the recipient, where costs and benefits are measured in terms of effects on expected lifetime reproductive success.

With these definitions and distinctions in hand, let us return to how Darwin dealt with problems like suicidal honeybees. Darwin understood that, somehow or another—and he was not quite sure just how—traits that affected reproductive success were passed from parents to offspring. Of course, he did not know about genes—Mendel was only just about to publish his results, and they remained unnoticed until 1900—but Darwin did sometimes discuss blood-borne particles he called "gemmules" that were passed from parents to offspring. What is more, he understood that all blood relatives—what we would call genetic relatives—resembled one another to varying degrees.

In a section of *The Origin* entitled "Objections to the Theory of Natural Selection as Applied to Instincts: Neuter and Sterile Insects," Darwin (1859) proposed that the riddle of altruism like that seen in honeybee workers "disappears when it is remembered that selection may be applied *to the family*, as well as the individual and may thus gain the desired end" (p. 237). Acts that benefit blood kin can be favored by natural selection. Even though a worker bee pays a huge cost by defending the hive, this cost is compensated by the benefits accrued by her genetic relatives at the hive. Kinship was the key to solving Darwin's problems with altruism.

In one sense, Darwin both posed and solved the conundrum of the evolution of altruism. The problem was confronted, and the remedy—what we would now call kin selection—was proposed. But in some important ways, Darwin failed to settle the issue. Without experiments or some sort of mathematical framework for his theory, he was never able to answer the questions his theory brought forth, namely, *precisely how does what we now call kin selection operate*? For example, *Just how does the* degree *of kinship affect the evolution of altruism*? Some blood kin like siblings are very closely related, but others, like second cousins, are less genetically related. *Does that matter, and if so, precisely how? Does it matter how costly the altruistic act is? Does it matter how great a benefit the recipient of altruism receives*? If so, *how are these costs and benefits to be measured, and how does ecology affect such costs and benefits*? For the most part, these questions lay unanswered for the next hundred or so years after publication of *The Origin of Species*, though there were certainly some interesting twists and turns during this time period (Dugatkin, 2007).

INCLUSIVE FITNESS THEORY AND THE EVOLUTION OF ALTRUISM

In 1963, William D. Hamilton published a paper entitled "The Evolution of Altruistic Behaviour" in *The American Naturalist* (Hamilton, 1963). He opened this paper as follows: "It is generally accepted that the behaviour characteristic of a species is just as much the product of evolution as the morphology" (p. 354). But, as Hamilton noted, there are some kinds of behavior that could not easily be explained by classic evolutionary thinking: "in particular . . . any case where an animal behaves in such a way as to promote the advantages of other members of the species not its direct descendants at the expense of its own" (p. 354). To address this issue, he built his own model examining the role of genetic relatedness in facilitating altruism.

Hamilton asked the reader to imagine a pair of genes—gene G, which codes for altruism, and gene g, which does not. G codes for an act that entails a cost to the actor but a benefit to others, while g codes for no such action. Under standard models, G is always at a selective disadvantage compared to g and hence should never increase in frequency. But, Hamilton argued, if the effects of genetic kinship were added to the standard model—creating an ***inclusive fitness*** model—altruism could evolve.

To build this more inclusive model, Hamilton used only three variables, labeled r, b, and c. Let us walk through each of these. The benefit that a recipient of an altruistic act obtained was denoted as b, and the cost paid by an altruist was labeled c. For example, imagine an altruist who brings food back to her siblings—here the benefit might be an extra chick surviving in the nest of the altruistic bird, while the cost might entail an increased risk of death during her foraging incursions.

To measure the genetic relatedness between altruists and those they aid, Hamilton used Sewall Wright's (1922) ***coefficient of relationship***, labeled r. In evolutionary terms, genetic relatedness centers on the probability that individuals share gene copies that they have inherited from a common ancestor or a set of common ancestors—parents, grandparents, and so on. Gene copies that are shared because of common ancestry are referred to as "identical by descent." For example, you and your siblings are genetic kin because you share some of the same gene copies that you inherited from your mother and father. Similarly, you and your first cousins are genetic relatives because you share a set of common ancestors—one set of grandparents.

If we can trace the common ancestor of two or more individuals, then we can calculate their relatedness (r). Consider the case of full siblings who have as their recent common ancestors their mother and father. There are only two ways that siblings can share a copy of gene X—via their mother or father. If sibling 1 has X, then there is a 50 percent chance she received it from her mother; if sibling 2 has X , there is a 50 percent chance that her mother passed this gene to sibling 2. So there is a 1 in 4 chance that the siblings share gene X through their mother. In a similar vein, there is a 1 in 4 probability that their father is the reason that the siblings share gene X. To calculate the chances that our siblings share gene X through *either* mother or father, we add the probabilities for each and obtain $1/4 + 1/4 = 1/2$, or 0.5. This r value can be calculated for any set of genetic relatives, no matter how distant. For example, the genetic relatedness between cousins is 1/8 (that is, $r = 0.125$), between grandparent and grandchild is 1/4 (that is, $r = 0.25$), and between aunts or uncles and their blood nieces and nephews is also 1/4 (that is, $r = 0.25$).

In Hamilton's model, natural selection favors the gene for altruism whenever $r \times b > c$. This equation has become known as Hamilton's rule, and it lies at the heart of inclusive fitness theory (also known as **kin selection**). If a gene for altruism is to evolve, then the cost (c) of altruism must be balanced by compensating benefits to the altruist. In Hamilton's inclusive model, the cost is balanced by the benefits (b) accrued by genetic relatives of the altruist because relatives *may* carry the gene for altruism as well. But relatives have only *some probability*—measured by r—of carrying the gene in question, and so the benefits received by the altruist must be devalued by that probability.

When we know the costs and benefits associated with an altruistic act, and we know the relatedness of the individuals involved, Hamilton's rule allows us to predict whether altruism will evolve for any value of r, b, and c. Because this rule is more easily satisfied when r is large, the more related individuals are, the more likely altruism is to evolve. In a similar vein, the greater the benefit/cost ratio for the altruist, the more likely altruism is to evolve (Hamilton, 1964, 1972; Grafen, 2007).

Hamilton's rule was first constructed for cases in which an altruist helps a *single* relative, but it can easily be modified to examine a single altruistic act that has effects on *many* blood relatives simultaneously. In that case, altruism evolves when $(\sum_{1}^{n} r \times b) > c$. In this equation, $\sum_{1}^{n}$ sums benefits across *all* genetic relatives that are helped by the act of an altruist. If an altruistic act affects *two* siblings ($n = 2$), it will be favored as long as the benefit received by the recipient is simply greater than the cost to the altruist—exactly half the benefit that is necessary for altruism to evolve when an altruistic act benefits a *single* sibling.

In 1977, Paul Sherman reported the results of his University of Michigan dissertation work on alarm calls and blood kinship in Belding's ground squirrels (*Spermophilus beldingi*) (Sherman, 1977). Though others had tested Hamilton's inclusive fitness theory before this paper, Sherman's publication was a watershed moment. He had spent more than 3,000 hours observing ground squirrel behavior in the Sierra Nevada mountains. Sherman, like many researchers before him, had noted that when a ground squirrel spots a predator, it sometimes stands ups and emits a piercing call. In response, other squirrels in the vicinity head for cover. But not all squirrels were equally likely to emit such **alarm calls**. *Why? Why were some individuals more likely than others to emit these calls—calls that drew attention to the caller but benefited others in the caller's group?* Sherman's answer would require him to integrate demography, natural history, and inclusive fitness theory (Sherman, 1981). Let us examine how.

When male Belding's ground squirrels reach maturity, they emigrate to new populations to find mates. Female squirrels, however, spend their whole lives in their **natal** (birth) population. What this means is that there are important differences in how adult males and females are related to others in their populations. By remaining in their natal populations, females—both young and old—are surrounded by genetic relatives. Mature males, who emigrate to new populations, however, are often living with individuals who are not their genetic kin.

Over the course of his three-year study, Sherman and his colleagues saw terrestrial predators—weasels, badgers (*Taxidea taxus*), dogs (*Canis lupus familiaris*), coyotes (*Canus latrans*), and pine martins *(Martes americana)*—and ground squirrels together on 102 occasions, and predators killed a total of six adult Belding's ground squirrels and three juveniles. Sherman found that when a terrestrial predator was spotted, female squirrels gave alarm calls much more often than expected by chance, but the converse was true for males. On average, when a predator was spotted by ground squirrels, 30 percent of the individuals present were adult females and 20 percent were adult males. However, when he analyzed which squirrel emitted the first altruistic alarm call, Sherman found that females did so about 65 percent of the time (more than twice the 30 percent that would be expected by chance alone), but males did so 2 to 3 percent of the time (about a fifth to a tenth of what would be expected by chance). Adult females, surrounded by genetic kin, were emitting the altruistic alarm calls; while adult males, who tend to be unrelated to those around them, kept quiet. Sherman, who had learned about Hamilton's rule in graduate school, realized that alarm callers were aiding their genetic relatives, as inclusive fitness theory predicts (Figure 11.1).

Further support for the hypothesis that kinship helps explain the distribution of alarm calls came when Sherman found that in rare instances when a female did leave her natal group and moved into a population of unrelated individuals, she gave alarm calls less frequently than did native females—when a female was not surrounded by her genetic relatives she dramatically reduced the frequency of her alarm calling.

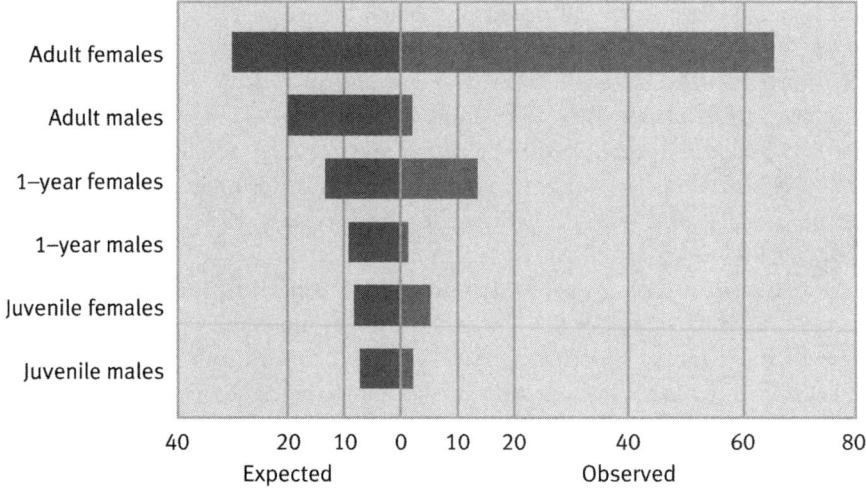

Figure 11.1. Alarm calls in Belding's ground squirrels (*Spermophilus beldingi*) (from Sherman, 1977, as modified in Dugatkin, 2013). Observed versus expected frequencies of alarm calls in Belding's ground squirrels. Females call much more often than expected by chance, while the opposite is true for males.

Work on the relation between altruism and kinship is not limited to vertebrates. Ever since Darwin first proposed that genetic relatedness may play a role in understanding the evolution of altruism, evolutionary and behavioral biologists have been interested in how kinship may have facilitated altruism in the social insects: the ants, bees, and wasps (which collectively make up the order Hymenoptera). In particular, social insect species display *eusociality*—a type of sociality that involves a reproductive division of labor, in which only a small fraction of individuals in a colony breed and others in the colony assist in cooperative rearing of offspring (Alexander et al., 1991).

Although eusociality has been uncovered in other species, it is most often associated with ants, bees, and wasps, where it has evolved independently at least eight times (Hughes et al., 2008). *Does inclusive fitness theory help us explain why eusociality exists in social insects?* And if so, *how?* To answer that question, we need to recognize that nests of eusocial species often contain thousands of individuals, many of whom are genetic relatives. What this means is that altruistic acts (for example, food sharing, cooperative nest defense) benefit not just one but many many genetic relatives. But this fact only gets us so far—such altruistic benefits are dispensed in *any* species in which relatives live near one another, not just in eusocial hymenoptera. *Why has eusociality evolved so many times in the hymenoptera, but not elsewhere?*

The answer to that question, in part, is tied to the unusual genetic system found in hymenopterans. Social insects are **haplodiploid**—that is, males are haploid, but females are diploid. In haplodiploid species, if a queen mates with just one male, sister workers in the colony are related to one another on average by a coefficient of relatedness (r) of 0.75. The upshot of this is that hymenopteran females are more related to their sisters than to their own offspring. So if kinship is important in driving altruism and other eusocial behaviors, we should see this primarily in female workers, not male drones. And that is what behavioral ecologists have found over and over: females raise up broods, defend the nest, and bring food back to the nest to share with others.

The importance of genetic relatedness in driving the evolution of eusociality in hymenoptera has also been examined using phylogenetic techniques. When queens mate with a single male (i.e., when queens are **monandrous**), genetic relatedness in colonies should be high. But when queens mate with many males (**polyandry**), the genetic relatedness in groups plummets. This difference generates a testable hypothesis: eusociality and monandry should be casually linked.

To test this hypothesis, William Hughes and his colleagues began with previously published information that eusociality has evolved nine times in hymenopterans (Hughes et al., 2008; Ratnieks & Helantera, 2009). Hughes and colleagues hypothesized that for eusociality to have been favored when it first appeared, the ancestral mating system in these cases should have been queens mating with a single male (monandry). Their analysis supports this prediction—in all the lineages they analyzed (lineages that included 267 different species), they found that, as predicted by inclusive fitness theory, monandry was the ancestral state.

It is important to understand that the mathematical models Hamilton and others have derived predict that natural selection will favor behavioral traits that increase inclusive fitness and that this will often be facilitated by remaining in the vicinity of kin. But that is not always the case. More specifically, individuals who have a higher inclusive fitness when remaining with their family should stay as part of the family unit, while those who have better opportunities for increasing their inclusive fitness elsewhere are predicted to depart and avail themselves of that opportunity. For example, consider Stephen Pruett-Jones's work with the superb fairy wren (*Malurus cyaneus*), an Australian bird.

In superb fairy wrens, a breeding pair is often helped by its young (but sexually mature) male offspring, who provide their siblings with additional food and protection (Pruett-Jones & Lewis, 1990). If there are no territories available for these young helpers to start their own families, and if female breeding partners are scarce, then remaining at their parent's nest and helping raise their siblings is the only thing young males can do to increase their own inclusive fitness. But *what happens when new territories open up in the vicinity of these sexually mature male helpers?* To find out, Pruett-Jones removed 29 superb fairy wren males from their territories, creating new breeding opportunities for male helpers in the vicinity of the removals. All but one of the 32 male helpers who could have dispersed to the newly opened territories did so, and they did so quickly—new territories were usually occupied by former male helpers within six hours. When a better way to increase their own inclusive fitness appeared, male helpers seized the opportunity for a breeding territory, even though this caused the disbanding of the family units these helpers were originally part of.

KIN RECOGNITION

Because interactions with genetic relatives will often favor the evolution of altruism, **behavioral ecologists** have studied whether, and by what means, individuals can recognize their kin (Holmes, 2004). Consider **kin recognition** seen in penguins. In some species of penguins, parents travel long distances to the sea to obtain food to bring back to inland areas where their chicks have hatched. When they return from their journey, parents must find their young among thousands of screaming, hungry chicks in a colony (Aubin & Jouventin, 2002; Brumm & Slabbekoorn, 2005). *How do parents know which chicks are their offspring?* For species like the king penguin (*Aptenodytes patagonicus*) and the emperor penguin (*Aptenodytes fosteri*) the answer appears to center on complex vocal cues that allow for kin recognition via **vocal signatures** emitted by the young (Aubin & Jouventin, 1998; Jouventin et al., 1999; Aubin et al., 2000; Lengagne et al., 2000).

Some penguin species are as not proficient as the king and emperor penguins at recognizing the vocal signatures of their offspring. Individuals who build nests, for example, are not as adept at recognizing the vocal calls of their young as are individuals who live in dense colonies and do not nest (Jouventin & Aubin, 2002; Searby et al., 2004). *Why?* Researchers hypothesize that this is due to the fact that parents in nest-building species can find their offspring by remembering the location of their nests. Any chick in their nest is likely their offspring, and hence natural selection to recognize offspring by vocal cues in these species is weak. Where the problem of kin recognition is more difficult—in dense colonies with no nests—natural selection favors the evolution of more complex vocal recognition systems.

A number of kin-recognition models center on an internal **template** against which genetic relatedness is gauged (Reeve, 1989). The basic idea is that individual A determines if individual B is kin or nonkin based on how closely individual B matches the internal template of individual A. The internal template may be generated genetically, via learning, or via social learning, but in all cases the animal infers the degree of kinship as some function of the extent to which others match its own template.

A fascinating case of template matching has been studied in the behavior of spadefoot toad tadpoles (*Scaphiopus bombifrons*) (Elgar & Crespi, 1992; Pfennig et al., 1993). Two feeding **morphs** of spadefoot toads are found in ponds: in one morph, juveniles feed on small, often drifting vegetative clumps of food (known as **detritus**) and typically develop into herbivores when they mature. In the second morph, juveniles feed on shrimp and tend to mature into carnivorous cannibals as they mature.

David Pfennig (1993) studied kin recognition in both herbivore and cannibal spadefoot morphs by testing morphs in the presence of either unfamiliar siblings or unfamiliar nonrelatives. Herbivores preferred associating with their siblings over unrelated individuals, presumably because of the inclusive fitness benefits associated with interactions with genetic kin. On the other hand, the carnivorous cannibal morphs spent more time near unrelated individuals, presumably to avoid the costs of killing their genetic kin.

Pfennig and his colleagues also offered carnivores a choice between unfamiliar siblings and unfamiliar nonrelatives in a protocol that allowed carnivores to actually eat other tadpoles. Carnivores were not only more likely to eat unrelated individuals, but they were able to distinguish between relatives and nonrelatives by taste. Pfennig found that carnivores were equally likely to suck relatives and nonrelatives into their mouths, but they released their relatives much more frequently than unrelated individuals.

CLOSING THOUGHT

The study of animal behavior was revolutionized by the introduction of inclusive fitness models of the evolution of altruism. This continues to be one of the most actively researched areas in ethology, with modern work employing molecular genetic and phylogenetic analyses to expand the frontiers of research in this area.

REFERENCES AND SUGGESTED READING

Alexander, R. D., K. M.Noonan, & B. J. Crespi. (1991). The evolution of eusociality. In P. Sherman, J. U. M. Jarvis, & R. D. Alexander (eds.), *The Biology of the Naked Mole-Rat* (pp. 3–44). Princeton, NJ: Princeton University Press.

Aubin, T. & P. Jouventin. (1998). Cocktail-party effect in king penguin colonies. *Proceedings of the Royal Society of London, B,* 265, 1665–1673.

Aubin, T., & P. Jouventin. (2002). How to vocally identify kin in a crowd: The penguin model. *Advances in the Study of Behavior,* 31, 243–277.

Aubin, T., P. Jouventin, & C. Hildebrand. (2000). Penguins use the two-voice system to recognize each other. *Proceedings of the Royal Society of London, B,* 267, 1081–1087.

Beecher, M. D., I. Beecher, & S. Hahn. (1981). Parent-offspring recognition in bank swallows: II. Development and acoustic basis. *Animal Behaviour,* 29, 95–101.

Beecher, M. D., I. Beecher, & S. Lumpkin. (1981). Parent-offspring recognition in bank swallows: I. Natural history. *Animal Behaviour,* 29, 86–94.

Beecher, M. D., B. Medvin, P. Stoddard, & P. Loesch. (1986). Acoustic adaptations for parent-offspring recognition in swallows. *Experimental Biology,* 45, 179–193.

Blaustein, A. (1983). Kin recognition mechanisms: phenotype matching or recognition alleles. *American Naturalist,* 121, 749–754.

Browne, J. (2002). *Charles Darwin: The Power of Place.* New York: Knopf.

Brumm, H., & H. Slabbekoorn. (2005). Acoustic communication in noise. *Advances in the Study of Behavior,* 35, 151–209.

Darwin, C. (1859). *On the Origin of Species.* London: J. Murray.

Dawkins, R. & J. Krebs. (1979). Arms races between and within species. *Proceedings of the Royal Society of London, B,* 205, 489–511.

Dugatkin, L. A. (2007). Inclusive fitness theory from Darwin to Hamilton. *Genetics,* 176, 1375–1380.

Dugatkin, L. A. (2013). *Principle of Animal Behavior.* Third edition. W. W. Norton: New York.

Elgar, M. & B. Crespi (eds.). (1992). *Cannibalism: Ecology and Evolution among Diverse Taxa.* Oxford, UK: Oxford University Press.

Frommen, J. G., C. Luz, & T. C. M. Bakker. (2007). Kin discrimination in sticklebacks is mediated by social learning rather than innate recognition. *Ethology,* 113, 276–282.

Grafen, A. (2007). Detecting kin selection at work using inclusive fitness. *Proceedings of the Royal Society of London, B,* 274, 713–719.

Hamilton, W. D. (1963). The evolution of altruistic behavior. *American Naturalist,* 97, 354–356.

Hamilton, W. D. (1964). The genetical evolution of social behaviour. I and II. *Journal of Theoretical Biology,* 7, 1–52.

Hamilton, W. D. (1972). Altruism and related phenomena, mainly in social insects. *Annual Review of Ecology and Systematics,* 3, 192–232.

Holmes, W. & P. W. Sherman. (1983). Kin recognition in animals. *American Scientist,* 71, 46–55.

Holmes, W. G. (2004). The early history of Hamiltonian-based research on kin recognition. *Annales Zoologici Fennici,* 41, 691–711.

Hoogland, J. & P. W. Sherman. (1976). Advantages and disadvantages of bank swallow coloniality. *Ecological Monographs*, 46, 33–58.

Hughes, W. O. H., B. P. Oldroyd, M. Beekman, & F. L. W. Ratnieks. (2008). Ancestral monogamy shows kin selection is key to the evolution of eusociality. *Science*, 320, 1213–1216.

Jouventin, P. & T. Aubin. (2002). Acoustic systems are adapted to breeding ecologies: Individual recognition in nesting penguins. *Animal Behaviour*, 64, 747–757.

Jouventin, P., T. Aubin, & T. Lengagne. (1999). Finding a parent in a king penguin colony: The acoustic system of individual recognition. *Animal Behaviour*, 57, 1175–1183.

Lengagne, T., T. Aubin, P. Jouventin, & J. Lauga. (2000). Perceptual salience of individually distinctive features in the calls of adult king penguins. *Journal of the Acoustical Society of America*, 107, 508–516.

Manning, C. J., D. A. Dewsbury, E. K. Wakeland, & W. K. Potts. (1995). Communal nesting and communal nursing in house mice, *Mus musculus domesticus*. *Animal Behaviour*, 50, 741–751.

Manning, C. J., E. K. Wakeland, & W. K. Potts. (1992). Communal nesting patterns in mice implicate MHC genes in kin recognition. *Nature*, 360, 581–583.

Ortega, C. (1998). *Cowbirds and Other Brood Parasites*. Tucson, AZ: University of Arizona Press.

Packer, C., S. Lewis, & A. Pusey. (1992). A comparative analysis of non-offspring nursing. *Animal Behaviour*, 43, 265–281.

Pfennig, D. W., H. K. Reeve, & P. W. Sherman. (1993). Kin recognition and cannibalism in spadefoot toads. *Animal Behaviour*, 46, 87–94.

Pruett-Jones, S. G. & M. J. Lewis. (1990). Sex ratio and habitat limitation promote delayed dispersal in superb fairy-wrens. *Nature* 348, 541–542.

Ratnieks, F. L. W. & H. Helantera. (2009). The evolution of extreme altruism and inequality in insect societies. *Philosophical Transactions of the Royal Society of London, B*, 364, 3169–3179.

Reeve, H. K. (1989). The evolution of conspecific acceptance thresholds. *American Naturalist*, 133, 407–435.

Searby, A., P. Jouventin, & T. Aubin. (2004). Acoustic recognition in macaroni penguins: An original signature system. *Animal Behaviour*, 67, 615–625.

Sherman, P. W. (1977). Nepotism and the evolution of alarm calls. *Science*, 197, 1246–1253.

Sherman, P. W. (1981). Kinship, demography, and Belding's ground squirrel nepotism. *Behavioral Ecology and Sociobiology*, 8, 251–259.

Sherman, P. W. & W. Holmes, (1985). Kin recognition: Issues and evidence. *Fortschritte der Zoologie*, 31, 437–460.

Waldman, B. (1987). Mechanisms of kin recognition. *Journal of Theoretical Biology*, 128, 159–185.

Wright, S. (1922). Coefficients of inbreeding and relationship. *American Naturalist*, 56, 330–338.

12
Evolutionary History of Behavior

Terry J. Ord

BEHAVIORAL ECOLOGY MEETS PALEONTOLOGY AND PHYLOGENETICS

Uncovering the past history of animal behavior seems impossible; unless we can observe behavior directly, it would appear there is no way to know what the behavior of an extinct animal might have been like. Direct observation seems an obvious prerequisite for the study of animal behavior, but *is it really necessary? Will we never know how the dinosaurs interacted with one another and their environment without a time machine or a way to resurrect them? Can we say nothing about the behavior of the evolutionary ancestors of animals that are alive today?*

Science is about **deduction** and **inference**. Direct observation of the phenomena under study is an advantage, but it is by no means a necessity. Granted, there are some things we will never know about the behavior of extinct animals without direct observation. But much can still be learned about the evolutionary history of animal behavior through some clever detective work and modern scientific methods.

There are two ways the history of behavior can be studied: combining fossil evidence with information from contemporary analogues of similar species alive today, and the reconstruction of past history using phylogenies and the comparative method. The first approach extrapolates the **behavioral ecology** of extinct species by exploiting what we know about the behavior of living species. Behavior is often reflected in the morphology of an animal and other characteristics that can leave traces in the fossil record (e.g., track marks; Mazin et al.,

2009). These preserved characteristics can therefore be used to similarly infer the probable behavior of prehistoric animals. The second approach is similar to the first in the sense that it also uses information on the behavior of living species, but it relies on *phylogenetics* rather than fossil evidence to infer the evolutionary history of behavior. By comparing the behavior of species that are phylogenetic close relatives it is possible to reconstruct the likely behavior of the common ancestor of those species. As a basic example, if a set of closely related species hunted in small groups, then it is reasonable to assume that the common evolutionary ancestor of those species also hunted in groups. This *phylogenetic comparative method* can be taken even further to investigate the factors that might have influenced the initial evolution of particular behaviors.

Combining fossils with information from living species and the application of the phylogenetic comparative method has led to many important discoveries about the evolutionary history of behavior. These discoveries have in turn helped behavioral scientists understand why animals in existence today behave the way they do and how animals have adapted to their environment. The latter is especially important for knowing how animals respond to environmental change, which has important implications for conservation. In the following sections, examples are presented that illustrate some of the ways scientists have studied the history of animal behavior and what has been learned about how animals respond to natural selection and other evolutionary forces.

LINKING FOSSILS TO THE BEHAVIOR OF LIVING SPECIES

Questions that often arise when contemplating the behavior of extinct species are how social they might have been and whether their behavioral ecology contributed to their ultimate demise. It is understandable why animal *sociality* is an especially evocative topic to us because we are a highly social species ourselves. Sociality implies richness in the lives of animals that in itself is fascinating. But the social behavior of animals is also important in shaping the evolution of species, from how they reproduce to how they find their food. Animal sociality is therefore a major focus of research in behavioral ecology. Knowing the constraints on an animal's behavior can also help us understand why some species and not others might have gone extinct. This is important to know so we can predict how animals today will cope with environmental change resulting from human impacts and global shifts in climate.

What evidence do we have for social behavior in extinct species, and how do we know what aspects of behavior might have contributed to species extinction?

Social Sabertooths and Migratory Mastodons

Sabertooth cats (*Smilodon fatalis*) were about the size of the largest tigers in existence today and roamed North America during the late Pleistocene (1.6 million to 10,000 years ago). Fossils reveal powerful forelimbs and massive canines, indicating a formable predator. Classically, sabertooths were believed to have been solitary, like today's cheetahs or American mountain lions. But the large number of fossils clumped together in the Californian tar seeps of Rancho La Brea is at odds with this "lone hunter" lifestyle. When fossils of the same species are found clustered together, it is often taken as evidence of sociality (Ladevéze et al., 2011). For the sabertooth, however, things were ambiguous. The consensus among paleontologists was that sabertooths were attracted to the tar seeps by the sounds of struggling prey caught in the tar and themselves ended up becoming trapped. Debate, however, centers on whether sabertooths were attracted separately or in groups. There were two competing hypotheses. The first hypothesis was that sabertooths were solitary and attracted individually to the tar seeps. The second hypothesis was that sabertooths were social—that is, hunted in packs—and were attracted to the tar seeps as groups.

Testing these hypotheses presented a challenge. The number of sabertooth cats recovered from the tar seeps was the only information ***paleontologists*** had of what occurred in the Pleistocene. By itself, this information gives only the final outcome, not how that outcome came about. The breakthrough came when paleontologists teamed up with several experts on the behavior of present-day large carnivores (Carbone et al., 2009). These behavioral ecologists had conducted many field surveys of African carnivores by "calling in" predators through playing audio recordings of distressed prey. The researchers argued that these playbacks, which were originally designed to estimate carnivore abundance, were analogous to the sounds of prey caught in the La Brea tar seeps (Van Valkenburgh et al., 2009). That is, the types of African carnivores attracted by the playbacks were, in effect, a real-world simulation of events at the La Brea tar seeps in the Pleistocence.

The behavioral ecologists knew from direct observation and other studies which carnivores attracted to playbacks were solitary hunters (e.g., cheetahs [*Acinonyx jubatus*] and leopards [*Panthera pardus*]) and which hunted in groups (e.g., wild dogs [*Lycaon pictus*] and lions [*Panthera leo*]). Their data showed clear differences in the proportion of solitary and social predators attracted to playbacks (Figure 12.1). With this information in hand, the paleontologists reevaluated the number of sabertooth cats attracted to the tar seeps relative to other types of animals. If sabertooth cats were solitary, as traditionally assumed, their numbers were vastly overrepresented in the La Brea tar

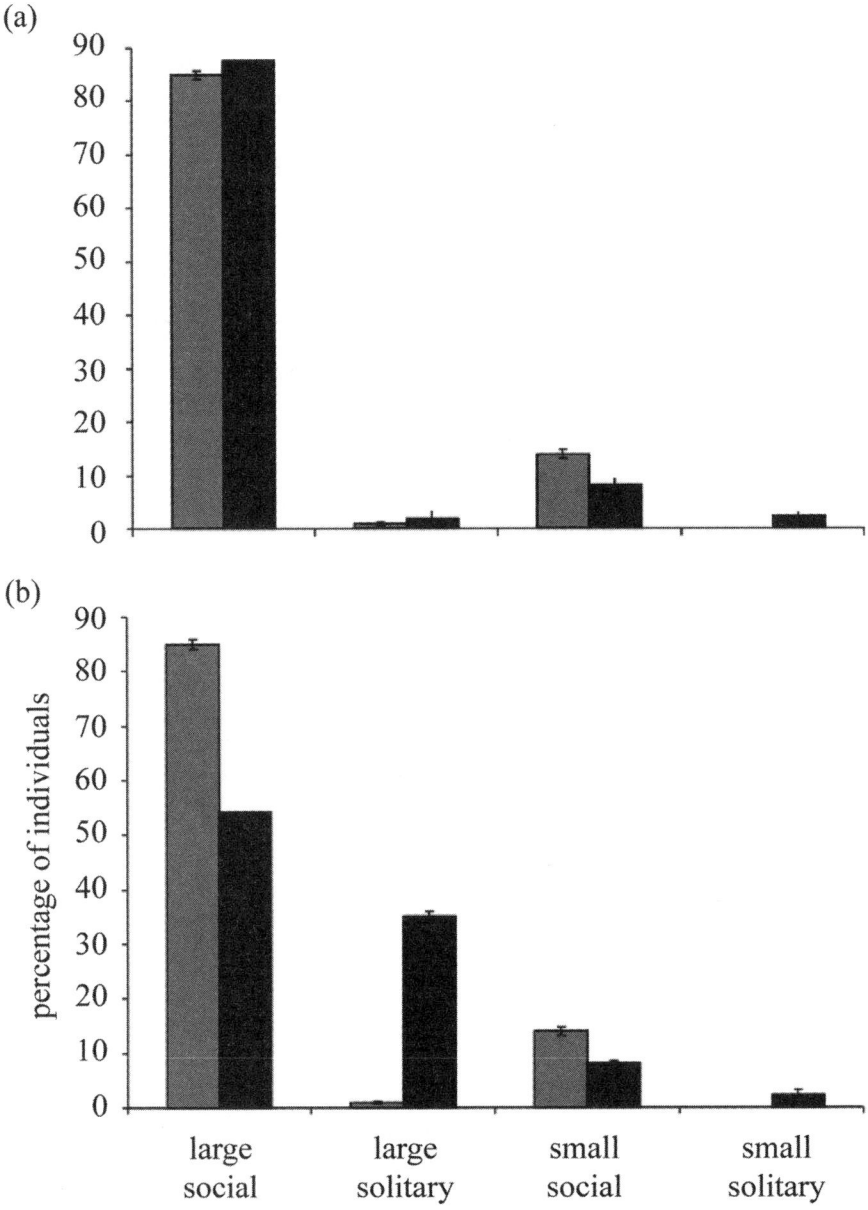

Figure 12.1. Shown in gray are the percentages of African carnivores, grouped by body size and sociality, attracted to audio playbacks of distressed prey animals, which were heavily skewed towards large, pack-hunting predators. In the upper panel (a) shown in black are the percentage of fossilized remains found in the La Brea tar seeps based on the assumption that sabertooth cats hunted in groups, while the lower panel (b) shows the percentage of fossils based on the assumption that sabertooth cats were solitary. Error bars are 95 percent confidence intervals. (This plot is reprinted from Carbone et al., 2009, by permission of Highwire Press.)

seeps according to the number of living solitary predators attracted to audio playbacks (Figure 12.1b). By contrast, if sabertooth cats were social, then their numbers were almost exactly the number predicted by the proportion of group-hunting predators attracted to playbacks (Figure 12.1a). While this data cannot provide conclusive proof, it does offer compelling support for the social hypothesis.

The study of sabertooth cats was possible because there were numerous surviving analogues—for instance, cheetahs, leopards, and lions—that differed in sociality and could be used to deduce the possible behavior of the sabertooth based purely on their abundances in the La Brea tar seeps. But *what if few (or no) contemporary analogues exist for comparison?* This was a problem faced by paleontologists interested in the behavior of mastodons. These massive creatures were related to present-day elephants in the sense that they both belong to the same order of mammals (Proboscidea) and share some similarities in morphology. But the similarities are generally few, and mastodons and elephants were certainly not as genetically related as the sabetrooth was to today's big cats (Janczewski et al., 1992; Rohland et al., 2010). Furthermore, the elephants represented the only living analogue of mastodons, so even if they were phylogenetic close relatives, the comparison is limited to one. So when paleontologists wished to investigate the probable migration patterns of mastodons, an alternative approach had to be found.

The question of mastodon migration was a topic of interest to scientists because it might reveal why these large mammals went extinct 11,500 years ago (Hoppe et al., 1999). Mastodons lived in North America alongside another iconic species, the mammoth, which were close relatives to elephants (Rohland et al., 2010). Several hypotheses had been proposed about why mastodons and mammoths suddenly disappeared after having existed for millions of years. One hypothesis related to the impacts of climate change on the environment and whether the behavioral ecology of mastodons and mammoths hindered their ability to cope with the accompanying ecological changes.

We know from many species today that migration allows animals to escape freezing winters or scorching summers and exploit seasonally fluctuating resources at different locations. A spectacular example is the annual migration of thousands of Serengeti wildebeest (*Connochaetes taurinus*) that walk hundreds of kilometers south at the onset of the dry season to follow rainfall and shifting food resources (Boone et al., 2006). To discover the annual movement of mastodons, paleontologists took a novel approach and examined the fossilized remains of their teeth.

In living animals, the isotope ratio in the chemical element strontium found in the enamel of teeth reflects the isotope signature of ingested food.

In the case of plant foods, strontium isotope ratios depend on soil type and vary geographically. Put simply, by measuring the strontium isotopes of mastodon teeth, it was possible to retrace where the animals had been foraging during their lifetime. If mastodons uncovered in southern Florida had migrated to escape the winter from as far north as the Appalachian Mountains in Georgia, then the isotope signatures of their teeth should reflect the differences in isotopes present in the vegetation from the two regions. This is precisely what the researchers found (Hoppe et al., 1999). Isotope signatures revealed frequent mastodon migration over hundreds of kilometers between the Appalachians and southern Florida. In contrast, the isotope signature of mammoth teeth also uncovered in Florida indicated that these animals did not range nearly as far and were not, as previously hypothesized, migratory (Hoppe et al., 1999).

While the migration of mastodon implies that these animals should have been able to buffer themselves against the ecological changes resulting from climate shifts better than mammoths, recent studies on the migration of living species suggested this might not have been the case. Even subtle regional changes in temperature have lead to the mismatch of migration events with seasonal fluctuations, placing the survival of some species in jeopardy (e.g., Saino et al., 2011). It is still unclear the extent to which migration in mastodons and its absence in mammoths contributed to their extinction, but new data on other species—both extinct and living—should help resolve this question. It appears, though, that climate change and whether or not species migrated was not the primary contributing factor for the extinction of the North American megafauna at the end of the Pleistocene (Ripple & Van Valkenburgh, 2010).

Fatherly, Musical Dinosaurs

The repeated discovery of the remains of adult dinosaurs alongside fossilized egg clutches gives a strong indication that parental care was an important component of the behavior of some dinosaurs (Norell et al., 1995). Whether animals provide parental care, and especially who provides parental care, reflects the mating system of species. In birds, males often provide all or part of the parental care (e.g., incubating eggs, feeding hatchings), whereas females are generally the sole caregivers in mammals and in the few reptiles like the crocodile that exhibit parental care. *Were dinosaurs like crocodiles, in which females provided all care, or more like birds, in which males provided care? If dinosaurs were like birds, was vocal communication an integral component of their social behavior as it is in birds? What did those calls sound like?*

In living species, the size of egg clutches reflects the type of parental care exhibited by species. When males provide all care, larger clutches are maintained than if only females provide care (Varricchio et al., 2008). This is presumably because males do not incur the considerable cost associated with egg production and can invest more in incubating eggs at the expense of feeding. Paleontologists used the relationship of parental care type and clutch size to sex the caregivers in the egg-nesting Cretaceous troodontid and oviraptorid dinosaurs. Based on the size of egg clutches, these dinosaurs were most consistent with a system of exclusive male parental care and not *biparental* or *maternal care* (Varricchio et al., 2008).

Further evidence for male care came from a closer examination of the bones associated with the fossilized egg nests. Female birds and crocodiles leach large amounts of calcium and phosphorus from their tissues during egg production, and this leaves telltale signs in the histology of their bones. Cross sections of fossilized bones confirmed the adult dinosaurs associated with nests were male; there was no evidence of calcium and phosphorus leaching (Varricchio et al., 2008). Taken together, not only were some dinosaurs building nests and incubating eggs (Norell et al., 1995), but paleontologists have been able to determine that parental care was most likely provided by males and not females, pushing back the origin of ***paternal care*** to before the evolution of birds.

Comparisons of behavior among living species have also been able to reconstruct whether dinosaurs communicated vocally and what those calls might have sounded like. The length of the cochlear—or inner ear—correlates closely with the hearing sensitivity of species and can be used to predict the frequency range, the mean frequency, and even the complexity of vocal calls produced by animals (Walsh et al., 2009). The cochlear has been preserved in several fossils, for example the *Archaeopteryx*, a bird-like precursor from the Late Jurassic. Using the equation derived from the statistical relationship between the length of the cochlear and the vocal characteristics of living animals, researchers have argued that *Archaeopteryx* had a vocal repertoire very similar to the present-day emu (*Dromaius novaehollandiae*) (Walsh et al., 2009). This also implies a reasonable complex social life for *Archaeopteryx* because the complexity of communication tends to reflect the level of social complexity in species (Freeberg et al., 2012).

Prehistoric Polygyny

It is common in nature for males to compete aggressively among themselves for access to females. And in aggressive competition, size matters. Large males win more fights, mate with more females, and subsequently produce more offspring. The selective advantage of large males in contests typically leads to the evolution of increasing ***sexual size dimorphism*** in species. That is, males become increasingly larger than females over evolutionary time. If body size is strongly skewed towards males in fossils, then this is a strong indication that males probably competed aggressive with one another in a ***polygamous*** mating system (***monogamous*** species are typically ***sexually monomorphic*** in size). For example, the sexual size dimorphism of fossils has been used to infer that strong male-male competition and ***polygyny*** existed in the extinct relatives of present-day marsupials from the early Palaeocene (Ladevéze et al., 2011). This implies that the solitary nature of many marsupials today has resulted from a loss of sociality rather than being ancestral as initially assumed.

Which male a female chooses to mate with contributes not only to the evolution of male size—females preferentially mate with large males over small males—but also the evolution of male ornamentation. ***Sexual ornaments*** in males are common in nature and provide valuable cues to females on the quality of a male as a potential mate. This is because only males in top condition can incur the energetic and developmental costs associated with the possession of a large ornament. Classic examples of female-driven male ornamentation are the elaborate plumages of many male birds (e.g., peacock [*Pavo cristatus*] trains). Others include conspicuous rostral appendages, such as fleshy horns, and large throat fans in lizards.

The presence of ornaments in an extinct animal, especially if that animal were confirmed as male, would imply a polygamous mating system. It would also provide strong evidence that females were highly selective about which males they chose to mate with. In present-day animals, ornaments are also often associated with elaborate ***courtship displays***. Determining whether extinct animals had ornaments is helped if the features preserved in fossils are similar in appearance to confirmed ornaments in species today. Unfortunately, common ornaments like feather plumages or fleshy structures are rarely left in the fossil record (*Archaeopteryx* provides an unusually clear example of feather ornamentation). But there are also other, often more bizarre structures preserved in fossils that paleontologists speculate might have also functioned as ornaments. Yet there are also frequently several plausible explanations these structures as well.

In some instances, it has been possible to test alternative hypothesis for the function of elaborate morphological structures (Tomkins et al., 2010). *Pteranodon* were large flying pterosaurs with wingspans of many meters found in North America in the Late Cretaceous, some 85 million years ago. Fossils show these pterosaurs had large, prominent crests on their head (Figure 12.2), which were especially exaggerated in males (i.e., the crests were sexually size dimorphic). Several hypotheses for what these crests might have been used for included a rudder to facilitate flying, a heat-dissipating or -absorbing device to aid thermoregulation, or an ornament used to attract mates. Stranger still

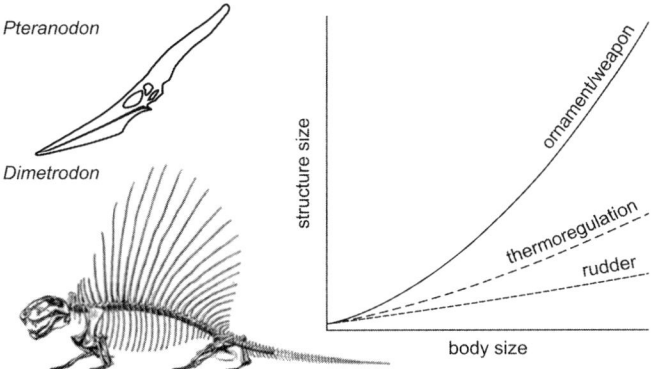

Figure 12.2. Illustrations of the head crests of the flying *Pteranodon* and dorsal sail of *Dimetrodon* alongside predicted allometric slopes if crests or sails functioned as ornaments, as devices for thermoregulation, or as rudders (in *Pteranodon*). The estimated allometry of head crests was $\times^{6.94}$, while for dorsal sails it was $\times^{1.73}$; these estimates were consistent with the allometry of ornaments. (Sketches of *Pteranodon* and *Dimetrodon* are reprinted from Tomkins et al., 2010, by permission of University of Chicago Press.)

were the massive sails found on the backs of *Dimetrodon* and *Edaphosaurus* dinosaurs (Figure 12.2; these animals provided the inspiration for many a 1950s Hollywood monster movie). These large predators lived during the late Carboniferous and early Permian around 300–260 million years ago in North America and Europe. The classic hypothesis was these large sails were structures used in thermoregulation. But it was also possible that they might have been ornaments.

Behavioral ecologists who have studied male ornamentation in living species have discovered the size of ornaments commonly exhibit **positive allometry** (Kodric-Brown et al., 2006). Positive allometry occurs when larger individuals have disproportionately larger structures—like ornaments—compared to smaller individuals. This allometric pattern is believed to happen whenever ornaments are costly to produce. Larger animals are better able to bear the costs of having a large ornament than smaller animals, leading to a disproportionate increase in ornament size with overall body size. Behavioral ecologists decided to use this phenomenon to test whether *Pteranodon* head crests and *Dimetrodon* and *Edaphosaurus* sails were ornaments (Tomkins et al., 2010). The researchers also tested the alternative hypotheses relating to rudders and thermoregulation, which biophysics predicted would have their own unique relationships with body size (Figure 12.2).

By comparing the size of head crests and sails in fossils for a range of individuals varying in body size, the rudder and thermoregulation hypotheses were rejected by the magnitude of the allometric slopes (it was too large; Figure 12.2). But these allometric slopes were well within the range for what would be expected for ornaments (Tomkins et al., 2010). By extension, these apparent ornaments—head crests and dorsal sails—implied that male *Pteranodon*, *Dimetrodon*, and *Edaphosaurus* dinosaurs competed among themselves for mating opportunities with females and that those females probably exerted a strong preference for males with the largest ornament.

COMPARE AND CONTRAST TODAY TO RECOVER BEHAVIORS PAST

Carl Linnaeus created a classification scheme based on a nested hierarchy of shared characteristics among organisms, a scheme that remains at the heart of modern taxonomy. This scheme exploited the common observation of naturalists that groups of species often shared physical characteristics. Darwin argued a hundred years after Linnaeus that such shared features reflected the shared ancestry among species (Darwin, 1859). He expanded on this idea to formulate the theory of evolution, encapsulated by the notion of "descent with modification": descendent species inherit features from evolutionary ancestors, with those features becoming modified over evolutionary time through natural selection. Today we understand evolution as a process that is more complex than Darwin could have appreciated in his day. But this idea of common ancestry and its influence on the features expressed by animals—and this includes an animal's behavior—also provides a powerful way to uncover the probable behavior of evolutionary ancestors.

By comparing the similarities and differences among phylogenetic close relatives, it becomes possible to map behaviors onto a phylogeny and retrace their likely origin. In this section, we review examples of some remarkable animal behavior and how comparative biologists have uncovered the evolutionary history of those behaviors with the aid of modern phylogenetic methods.

Ancient Squirrels Exploited Smelly Predators

For a number of North American ground squirrel species, rattlesnakes pose an acute threat to the survival of offspring. At dusk, rattlesnakes use smell to locate burrows in which squirrel pups are sheltering. Adult ground squirrels have evolved a number of strategies to reduce the likelihood of their pups being eaten, such as mobbing a snake before it enters a burrow. Mobbing is meant to harass the snake so much that it is forced to give up hunting and leave the area. It is not always successful.

Some ground squirrel species, and in particular the females and juveniles of those species, add another defense: they smear themselves with the chewed-up remains of shed rattlesnake skins (rattlesnakes molt frequently). This behavior was puzzling to researchers at first. It seemed that the squirrels were willingly covering themselves with the scent of their main predator. Experiments later revealed that this anointing behavior helped to minimize predation. Rattlesnakes avoided the burrows of squirrels that had recently anointed themselves with rattlesnake sheds (Clucas et al., 2008). The scent-application behavior was a novel antipredator strategy. The scent of the rattlesnake not only hid the squirrel's own smell but also gave the false impression that a squirrel's burrow had already been visited by a rattlesnake.

But *what was the evolutionary history of this extraordinary anointing behavior? Did anointing evolve multiple times independently in each squirrel species or just once early in the history of the squirrel family and was retained in those species performing the behavior today?*

An extensive study of different squirrel species throughout North America and Mexico was undertaken to determine which species performed the anointing behavior and which species did not (Clucas et al., 2010). This survey in itself led to some fascinating findings. It revealed that squirrels not only anointed with rattlesnake scent but weasel scent as well; weasels were another key predator of ground squirrel pups in some areas of the Americas. That is, the behavior was not specific to rattlesnakes but general to predators using smell to hunt for squirrel pups.

The researchers then used the phylogenetic relationships among squirrel species to reconstruct the evolutionary history of predator scent application. The behavior was ancient, evolving once some 28 million years ago (Figure 12.3) and possibly even as far back as 75 million years ago (Clucas et al., 2010). Yet, in another twist to the story, the earliest fossils of rattlesnakes and weasels were roughly 15 million years ago. This meant that scent application evolved well before rattlesnakes and weasels even existed. It must therefore have evolved as an antipredator strategy to some other, now extinct predator that also relied on smell to hunt for prey. A good candidate was the ancient North American boa snake that existed during the Oligocene (Clucas et al., 2010).

Further analysis of the relationship between the overlap of rattlesnakes and weasels with living ground squirrels gave other insights as well. Squirrel species that no longer faced predation from rattlesnakes had lost the scent-application behavior. In a couple of cases, squirrels have since come back into contact with rattlesnakes but have not reevolved the behavior. It seems then that applying predator scent was easily lost in the absence of predation, was a highly unusual event in the prehistory of the squirrel family, and has not been repeated since.

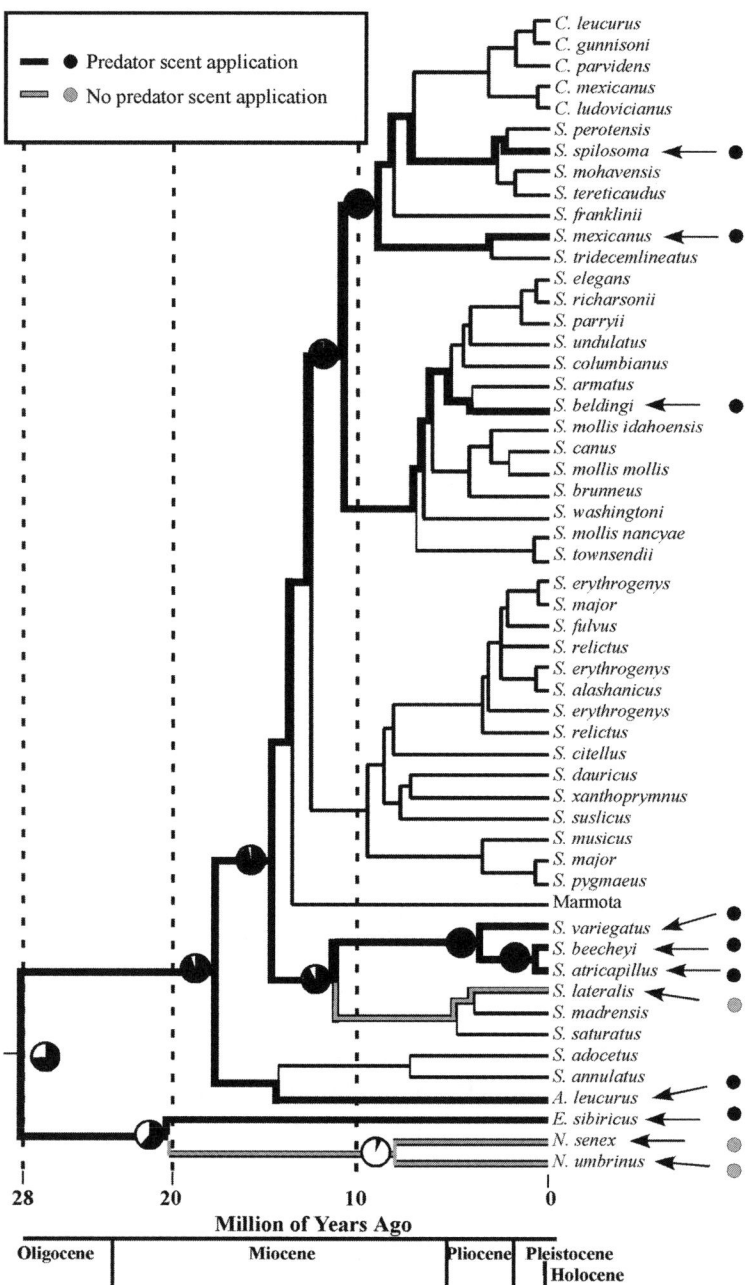

Figure 12.3. Ancestor reconstructions of the predator-scent-application behavior of North American ground squirrels. Filled lines on the phylogeny depict reconstructions using parsimony; pie charts depict reconstructions using maximum likelihood. The phylogeny of scent application indicates that the behavior evolved at least 28 million years ago. Dots at the tips of the phylogeny indicate which species did and did not anoint themselves with predator scent when researchers placed scent out in the environment for squirrels to inspect. (This figure is reprinted from Clucas et al., 2010, by permission of John Wiley & Sons.)

The ground squirrels are a wonderful example of how field studies of behavior on a select number of species and experiments in the laboratory documenting the adaptive significance of initially puzzling behavior can be integrated with evidence from phylogeny and paleontology to show the history of animal behavior. In doing so, researchers were able to not only date the origin of behavior but also show how unusual the evolution of such an odd behavior like predator-scent application might have been.

Reconstructing Past Mating Calls

What the vocal communication of extinct species might have sounded like, such as the calls of the *Archaeopteryx*, has been inferred from the shape of the cochlear preserved in fossils (see "Fatherly, Musical Dinosaurs" above). But there is another way scientists can reconstruct the calls of ancestral species. By exploiting detailed information on the call structure of living species and the phylogenetic relationships of those species, it has been possible to recreate the calls of evolutionary ancestors.

In a remarkable study on Central American túngara frogs (*Physalaemus* spp.) (Ryan & Rand, 1995), phylogenetic methods were used to reconstruct the evolutionary history of mating calls (Figure 12.4). Male frogs in this species group produce a whine to attract females. Each species has a variant on this call, and it has been assumed from this species variation that the whine was also important in species recognition. A female needs to recognize males of her own species as potential mates; otherwise she is in danger of wasting her efforts with a male from the wrong species.

Once researchers had reconstructed the ancestor calls, they synthesized the calls on a computer and played them back to females of living species in mate-choice experiments. The researchers wanted to test how female responses to male calls had changed over evolutionary time.

Females responded to both the calls of conspecific males and those of males from their immediate evolutionary ancestors. This showed that female preferences were not especially tuned to the specific call of males from their own species. Female responses did drop off, however, as evolutionary ancestors became older; the longer females had been separated from ancestors, the longer their preferences have had time to change.

The study indicated in a novel way that changes in male calls were generally accompanied by shifts in female preferences. The match was not perfect, and this was interesting. It revealed that female responses were not the driving force behind changes in the structure of male calls, as would be expected if the differences in male whines among species today were the product of the need for accurate species recognition by females. Rather, shifts in the structure

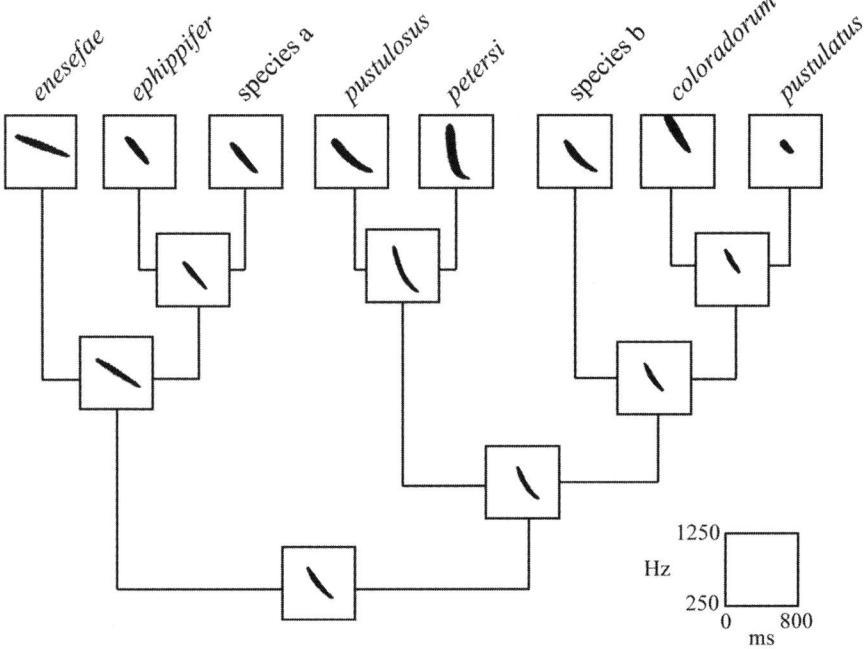

Figure 12.4. Ancestor reconstructions of the mating call of male túngara frogs of the genus Physalaemus. Shown are the sonograms of calls; those reconstructed at phylogenetic nodes were synthesized by researchers and played back to females of living species. (This figure is reprinted from Ryan and Rand, 1995, by permission of the American Association for the Advancement of Science.)

of the male whine were either generated from random mutation, or **genetic drift**, or in response to some selection pressure other than female preference (e.g., properties of the acoustic environment, male competition, or predation). The fact that female preferences did generally track the evolution of male calls showed that the **coevolution** of male calls and female preferences has still been an important aspect of how communication has evolved in the group but just that female preference has not been the engine of change in male calls.

Follow-up studies have since shown this coevolution between signal and receiver extends beyond the whine component of the mating call. It includes the elaboration of this call with a series of chucks added to the end of the call. Not all species add these chucks, but the more complex call that it creates is strongly preferred by females over just the whine by itself. Early reports using phylogenetic reconstructions suggested that the preference for chucks predated the evolution of the chuck (Ryan & Rand, 1993), perhaps because of a latent sensory bias for complex acoustic stimuli rather than a specific preference for the chucks themselves. This was an exciting prospect because it suggested that sensory biases in receivers could have had an important influence on the evolution of the chuck component in the mating call. There has since been a reexamination of this sensory bias hypothesis using a more detailed phylogeny and broader species survey of female responses to calls that included chucks (Ron, 2008). This later study found no evidence for females preferring calls with chucks before the chuck evolved. Instead, the chuck produced by males and the preference for it in females had coevolved in a similar manner as the whine component of the call (Ryan & Rand, 1995).

This provides an important cautionary note: methods used to extrapolate the evolutionary history of behavior are dependent on the data available at the time the analysis is done. If these data are updated, for example if new information comes to light, conclusions can and should be revised. This should always be kept in mind when considering evidence from fossils or phylogenetic comparative studies.

The Origin of Caterpillar Communication

While Charles Darwin is best known for his ideas on evolution outlined in what are now classic works such as *The Origin of Species* (1859), he was also interested in the origin of social communication. In his book *The Expression of the Emotions in Man and the Animals* (1872) he pondered how some forms of communication might have originally evolved from the grunts of physical exertions or posturings that occurred during aggressive contests among animals. Early ethologists like Niko Tinbergen (1952) and Konrad Lorenz (1966) explored this idea further and postulated that activities not initially associated with communication become social signals through a process of ritualization. For example, the aggressive head-bob displays of territorial lizards might have initially evolved from moving the head up and down to aid depth perception and help gauge the distance of territory intruders before launching an attack. Behaviors like head-bobs that provide cues on an animal's intentions were hypothesized to evolve into social signals through a process in which the behavior becomes simplified and exaggerated in structure and then repeated in a stereotyped sequence. Testing whether this process of ***ritualization*** explains the evolution of communication has proven difficult. It requires that the initial precursors of a communicative behavior still be in existence today alongside the very signals they are believed to have evolved into. Opportunities to study ritualization are therefore quite limited.

One of the best examples is the territorial vibration signal of caterpillars (Scott et al., 2010). Many caterpillars build shelters out of silk and leaves.

Individuals defend these shelters against interlopers that have either lost or failed to build shelters of their own. In some species, individuals advertise ownership of a leaf shelter using complex vibration signals. The signals consist of an elaborate sequence of leaf scraping using a specialized hardened "oar" on the abdomen and rapid drumming of the mandibles against the leaf surface (Scott et al., 2010). On closer examination, researchers found the abdomen-scraping part of the signal had the same sequence of movements as the crawling cycle of the caterpillar along the leaf. This suggested this part of the signal was derived from crawling behavior.

Researchers then looked at caterpillar species that did not perform vibration signals (Scott et al., 2010). In these nonsignaling caterpillar species, shelter owners are on alert for the vibrations of a potential intruder crawling along the branch. When detected, shelter owners start whipping violently backwards and forwards across the leaf in an attempt to knock the intruder off the leaf. The intruder can feel the vibrations of this flaying behavior and sometimes retreat before reaching the shelter. In many cases, this does not happen, and shelter owners must fight it out with the intruder. (In signaling caterpillars, individuals almost never came to blows, and disputes were resolved through the exchange of vibration signals.) Through careful analysis of the vibrations of the flaying behavior and the vibrations produced by territorial signaling, the researchers found the stylized mandible drumming of the leaf surface in the signal had striking similarities with the flaying defensive behavior.

Through phylogenetic reconstructions and detail species comparisons, it became apparent there had been a progressive ritualization of the crawling and flaying behavior over evolutionary time to produce the synchronized vibrations used by signaling species today. Signal movements were repeated in long bouts, highly stereotyped, and far simpler and more exaggerated in structure than the sporadic movements associated with crawling and flaying defensive behavior (Figure 12.5). This met all the requirements of a ritualized behavior: simplification, exaggeration, repetition, and stereotypy.

There are other possible examples of the ritualization of noncommunicative behaviors into social signals, such as the foot-drumming displays of some mammals, but formal tests are few and difficult to perform. The study of caterpillar communication represents a particularly elegant and rare confirmation of the ritualized origins of a complex communication behavior.

THE DETECTIVE WORK OF EVOLUTIONARY HISTORY

A common theme in evolutionary research is the integration of multiple methods from a range of different disciplines. This integrative study is a general attribute of animal behavior research, but it is especially conspicuous in

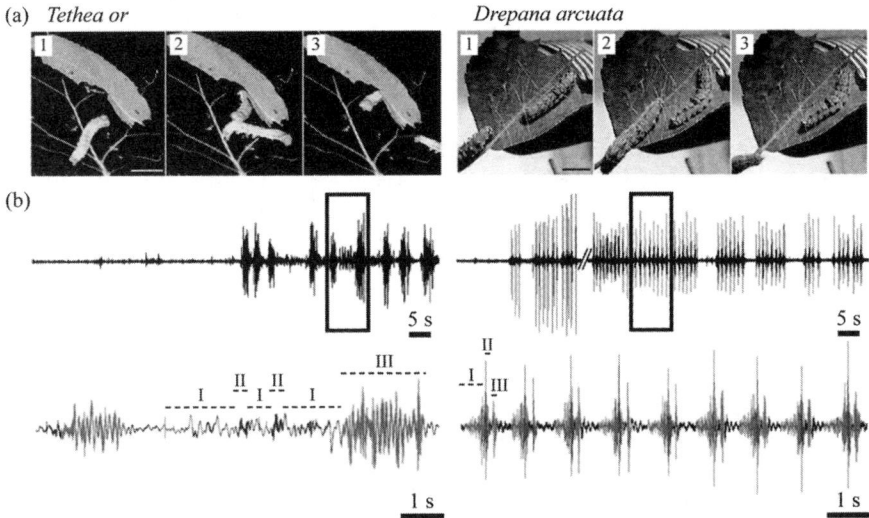

Figure 12.5. Shown in the upper panel (a) are video frames showing an encounter between a shelter owner and an intruder. On the left is a species that does not use territorial vibration signals, *Tethea or*, while on the right is a species that uses vibration signals, *Drepana arcuata*. In the lower panel (b), traces from a laser vibrometer show the vibrations produced by the flaying defensive behavior of the nonsignalling species on the left and the vibrations of the territorial signal from the signaling species on the right. The boxes show areas of the trace that are enlarged to show roman numerals corresponding to components that researchers found to be modified into the ritualized signal. (This figure is reprinted from Scott et al., 2010.)

evolutionary studies of behavior. This is because we are unable to observe the behavior of prehistoric animals directly, so we must infer it by combining various sources of evidence. This evidence comes in a variety of forms. The examples discussed in this chapter used information on morphological characteristics, the histology of bones, allometry, stable isotope signatures, the study of living animals, and statistical analyses that reconstructed changes in behavior into a phylogeny.

With careful consideration of the data that are available, it is possible to study the prehistory of behavior indirectly. Evolution on a smaller scale can also be studied using breeding experiments to identify the genes that regulate behavior and how they are inherited from one generation to the next. Other studies adopt the comparative approach among closely related species to investigate correlated evolutionary changes in behavior with the environment. For example, in the scent-applying squirrels, the hypothesis was that predation by rattlesnakes led to the evolution of squirrels anointing themselves with scent from shed rattlesnake skins. Another way the researchers examined this hypothesis was by testing whether there was a correlation between squirrel species that applied scent and geographic overlap with rattlesnakes. That is, squirrels that overlapped with rattlesnake ranges should apply scent, while squirrels that did not overlap with rattlesnakes, and therefore did not suffer rattlesnake predation, should not apply scent (given the opportunity when researchers artificially placed shed skins out in the environment for squirrels to inspect). A strong correlation was found and provided yet another piece of evidence that anointing behavior in squirrels was an adaptation to predation (Clucas et al., 2010).

To conclude, the imprint of evolutionary history is apparent in all types of behavior to a lesser or greater degree. By placing animal behavior in a historical context—whether it is through the study of fossils or comparisons among living species—scientists can understand the origin of behavior and in turn better interpret its present-day function (e.g., predator-scent application by squirrels or the role of frog calls in species recognition). This chapter has only briefly touched on the ways in which scientists have studied the evolution of animal behavior. More detailed reviews can be found in Ord (2010) and Ord and Martins (2010) for those readers who would like to delve deeper into this exciting area of research.

REFERENCES

Boone, R. B., S. J. Thirgood, & J. G. C. Hopcraft. (2006). Serengeti wildebeest migratory patterns modeled from rainfall and new vegetation growth. *Ecology*, 87, 1987–1994.

Carbone, C., T. Maddox, P. J. Funston, M. G. L. Mills, G. F. Grether, & B. Van Valkenburgh. (2009). Parallels between playbacks and Pleistocene tar seeps suggest sociality in an extinct sabretooth cat, *Smilodon*. *Biology Letters* 5, 81–85.

Clucas, B., T. J. Ord, & D. H. Owings. (2010). Fossils and phylogeny uncover the evolutionary history of a unique antipredator behaviour. *Journal of Evolutionary Biology*, 23, 2197–2211.

Clucas, B., D. H. Owings, & M. P. Rowe. (2008). Donning your enemy's cloak: Ground squirrels exploit rattlesnake scent to reduce predation risk. *Proceedings of the Royal Society of London, B*, 275, 847–852.

Darwin, C. (1859). *The Origin of Species*. London: Penguin Books.

Darwin, C. (1872). *The Expression of the Emotions in Man and the Animals*. London: John Murray.

Freeberg, T. M., R. I. M. Dunbar, & T. J. Ord. (2012). Social complexity as a proximate and ultimate factor in communicative complexity. *Philosophical Transactions of the Royal Society of London, B*, 367, 1785–1801.

Hoppe, K. A., P. L. Koch, R. W. Carlson, & S. D. Webb. (1999). Tracking mammoths and mastodons: Reconstruction of migratory behavior using strontium isotope ratios. *Geology*, 27, 439–442.
Janczewski, D. N., N. Yuhki, D. A. Gilbert, G. T. Jefferson, & S. J. O'Brien. (1992). Molecular phylogenetic inference from saber-toothed cat fossils of Rancho La Brea. *Proceedings of the National Academy of Sciences, USA*, 89, 9769–9773.
Kodric-Brown, A., R. M. Sibly, & J. H. Brown. (2006). The allometry of ornaments and weapons. *Proceedings of the National Academy of Sciences, USA*, 103, 8733–8738.
Ladevèze, S., C. de Muizon, R. M. D. Beck, D. Germain, & R. Cespedes-Paz. (2011). Earliest evidence of mammalian social behaviour in the basal Tertiary of Bolivia. *Nature*, 474, 83–86.
Lorenz, K. Z. (1966). Evolution of ritualization of behaviour in animals and man: Introduction. *Philosophical Transactions of the Royal Society of London, B*, 251, 273–284.
Mazin, J.-M., J.-P. Billon-Bruyat, & K. Padian. (2009). First record of a pterosaur landing trackway. *Proceedings of the Royal Society London, B*, 276, 3881–3886.
Norell, M. A., J. M. Clark, L. M. Chiappe, & D. Dashzeveg. (1995). A nesting dinosaur. *Nature*, 378, 774–776.
Ord, T. J. (2010). Phylogeny and the evolution of communication. In M. D. Breed & J. Moore (eds.), *Encyclopedia of Animal Behavior* (pp. 652–660). Oxford, UK: Academic Press.
Ord, T. J. & E. P. Martins. (2010). The evolution of behavior: Phylogeny and the origin of present-day diversity. In D. F. Westneat & C. W. Fox (eds.), *Evolutionary Behavioral Ecology* (pp. 108–128). New York: Oxford University Press.
Ripple, W. J., & B. Van Valkenburgh. (2010). Linking top-down forces to the Pleistocene megafaunal extinction. *BioScience*, 60, 516–526.
Rohland, N., D. Reich, S. Mallick, M. Meyer, R. E. Green, N. J. Georgiadis, A. Roca, & M. Hofreiter. (2010). Genomic DNA sequences from mastodon and woolly mammoth reveal deep speciation of forest and savanna elephants. *PLoS Biology*, 8, e1000564.
Ron, S. R. (2008). The evolution of female mate choice for complex calls in tungara frogs. *Animal Behaviour*, 76, 1783–1794.
Ryan, M. J. & A. S. Rand. (1993). Sexual selection and signal evolution: The ghost of biases past. *Philosophical Transactions of the Royal Society of London, B*, 340, 187–195.
Ryan, M. J. & A. S. Rand. (1995). Female responses to ancestral advertisement calls in tungara frogs. *Science*, 269, 390–392.
Saino, N., R. Ambrosini, D. Rubolini, J. von Hardenberg, A. Provenzale, K. Hüppop, O. Hüppop, A. Lehikoinen, E. Lehikoinen, K. Rainio, M. Romano, & L. Sokolov. (2011). Climate warming, ecological mismatch at arrival and population decline in migratory birds. *Proceedings of the Royal Society London, B*, 278, 835–842.

Scott, J. L., A. Y. Kawahara, J. H. Skevington, S.-H. Yen, A. Sami, M. L. Smith, & J. E. Yack. (2010). The evolutionary origins of ritualized acoustic signals in caterpillars. *Nature Communications*, 1, 4.

Tinbergen, N. (1952). "Derived" activities: Their causation, biological significance, origin, and emancipation during evolution. *Quarterly Review of Biology*, 27, 1–23.

Tomkins, J. L., N. R. LeBas, M. P. Witton, D. M. Martill, & S. Humphries. (2010). Positive allometry and the prehistory of sexual selection. *American Naturalist*, 176, 141–148.

Van Valkenburgh, B., T. Maddox, P. J. Funston, M. G. L. Mills, G. F. Grether, & C. Carbone. (2009). Sociality in Rancho La Brea *Smilodon*: Arguments favour "evidence" over "coincidence." *Biology Letters*, 5, 563–564.

Varricchio, D. J., J. R. Moore, G. M. Erickson, M. A. Norell, F. D. Jackson, & J. J. Borkowski. (2008). Avian paternal care had dinosaur origin. *Science*, 322, 1826–1828.

Walsh, S. A., P. M. Barrett, A. C. Milner, G. Manley, & L. M. Witmer. (2009). Inner ear anatomy is a proxy for deducing auditory capability and behaviour in reptiles and birds. *Proceedings of the Royal Society of London, B*, 276, 1355–1360.

Glossary

Abiotic—Nonliving chemical and physical factors in the environment that affect ecosystems.

Adaptation—The process of adjustment of an organism to its environment. The process of evolutionary modification that improves an organism's survival and reproductive success. A trait that evolved by natural selection for its current functional role.

Adaptationist—An unsubstantiated assumption that all or most traits are optimal adaptations.

Adaptive mate choice—Explanations in which ornament expression is associated with characteristics of prospective mates that are beneficial outside of a mate-selection context.

Aggregation—A group or collection of individuals.

Aggregation pheromone—A secreted or excreted chemical factor that attracts large numbers of conspecifics.

Aggression—Behavior used to subdue or repel another individual.

Aggressive mimicry—A predator mimicking some signal that is benign or attractive to the prey.

Agonistic—Relating to the range of activities associated with aggressive encounters between members of the same species, including threat, attack, appeasement, or retreat.

Agonistic character displacement—A specialized case of character displacement in which divergence occurs in traits that affect the rate, intensity, or outcome of competitive interactions.

Alarm call—A signal produced by an animal in response to danger.

Alarm pheromone—A chemical secreted by an animal that alerts others of the same species to the presence of danger.

Alarm substance—A chemical released by an injured individual that causes others to withdraw.

Allee effect—The reproductive rate of a population declining with declining population density.

Allogroom—To groom another individual.

Allospecific—A member of another species.

Alternative reproductive tactics—Different behavior patterns of one sex of a species to enhance reproductive success.

Altruism—An act that is costly to the donor and beneficial to the recipient, where costs and benefits are measured in terms of effects on expected lifetime reproductive success.

Altruistic punishment—Individuals punishing noncooperators at a cost to themselves.

Ambivalent behavior—Behavior that appears to result from conflicting motivations.

Amplified fragment length polymorphism (AFLP)—A highly sensitive molecular technique for detecting polymorphisms in DNA.

Anisogamy—Gametes differ is size or form.

Antithetical—Referring to the opposite.

Aposematism—Warning coloration.

Arbitrary—Traits that do not directly promote or are not tightly associated with survival and fecundity.

Armaments—A trait that can serve as an aid in contests with members of the same species (often of the same sex) but that is not required for foraging or protection from predators.

Assessment/communication phase—A stage of an encounter between a predator and prey when the prey judges how much of a risk this predator poses to its safety, when the predator weighs the costs and benefits of attack and the likelihood of success, and when there is overt exchange of signals or cues between the parties that allow them to assess each other more rapidly.

Associated social monogamy—A male and female of the same species spend most of their time together and coordinate their behavior.

Association hypothesis—An explanation for maternal care proposing that because internally fertilizing males release their gametes first, they are free to desert immediately, forcing females to choose between providing care alone or abandoning the breeding attempt.

Attack/flight phase—A stage of an encounter between a predator and prey when the predator decides whether to attack and the prey decides whether to flee.

Autonomic behavior—Automatic physiological responses that are not under the direct control of the animal.

Autotomy—The spontaneous casting off of a body part when the organism is injured or under attack.

Background matching—A camouflage strategy in which coloration blends in with the background habitat.

Back-up food supply hypothesis—An explanation proposing that some offspring are produced as food for the dominant offspring.

Bateman's principle—The view that females almost always invest more energy in offspring production than males, making females a limiting resource over which males will compete.

Beater effect—Prey flushed out by group activity becoming easy to capture.

Begging—Behavior by which depending young solicit care from their parents and other caregivers.

Behavioral—Referring to movements, postures, or displays produced by an animal.

Behavioral ecology—An approach to investigations of how different ecological circumstances affect animal behavior and the evolution of traits with functions that fit these different environments.

Behavioral (ethological) isolating mechanism—Behavior that prevents members of different species from reproducing with each other.

Behavioral syndrome—Correlation of rank-order differences among individuals through time or across situations.

Biotic—Any living component that affects other organisms in an ecosystem.

Biparental care—Parental care provided by both the father and the mother.

Bonanza resource—A resource, especially food, that is present in vast amounts but often for a short time.

Bower—A structure built by a male bowerbird to attract mates.

Breeding bout—The amount of time required for a male and female to produce viable offspring.

Broadcast signal—A signal that is propagated throughout the environment without a specific receiver.

Brood parasitism—Use of a host of the same species (intraspecific brood parasitism) or a different species (interspecific brood parasitism) to care for the parasite's offspring.

Capture/escape phase—A stage of an encounter between a predator and prey when the predator decides whether to attempt capture and the prey decides whether to attempt to escape.

Caste—A distinct category of individuals that provide particular services to a colony.

Central-place forager—An animal that forages from a fixed home location such as a nest.

Character displacement—Divergence in traits of species with overlapping distributions caused by competition between them.

Chase-away model—An explanation for ornaments that is based upon the fundamental conflict of interest between males and females in sexually reproducing organisms. Male ornaments are proposed to stimulate females to mate in a manner that benefits males but is not beneficial for females. The male ornament is favored because it enhances male mating success, but because being attracted to the ornament is bad for females, they evolve diminished response systems, which produces selection on males for any new traits that enhance or expand the stimulatory effect of the ornament.

Cheating—Use of a "dishonest" (unreliable) signal to manipulate a receiver.

Circadian rhythm—An endogenous activity rhythm with a period of about 24 hours.

Coefficient of relationship—A measure for the level of consanguinity (degree to which they are descended from a common ancestor) between two given individuals.

Coevolution—The evolution of two or more species, each adapting to changes in the other.

Colony—A dense cluster of breeding territories that do not contain only breeding sites.

Communal nesting—More than one breeding female occupying a single nest or nesting chamber within a burrow.

Communication—The behavior of one individual affecting the behavior of another individual.

Communication signals—Traits (structures and behavior) that have evolved specifically because they change the behavior of receivers in ways that benefit the signaler.

Compartmentalization—The extent to which subgroups in a society operate as discrete units.

Competitor—An individual of the same (intraspecific) or different (interspecific) species that vies for the same resource (e.g., food or living space) in an ecosystem.

Condition-dependent model—An explanation that proposes that ornaments are honest signals of condition because only individuals in good condition could produce a big ornament.

Confusion effect—A large number of prey escaping from predation at the same time making it difficult for the predator to pursue a specific individual prey animal.

Conspecific—A member of the same species.

Constraints—Factors that act to limit response or performance.

Contact call—A signal that functions to keep an animal in touch with a group and allow other group members to determine the location of the caller.

Contrast effects—Phenomena that occur when animals experience a contrast between what they have experienced in the past and what they are experiencing now.

Conventional sex roles—Males competing more intensely than females for access to mates.

Cooperative—Interaction of two or more individuals that is mutually beneficial, or beneficial to one and not beneficial to the other.

Cooperative breeding—A social system in which individuals contribute care to offspring that are not their own at the expense of their own reproduction.

Cooperative polygamy (mate-sharing)—Mating system in which multiple cobreeders of one or both sexes share mates of the opposite sex.

Cooperative polygynandry—Mating system in which multiple males share and mate with more than one female.

Copulation solicitation display—A female courtship display given before and during copulation in many species of birds.

Core—The average number of offspring reared to independence.

Cost/benefit analysis—Scoring the potential outcomes of the different options in terms of positive and negative effects on fitness.

Counteradaptation—Evolutionary responses of one species (or sex) in response to adverse adapations of another species (or the other sex).

Countershading—A camouflage strategy in which the back is darker in color than the belly.

Courtship display—A stereotyped pattern of behavior in animals that functions to attract and arouse a prospective mate.

Cross-fostering experiment—An experimental technique in which members of one brood are exchanged with those of another, producing a mix of genetic and foster offspring in each brood.

Crypsis—The ability of an organism to avoid observation or detection by other organisms.

Cryptic female choice—The ability of females to manipulate paternity by choosing which sperm fertilize their eggs.

Cuckold—A male whose mate engages in extra-pair copulation.

Cue—Circumstances that predict the likelihood of a subsequent event.

Culture—Sharing of behavior by a group of animals through social transmission among peers and between generations.

Currency—The means by which costs and benefits are measured.

Decision rules—The cognitive mechanisms that animals use to make decisions.

Decreasing marginal returns—Situation in which it becomes harder and harder for a forager to extract food in a particular circumstance or location.

Deduction—Reasoning from the general to the particular.

Defense—A stage of an encounter between a predator and prey in which the prey can employ behavioral or morphological defenses with which to deter or repel predators.

Detection/recognition phase—The stage during an encounter between predator and prey when both parties can discover then determine whether the other is a potential threat or a potential meal.

Detritus—The organic debris formed from the decay of organisms.

Developmental constraint—A bias on the development of behavior or a limitation on phenotypic variability caused by the structure, character, composition, or dynamics of the developmental system.

Dewlap—A longitudinal flap of skin that hangs beneath the lower jaw or neck of many vertebrates.

Diel vertical migration (DVM)—A pattern of up-and-down movement within the water column that some aquatic organisms undertake each day.

Diet selection—How animals decide whether to use a particular food resource.

Dilution effect—An individual's risk of predation decreasing as group size increases because the predator can only catch or eat a certain number of prey.

Diploid—A cell or an organism with a full set of paired chromosomes.

Dispersed social monogamy—A male and female of the same species sharing a territory but foraging and sleeping alone.

Displacement activity—A seemingly inappropriate behavior thought to result from two conflicting motivations in a particular situation.

Display—A stereotyped behavior that has evolved to function as a signal in communication.

Disruptive coloration—A camouflage strategy in which bold blocks of color or projections of the body break up the body outline against the background environment.

Diurnality—The behavior of animals that are active in the daytime.

Divergence—Acquisition of dissimilar traits by previously similar, related organisms.

DNA fingerprinting—A molecular technique that uses highly variable regions of DNA to identify individuals.

Dominance—The effect of one allele at a locus being partially or entirely hidden by another allele at the same locus. A social relationship in which one individual has power or priority of access over another.

Dynamic—Changing.

Eavesdropping—Observing the contests, behavior, or signaling of others.

Economically defendable—The benefits of the behaviors associated with having exclusive access to the area outweighing the costs of maintaining and defending it from conspecifics.

Ectospermalege—An adaption in female bed bugs that acts as a mating guide for the male to minimize the long-term damage of traumatic insemination of the female.

Emancipated—Freed from its original function and motivation.

Encounter—Predator and prey coming near enough to each other that detection is possible.

Environmental potential for polygamy—The degree to which the social and ecological environment allows one sex to monopolize the other sex as mates.

Eusocial—An advanced level of social organization, in which a single female or caste produces the offspring and nonreproductive individuals cooperate in caring for the young.

Evolution—Descent with modification.

Evolutionary ecology—An approach to the study of ecology that explicitly considers the evolutionary histories of species and the interactions between them.

Evolutionary tug-of-war—The evolutionary response to conflicting selection on a single trait.

Experimental evolution—The study of evolution using experimental treatments in the laboratory and measurement of responses of organisms over time.

Extrafloral nectaries—Nectar-producing structures not part of the flower.

Extra-pair copulation—Mating with partners other than the social mate.

Extra-pair paternity—Cases in which the social mate is not the father of particular offspring.

Facultative—An organism that does something (e.g., act as a parasite) as an option.

Feedback mechanism—A mechanism that tends to accelerate (positive) or to inhibit (negative) a process.

Female (harem) defense polygyny—Aggregation of females, often because of clumped food or nest sites, providing dominant males the opportunity to defend them and thereby gain multiple mating opportunities.

Filial cannibalism—A form of infanticide in which adults eat their own offspring.

Fisher's fundamental theorem—Any alleles that improve fitness will increase in frequency, reaching fixation in just a few generations.

Fitness—The ability to survive and reproduce; the contribution to the gene pool of the next generation made by a specified genotype or phenotype.

Fitness proxies—The currencies that animal behavior is adapted to maximize.

Foraging—Finding and collecting resources.

Forced copulation—A tactic in which male ducks attempting to coerce the females into mating so violently they will on occasion drown them.

Game theory—A mathematical method for calculating success based on the choices of others.

Generalist forager—An animal that uses many kinds of food.

Genetic bottleneck—A sharp reduction in population size that drastically reduces the genetic variation in a population or species.

Genetic covariance—The extent to which the genetic effects for two traits vary together.

Genetic drift—Random fluctuations in the frequency of a gene, usually in a small, isolated population, as a result of chance.

Genetic mating system—The characterization of mating relationships based on genetic analysis of maternity and paternity.

Genetic relatedness—A quantitative estimate of the proportion of genes shared between the genomes of any two individuals.

Genic capture model—An explanation that proposes that elaborated ornaments inevitably become associated with condition (become condition dependent).

Gestational diabetes—High blood sugar during pregnancy.

Good genes hypothesis—A proposed explanation for mate choice based on traits that reflect an individual's genetic quality.

Good parent model—Females benefit from a preference for males with large ornaments because they provide superior resources.

Handicap principle—An explanation first proposed by Zahavi in which ornamental traits serve as handicaps to survival such that only robust, strong, and healthy individuals could bear the burden of a big ornament.

Haplodiploid—A system in which unfertilized (haploid) eggs become males and fertilized (diploid) eggs become females.

Haploid—A cell or organism with a set of unpaired chromosomes.

Hatching asynchrony—A pattern in which eggs in a brood do not hatch together.

Heritability—The proportion of total variance in a trait that can be attributed to additive genetic variance in that trait.

Heterospecific—A member of a different species.

Home range—An area in which an animal or group of animals regularly travels in search of food or mates.

Honesty—When signals reliably predict something of use to the receiver.

Incest avoidance—Individuals not attempting to breed with close relatives.

Inclusive fitness—The ability of an individual to contribute its genes to the next generation, taking into account the shared genes passed on by the individual's close relatives.

Index of resource monopolization (Q)—A variance-based estimate of the potential for sexual selection.

Indicator models—Proposed explanations of mate choice based on traits that are condition-dependent and have associated costs, and thereby reliably indicate an individual's genetic quality.

Indicator trait—An ornamental trait that is reliably correlated with important qualities of a prospective mate.

Indirect genetic effects—Influences of genotypes of other individuals in the population on a focal individual.

Indirect reciprocity—Reciprocity by individuals other than those originally helped.

Individual recognition—The ability to discriminate among individuals using their identity cues.

Infanticide—A tactic by which new males increase their mating success by killing the existing offspring.

Inference—Deriving logical conclusions from premises known or assumed to be true. Reasoning from evidence.

Information—A reduction in uncertainty.

Information center—A gathering of individuals at which information about the location of food can be exchanged.

Information content (message)—How a receiver behaves in response to a signal or other stimulus.

Infrasonic—Very-low-frequency sounds that are inaudible to humans.

Insurance hypothesis—An explanation proposing that parents produce back-up offspring in case something happens to the first ones.

Intention movement—Behavior that occurs in preparation for another behavior.

Interlocus sexual conflict—The conflict between males and females mediated by different genes and traits.

Intersexual selection—Differential reproductive success that results from individuals of one sex choosing mates from members of the other sex.

Interspecific communication—Communication that occurs between or among members of different species.

Intralocus sexual conflict—A form of conflict in which the same genes are expressed in both males and females but are selected in opposite directions.

Intrasexual competition—Competition among members of one sex for mating access to members of the other sex.

Intrasexual selection—Differential reproductive success that results from intrasexual competition.

Intraspecific communication—Communication that occurs between or among members of the same species.

Kin recognition—The ability to distinguish between close genetic kin and nonkin.

Kin selection—A mechanism of evolution in which characteristics favor the survival of close relatives of an individual.

Kinship—The degree of genetic relatedness between individuals of a species.

Kleptoparasitism—Food stealing by members of the same species.

Lek—An assembly area where animals display and court potential mates.

Lek paradox—The situation that the only benefit for female mate choice when males provide no resources is good genes, but good genes are theoretically impossible.

Life history—The timing of key events in an organism's lifetime, as shaped by natural or sexual selection.

Life history traits—Characteristics that affect an organism's investments in growth, reproduction, and survivorship.

Lifelong monogamy—A form of social behavior in which individuals have only one mate for life.

Linkage disequilibrium—The nonrandom association of alleles at two or more loci, not necessarily on the same chromosome.

Lordosis—A female courtship display given before and during copulation in many species of mammals.

Major histocompatibility complex (MHC)—A cell surface protein molecule that helps the vertebrate immune system recognize foreign substances and that is encoded by a large gene family.

Many-eyes effect—Increased antipredator vigilance ability resulting from large group size.

Marginal—Offspring that exceed the average number that can be reared to independence.

Masquerade—A camouflage strategy in which an organism is detectable but resembles an inedible object.

Mate guarding—A tactic by which a male attempts to prevent the female from mating with other males.

Maternal care—Parental care provided by the mother.

Mating system—A behavioral strategy employed by a population of animals to obtain mates.

Maximum standardized selection differential—A method of estimating the maximum strength of sexual selection in units of phenotypic standard deviations.

Mean crowding—A quantitative measure of density dependence.

Mean spatial crowding (m)—A quantitative measure of the number of potential encounters that is based on the distribution of individuals in space.

Mean temporal crowding (t)—A quantitative measure of the number of potential encounters that is based on the distribution of individuals in time.

Mesopredator—Midlevel predator.

Meta-analysis—A statistical method that combines results of several studies that address related research hypotheses.

Metacommunication—Communication about communication.

Microsatellite DNA—Repeating sequences of two to six base pairs of DNA; also called simple sequence repeats (SSRs) or short tandem repeats (STRs).

Mobbing—An antipredator behavior in which prey individuals collectively attack or harass a predator.

Monandry—A form of social behavior in which a female mates with only one male.

Monogamy—A form of social behavior in which an individual has only one mate at any one time.

Monogyny—A form of social behavior in which a male mates with only one female.

Morisita index (I_δ)—A variance-based estimate of the potential for sexual selection that is based on the value of mating or reproductive success to each individual.

Morph—Each of several variant forms of a species of animal.

Morphological—Referring to physical characteristics of the animal.

Mutualism—Both individuals benefiting from interactions between them.

Natal—Relating to the place of an individual's birth or hatching.

Natural selection—The nonrandom process by which traits become either more or less common in a population as a function of differential survival or reproduction of their bearers.

Neuropeptide—A polypeptide that acts as a neurotransmitter or hormone.

Nocturnality—The behavior of animals being active in the nighttime.

Nomadic—An organism that moves or changes location frequently.

Nuptial gift—A food offering males provide to females during courtship.

Nurse ants—Individuals that clean and feed the begging larvae of the brood.

Obligate—An organism that does something (e.g., act as a parasite) by necessity.

Olfactory cue—A source of information based on odors.

Ontogenetic—Relating to the origin and development of individual animals.

Operational sex ratio—The relative number of sexually competing males that are ready to mate versus sexually competing females that are ready to mate.

Opportunity for selection (I)—An index of the amount of variation that could be subject to selection in a population.

Opportunity for sexual selection (I_s)—An index of the amount of variation in mating success that could be subject to sexual selection in a population.

Optimal foraging—Gathering resources to maximize the net energy intake (or other foraging currency) per unit time.

Ornaments—A trait that enhances the appearance of an animal, sometimes to the detriment of its survival.

Oviposition—The process of laying eggs by egg-laying animals.

Pair bond—An enduring preferential association between two sexually mature adults that is characterized by selective affiliation, contact, and copulation of the partners.

Paleontologist—A scientist who studies the fossilized remains of life.

Parasite—An organism that lives on or in an individual of another species (the host) from which it obtains nutrients.

Parasocial—Females of the same generation assisting one another, in some cases cooperating in nest construction but otherwise rearing their broods separately, and in other cases attending the broods cooperatively, although each female may still reproduce.

Parental care—Any behavior by a father that enhances the reproductive success (fitness) of offspring.

Parental investment—Any behavior that increases offspring's fitness but decreases the parents' ability to invest in other offspring.

Parental investment theory—An explanation proposed by Robert Trivers suggesting that the sex that invests more in parental care would ultimately control reproduction.

Parental table—A method of characterizing mating relationships that is based on a summary of the genetic parentage of all progeny in a breeding population regardless of whether the offspring are produced by selfing or outcrossing.

Parent-offspring conflict—The evolutionary conflict between parents and their offspring in optimal parental investment.

Paternal care—Care provided by a parent that increases the fitness of offspring.

Peak shift—Foragers rejecting items that appear too similar to unwanted items, even to the point of rejecting many suitable items.

Permeability—The degree to which societies interact with one another.

Personality variation (personalities)—Consistent differences in behavior among individuals.

Phenotype matching—Matching the look or smell of an individual to an internal template independent of environmental or other external cues.

Pheromone—A secreted or excreted substance that elicits a social response in members of the same species.

Phylogenetic—Relating to the evolutionary history of animals or behavior.

Phylogenetic comparative method—Use of information on the evolutionary relationships of organisms (phylogenies) to compare species.

Piscivorous—Fish-eating.

Plugs—A tactic by which a male blocks the female's reproductive tract.

Polyandry—A form of social behavior in which a female has more than one mate; females mate multiply, resulting in stronger sexual selection on females.

Polygamy—A form of social behavior in which individuals have more than one mate.

Polygynandry—Both sexes mate multiply, resulting in variable strength of sexual selection in both sexes.

Polygyny—A form of social behavior in which males have more than one mate; males mate multiply, resulting in stronger sexual selection on males.

Population biology—The study of populations of animals, especially the regulation of population size, life history traits, and extinction.

Positive allometry—Larger individuals have disproportionately larger structures compared to smaller individuals.

Postcopulatory sexual selection—Selection that results from competition within a female's reproductive tract between sperm of different males. Involves both the traits of the sperm and of the female.

Potential reproductive rate—The maximum number of independent offspring that parents can produce per unit of time.

Predator—An organism that eats another (prey) species.

Preeclampsia—A condition in which a pregnant woman develops high blood pressure.

Prenuptial molt—Replacement of all or some of a bird's feathers prior to the breeding season.

Private information—Information acquired from an individual's own experience.

Promiscuity—A social behavior in which both males and females mate with multiple partners.

Protean behavior—Fast, erratic turns to escape a predator.

Provisioning—Storing food or providing food to others.

Proximate—Referring to the immediate trigger or mechanism that controls the performance of the behavior.

Public information—Social or nonsocial information that is accessible to others.

Pursuit deterrence—Communicating to a predator the ability to evade capture and thereby causing the predator to forgo pursuit.

Random distribution—The position of one individual being independent of the position of another individual (i.e., individuals do not attract or repel each other).

Random walk—A path that consists of a succession of random steps.

Ratchet model—A mechanism by which selection on an indicator trait favors males that display the trait independent of their quality, producing selection on females to shift preferences to ornamentation that restores the association with male quality. The ensuing coevolutionary cycle. in which males produce ornaments without costs and females select for more and elaborate and costly forms of ornaments, results in large, costly, and elaborate traits

Receiver—A recipient of a signal.

Reciprocal altruism—An act by one animal that temporarily reduces its fitness while increasing another animal's fitness, with the likelihood that the other animal will subsequently respond in kind.

Recognition—A stage of an encounter between a predator and prey in which the other species is determined to be potential prey, irrelevant, a conspecific, or a potential threat.

Redirected behavior—The shifting of some behavior (e.g., aggression) away from the primary target and toward another, less appropriate object.

Reliable—A signal that accurately predicts something useful to the receiver.

Reproductive competition—Individuals are reproductively suppressed by more dominant individuals.

Reproductive effort—The proportion of its resources that an organism expends on reproduction.

Reproductive skew—Variance in reproductive success of each sex that contributes to variance in fitness, thus affecting the strength of selection. The degree to which reproduction is monopolized by one or a few individuals.

Reproductive success—The production of fertile offspring, often measured as the number of offspring produced by an individual.

Resource defense—Behavior by an animal that prevents another animal from gaining access to an important resource.

Resource-defense social mating system—A form of social behavior in which mating relationships are dependent on possession of critical resources by members of one sex.

Resource-tracking hypothesis—An explanation suggesting that parents produce the number of young that can be raised in a good year and sacrifice some if resources turn out to be insufficient.

Risk averse—Reluctant to accept a high but uncertain payoff.

Risk prone—A behavioral strategy that favors high risk in return for potentially high reward.

Ritualization—The evolution of signals through a process in which the behavior becomes simplified and exaggerated in structure and then repeated in a stereotyped sequence.

Ritualized displays—Stereotyped signals that evolved by ritualization.

Runaway model—A proposed explanation for sexual selection first proposed by R. A. Fisher in which sexual selection of traits is based upon a "runaway" positive feedback mechanism.

Satisficing—Meeting a minimum food requirement.

Scatter-hoarding—A behavior in which an animal stores (caches) food in times of surplus for times when food is less plentiful.

Schreckstoff—Olfactory cues in the form of alarm odors of depredated conspecifics.

Search image—A mechanism by which predators learn to focus attention on the cryptic prey type most frequently encountered during recent searching.

Secondary sexual traits—Characteristics that distinguish males and females of a species but that are not directly part of the reproductive system.

Sedentary—An organism that is attached to the substrate or does not normally move long distances.

Seismic communication—Signals that are propagated through the ground.

Selfish genetic element (SGE)—Segments of DNA that promote their own transfer to the next generation at the expense of other genes.

Selfish herd—The reduction of an individual's risk of predation by placing others in the herd at risk.

Selfishness—The individual performing a behavior benefits at the expense of the recipient.

Seminal fluid proteins (SFPs)—Proteins in the seminal fluid that are transferred to the female during mating.

Sender—The originator of a signal.

Sensory bias—An explanation for the evolution of novel traits as a result of inherent biases in the sensory system and brain.

Sensory drive—A model for signal evolution in which the environment affects signal intensity and fidelity, the sensory capabilities of the receiver evolve to achieve the greatest sensitivity at the region of highest environmental transmission, and signal properties then evolve to match this sensory system.

Sensory exploitation model—An explanation for ornament evolution that is based upon an arbitrary trait's ability to exploit sensory-response systems that function in contexts other than mate choice.

Sensory trap—Males attracting females for mating by producing signals that mimic other stimuli occurring in different contexts and to which females are adapted to respond.

Sequential choice theory—The view that choice depends on what other items a forager is likely to encounter later.

Serial polygamy—A form of social behavior in which an individual has a series of mates.

Sex difference in sexual selection (I_{mates})—A measure of the strength of sexual selection that is based on the source of variation in reproductive success of males under a strictly polygynous mating system.

Sex pheromone—A secreted or excreted substance that attracts a mate and elicits sexual behavior.

Sex-role reversal—Females competing more intensely than males for access to mates.

Sexual conflict—The conflict in evolutionary interests between males and females.

Sexual dimorphism—A distinct difference between males and females of the same species.

Sexual excitant (aphrodisiac)—A secreted or excreted substance that enhances sexual behavior.

Sexually antagonistic coevolution (SAC)—Cyclical coevolution between the male and female of a species resulting from a sequence of adapations and counteradaptations between them.

Sexual monomorphism—Males and females of a species being indistinguishable.

Sexual ornament—A conspicuous structure of an animal that is used in displays to attract mates or intimidate rivals.

Sexual selection—Differential reproductive success caused by competition for mates.

Sexual selection gradient (Bateman's gradient)—The relationship between mating success and reproductive success that is used to quantify the strength of sexual selection.

Sexual size dimorphism—A difference in size between males and females of a species.

Siblicide—The killing of a young individual by its sibling.

Signal—A characteristic that has evolved by natural selection to convey information to other organisms.

Signal detection theory—A means of quantifying the ability to discern between information-bearing energy patterns (signals) and random energy patterns that distract from the information (noise).

Skewed sex ratio—Numerically greater representation of one sex over the other.

Snake-harassment behavior—Prey sequentially approach and retreat from snakes while kicking substrate at them and even biting in order to deter attack.

Social behavior—The interactions among animals, usually of the same species, ranging from simple attraction between individuals, to temporary feeding aggregations and mating swarms, to multigenerational family groups with cooperative brood care.

Sociality—The tendency to form social groups.

Social learning—Facilitation of the acquisition of adaptive patterns of behavior by observing the behavior of others.

Social mating system—The characterization of mating relationships based on social interactions and parental care strategies.

Social mediation—A mechanism that can maintain signal honesty by testing of signals and punishment of unreliable ("dishonest") signalers.

Socioecology—The study of interactions among the members of a species and between them and the environment.

Soldiers—Individuals that possess weapons used to defend the family.

Spatial and temporal mean crowding—A method of characterizing mating relationships that is based on a mathematical interpretation of distributions in space and time.

Specialist forager (specialization)—An animal that uses one or a few kinds of food.

Speciation—The evolutionary process by which new biological species arise.

Species recognition hypothesis—An explanation, first proposed A. R. Wallace, arguing that arbitrary traits evolved to signal species identity.

Sperm competition—The process by which sperm cells of two or more different males compete to fertilize an egg of a female.

Spite—One individual performing a costly behavior that hurts the recipient, so both pay a cost.

Stable equilibrium—An equilibrium in which a disturbance results in a return to the equilibrium condition.

Static—Not changing.

Stereotyped—Performed with little variation.

Stillness—Remaining motionless to avoid detection.

Stotting (spronking)—Bounding up and down while fleeing a predator.

Subsocial—Females associating to a varying degree with their offspring, ranging from building a nest and laying an egg on a prey item that is sufficient to allow the larva to develop into an adult, to providing direct care for prolonged period of time but then departing before the young emerge as adults, to the situation where mothers stay at the nest until offspring emerge and those newly emerged offspring then assist their mother in rearing of additional broods.

Tail-flagging—Prey wagging their tails back and forth as a signal to a predator.

Teaching—Behavior that facilitates learning of a naïve individual at some cost to the teaching individual.

Template—A pattern, mold, or structure that guides the formation of a duplicate copy. An auditory template is a memory used to guide the development of bird song.

Territorial behavior—The acts by which an animal (or group of animals) protects an area from trespassing by others of the same species.

Territory—A fixed area containing specific resources such as food, nest sites, or mates from which one or more individuals exclude other members of the same species.

Testosterone—A steroid hormone in the androgen group, found in vertebrates.

Time in—The amount of time that an adult is sexually active or capable of mating.

Time out—The amount of time that an adult spends resting or preparing for mating.

Trade-off—A situation that involves losing one aspect of something in return for gaining another aspect.

Transgenic—Pertaining to a gene or genes transferred from another species, or a species that possesses a gene or genes transferred from another species.

Transparency—A camouflage strategy in which the body is clear.

Trapline—A circuit of specific foraging sites, much like a line of traps used by trappers.

Ultimate—Referring to how a behavior affects the survival and reproductive success of animals performing the behavior.

Ultrasonic—Above the range of human hearing.

Uniform spacing—A distribution of individuals or territories with relatively constant distances between neighbors.

Unreliable—A signal that does not accurately predict something useful to the receiver.

Variance—A measure of the spread of values on either side of the center.

Várzea—Seasonally flooded forest or woodland along rivers in the Amazon.

Vasopressin—A peptide hormone that plays a key role in regulating water, glucose, and salts in the blood as well as social behavior and bonding.

Vocal signature—Distinctive cues that allow individual recognition by voice.

Waggle dance—A behavior performed by a scout honey bee to indicate the direction and distance to food.

Workers—Individuals that perform household duties and care for young.

About the Editor and Contributors

EDITOR

KEN YASUKAWA is Mead Family Professor of Biology at Beloit College and Honorary Fellow in Zoology at the University of Wisconsin at Madison. Using Niko Tinbergen as his model, he has studied the behavior and ecology of birds and has focused primarily on the red-winged blackbird since 1973.

CONTRIBUTORS

JANIS L. DICKINSON is a Professor in Natural Resources and Arthur A. Allen Director of Citizen Science at the Lab of Ornithology at Cornell University. Her interests are in cooperative breeding and other forms of social behavior at both proximate and ultimate levels and in research on socially networked citizen science as a collective action.

ANNA DORNHAUS is an Associate Professor in the Department of Ecology and Evolutionary Biology at the University of Arizona. She is interested in the emergence of complexity and increased efficiency through collective behavior, the effects of scaling in complex systems, and the role of learning and individual variability for collective success in social insects. She has also studied many aspects of foraging, such as search strategies, learning, optimality, communication, and recruitment.

RENÉE A. DUCKWORTH is an Assistant Professor in the Department of Ecology and Evolutionary Biology at the University of Arizona. She is interested in ecological cycles, hybridization, dispersal, colonization of novel environments, maternal effects, and the genetics of personality.

LEE ALAN DUGATKIN is Professor and Distinguished University Scholar at the University of Louisville. He is currently studying the evolution of cooperation, the evolution of aggression, the interaction between genetic and cultural evolution, the evolution of antibiotic resistance, and the evolution of risk-taking behavior and has various projects dealing with the history of science.

GEOFFREY E. HILL is Professor and Curator of Birds in the Department of Biological Sciences at Auburn University. His research focuses on the function and evolution of ornamental traits in birds and on the coevolution of hosts and pathogens using approaches ranging from behavioral ecology and phylogenetic reconstruction to immunology, parasitology, proteomics, and functional genomics.

BRIAN KEANE is Associate Professor in the Department of Zoology at Miami University. His research combines the use of molecular genetic techniques with field studies on natural populations to address questions in evolutionary biology and behavioral ecology.

WALTER D. KOENIG is a Senior Scientist at the Lab of Ornithology and Department of Neurobiology and Behavior at Cornell University. He is interested in animal social behavior and behavioral ecology.

ZENOBIA LEWIS is a Lecturer at the Institute of Integrative Biology at the University of Liverpool. She is an evolutionary biologist and behavioral ecologist who uses arthropods to investigate questions of sexual selection, sexual conflict, and reproductive biology.

KENYON B. MOBLEY is a Postdoctoral Fellow at the Max Planck Institute for Evolutionary Biology who is interested in sexual selection, mating systems, and molecular evolution. His research seeks to understand how behavior, the environment, and evolutionary history interplay to shape contemporary populations in a variety of fish species including pipefishes, gobies, and sticklebacks.

TERRY J. ORD is a Lecturer in Evolution at the University of New South Wales in Sydney, Australia. He is interested in how phylogeny, natural selection, and sexual selection contribute to the evolution of behavioral diversity.

MICHELLE PELLISSIER SCOTT is Emerita Professor of Biological Sciences at the University of New Hampshire. Her research centers on the use of insects as model systems to understand the costs and benefits of reproductive strategies. She focuses on silphid beetles, a diverse group of insects that use carrion as a resource for reproduction, because they can be studied in the field and in the laboratory while retaining an important element of naturalism.

NANCY G. SOLOMON is Professor of Zoology at Miami University. She is a behavioral ecologist whose research interests focus on the ecology, reproduction, and behavior of small mammals and on sociality and mating patterns.

THEODORE STANKOWICH is an Assistant Professor in the Department of Biological Sciences at California State University–Long Beach. He studies the evolutionary ecology of antipredator defenses in a wide variety of taxa, particularly mammals. He has worked on risk assessment during predatory encounters, the effects of relaxed selection on predator recognition, and the evolution of defensive weaponry and aposematism.

ZULEYMA TANG-MARTÍNEZ is Founders Professor of Biology at the University of Missouri–St. Louis. She is interested in animal communication and focuses on chemical communication.

About the Editorial Board

Daniel T. Blumstein, Department of Ecology and Evolutionary Biology, University of California–Los Angeles, Los Angeles, California

Dorothy L. Cheney, Department of Biology, University of Pennsylvania, Philadelphia, Pennsylvania

David L. Clark, Department of Biology, Alma College, Alma, Michigan

Ellen S. Davis, Department of Biological Sciences, University of Wisconsin–Whitewater, Whitewater, Wisconsin

Bennett G. Galef Jr., Department of Psychology, Neuroscience, and Behaviour, McMaster University, Hamilton, Ontario, Canada

James C. Ha, Department of Psychology, University of Washington, Seattle, Washington

Kim L. Hoke, Department of Biology, Colorado State University, Ft. Collins, Colorado

Robert L. Jeanne, Department of Entomology, University of Wisconsin–Madison, Madison, Wisconsin

Patricia B. McConnell, Department of Zoology, University of Wisconsin–Madison, Madison, Wisconsin

Douglas W. Mock, Department of Biology, University of Oklahoma, Norman, Oklahoma

Michael Noonan, Department of Biology, Canisius College, Buffalo, New York

Sarah R. Partan, School of Cognitive Science, Hampshire College, Amherst, Massachusetts

Robert M. Seyfarth, Department of Psychology, University of Pennsylvania, Philadelphia, Pennsylvania

Zuleyma Tang-Martínez, Department of Biology, University of Missouri–St. Louis, St. Louis, Missouri

Index

Abiotic habitat, 233
Abiotic variables, 133
Acorn woodpeckers, 258, 277–78
Acoustic communication, 56–57
Adams, Dean, 316
Adaptations: antipredator, 25, 27; behavioral, 30, 42, 43; evolution of, 199; sexual conflict and, 179
Adaptionists, 3–4
Adaptive alert responses, 33
Adaptive hypothesis, 299–300
Adaptive mate choice, 161–62, 166–67
AFLP (Amplified fragment length polymorphism), 104
African carnivores, 341–42
African elephants, 205
African starlings, 277
African wild dogs, 242
Aggregation, 226–28, 236–37, 268–69
Aggregation pheromones, 58
Aggression: aggressive personality variation, 296–99; agonistic behavior, 62–63; body size, 302; consistency in expression of, 299–301; correlations with other traits, 301–10; definition of, 295–96; dominance, 303–6; ecological consequences of, 310–15; evolutionary consequences of, 315–16; inconsistency of correlations with other traits, 309–10; individual spacing and population cycles, 310–12; introduction to, 295–96; invasion and range limits, 312–15; parental care, 306–8; population cycles, 310–12; range limits, 312–15; spacing behavior and, 239–40; summary, 316–17; temperament traits, 308–9
Aggression-boldness-explorativeness syndrome, 309
Aggressive mimicry, 32
Aggressive personality variation, 296–99
Agonistic behavior, 62–63, 69–70, 71, 75
Agonistic character displacement, 315–16
Akimoto, Shin-ichi, 186
Alarm calls, 53–54, 64, 84, 272, 331–32
Alarm odor, 53
Alarm pheromone, 53, 58, 67
Alarm substance, 67
Allee effect, 241
Allocation trade-offs, 163
Allogrooming, 58, 59
Allospecific songs, 263
Alternative reproductive tactics, 234
Altruism: alarm calls as, 272; causes of, 266–67; definition of, 327–28; inclusive

fitness theory, 329–34; introduction to, 327–28; kin recognition, 334–35; reciprocal, 266–67
Altruistic punishment, 267
Altruistic workers, 204
Amazonian manatees, 28
Ambivalent behaviors, 71, 73
The American Naturalist, 329
Amoeba, 262
Amplified fragment length polymorphism (AFLP), 104
Andersson, Malte, 122
Androgens, 236
Anisogamy, theory of, 101, 122
Antelope, 70
Anthropogenic pollution, 133
Antipredator adaptations, 25, 27, 44, 46
Antithetical displays, 63, 75
Ants, 6, 64, 65, 204, 261–62, 309, 312–13
Aposematism, 33, 36–39, 272–73
Arbitrary, ornamental traits as, 153
Arctic foxes, 314
Arctic skuas, 231
Argentine ant, 312–13
Arginine vasopressin, 235, 236
Armaments. *See* Ornaments and armaments
Armor, 41–42
Arnold, Stevan, 122, 162
Arnqvist, Göran, 184, 185
Arraunt, Eduardo, 28
Artificial selection experiments, 304–5
Assessment, 34–35
Assessment/communication phase, 26, 34–39
Associated social monogamy, 225
Association hypothesis, 202
Atlantic salmon, 234
Attack, 24, 40–41
Attack/flight phase, 26, 39–41
Australian fairy-wrens. *See* Superb fairy wren
Australian sawfly, 271
Autonomic behaviors, 71, 72

Baboons, 260, 266, 275
Baby birds, 53
Background matching, 31
Background noise, 82

Back-up food supply, 209–10
Badgers, 230–31
Banded mongooses, 228
Bank voles, 241, 303
Banner-tailed kangaroo rats, 85
Barbour, Matthew, 85
Bark beetles, 58, 64
Barred owls, 313
Bateman, Angus J., 103, 119
Bateman gradient, 111, 120, 121, 122–24
Bateman's principles, 103, 119–20
Bats, 15, 54, 264, 266
Beater effect, 273
Bed bugs, 178–79
Beetles, 58, 64, 181, 199–200, 210, 269
Begging behavior, 210–12
Behavior, in aggregations, 226–27
Behavioral adaptations, 30, 42, 43
Behavioral ecologists, 334
Behavioral ecology, 1, 15, 145, 214, 339
Behavioral isolating mechanisms, 68
Behavioral messages, 53
Behavioral signals, 55
Behavioral syndrome, 14, 308
Behavior ecology, 339–40
Belding's ground squirrels, 265, 272, 330–32
Berglund, Anders, 167–68
Bighorn sheep, 151
Biotic variables, 133
Bird flocks, 226–27
Birds: chemical signals, 59; diversity in behavior, 15; parental care, 198–99; parent-offspring relations, 215; sexual conflict in, 185; siblicide, 208; vision systems, 30–31
Bird song, 53, 262–64
Black-tailed deer, 34, 43
Bluebirds, 311–12. *See also* Mountain bluebirds; Western bluebirds
Blue-footed boobies, 209
Blue tits, 205
Blumstein, Dan, 34
Body size, and aggression, 302
Bonanza resources, 198
Bonobos, 2
Boonstra, Rudy, 235–36
Bowerbirds, 56, 60

Bowers, 56
Breeding bout, 106, 107
Broadcast signal, 53–54
Broad-nosed pipefish, 123, 128, 129–30, 134–35
Brockmann, Axel, 66, 67
Brood parasitism, 268
Brook sticklebacks, 316
Brook trout, 44, 45
Brown, Grant, 44
Browne, Janet, 327
Brown-headed cowbirds, 268
Brown recluse spiders, 59–60
Brown trout, 237
Brumm, Henrik, 82
Bumble bees, 7
Burying beetles, 181, 199–200, 210
Butler, Mark, 237

California ground squirrels, 36, 37, 43, 85
California mice, 236, 307–8
California towhees, 31
California voles, 229
Camouflage, 31–33
Campbell's monkeys, 84
Cannibalism, 206–7
Cape porcupine, 225
Capture/escape, 24
Carazo, Pau, 81
Carcasses, 198–99
Cardinalfish, 207
Care givers, 200–203
Carnivores, 1–2, 12
Caro, T., 39
Casas, Jérôme, 41
Castes, 259, 280
Caterpillar communication, 353–54
Cattle egrets, 209
Cavity nesters. See Mountain bluebirds; Western bluebirds
Central-place foragers, 3, 11
Chacma baboons, 260
Chapman, Tracey, 183–84
Character displacement, 315–16
Chase-away model, 156–60, 167
Cheating, 151
Chemical communication, 57–59, 64

Chemical weapons, 41, 42
Cheney, Dorothy, 80
Chimpanzees, 2, 15, 56
Chipmunks, 314
Chitty, Dennis, 310, 312
Chivers, Doug, 44
Chucks, 352–53
Cichlid fishes, 63–64, 309
Circadian rhythms, foraging and, 6
Clark, Rulon, 85
Cleaner fish, 86, 266–67
Cliff swallows, 12, 269–70
Climate, 133
Clumped resources, access to, 275–76
Clutton-Brock, Tim, 116
Cochlear, 345, 351
Cockroaches, 11, 42–43, 265
Coefficient of relationship, 329
Coevolution, 352
Collared flycatchers, 202
Collective defense, 227
Coloniality, 227–28, 231, 269–70
Coloration, 33, 36–39
Combat weapons, 41, 42
Comb-footed spiders, 309
Comeault, Aaron, 37–39
Communal defense, 231
Communal nesting, 228
Communication: acoustic, 56–57; agonistic behavior, 62–63, 69–70, 71, 75; chemical, 57–59, 64; contexts of, 61–67; definition of, 52; in families, 210–12; food and predators, 64–67; group coordination, 64; honesty of, 76–79, 150–51, 162–64, 211–12; information content, 53, 67–71; interspecific, 51, 83–86; introduction to, 51–54; manipulation, 76; modalities of, 54–60; multiple modalities, 60; parental care, 63–64; predation and antipredator behavior, 35–36, 64–67; with predators, 35–36, 64–67; ritualization and origins of displays, 71–74; sexual behavior, 61–62; signals conveying information, 79–83, 210–12; summary, 86; tactile and vibrational, 59–60; typical- and variable-intensity displays, 74–76; vibrational,

59–60; visual, 55–56; vocal, 345, 351–53
Communication signals, 12–13, 52, 79–83, 210–12
Compartmentalization, 259
Competition, vs. mate choice, 149–50
Competitors, 199
Competitor-to-resource ration (CRR), 115
Condition-dependent model, 163
Conflict. *See* Families, conflict and cooperation in
Confusion effect, 227
Conservation, effects on intraspecific variation in spacing behavior, 241–42
Conspecific cues, 228
Conspecifics, 222
Conspecific tutor songs, 263
Conspicuousness, 39
Constraint hypothesis, 299–300
Constraints, 4
Contact calls, 64
Contrast effects, 9
Conventional sex roles, 109
Cooper, William, 39–40
Cooperation. *See* Families, conflict and cooperation in
Cooperative breeding: based on limited resources, 275–76; eusociality, 281; extended families, 203; reproductive skew, 276–80; social behavior, 276–80
Cooperative defense, 271–73
Cooperative polygamy, 277–78
Cooperative polygynandry, 277, 279
Cooper's hawks, 31
Copulation solicitation display, 69
Coral reef fish, 298
Core offspring, 208
Coss, Richard, 33
Cost/benefit analysis, 197, 203, 213
Costs of providing care, 200
Costs of remaining, 39
Cougars, 23–24
Counteradaptation, 179
Countershading, 32
Courtship behavior: in dinosaurs, 346–48; honesty in communication and, 76–77; information content in, 69; intention movements, 72; sexual behavior, 61–62; sexual conflict in, 176; tactile and vibrational communication, 59–60
Courtship displays, 346–48
Cowbirds, 268
Cox, Matt, 39
Coyotes, 23–24, 231
Crabs, 306
Cresswell, Will, 40
Crickets, 61
Cross-fostering experiments, 214
Crow, James F., 105
Crowned eagles, 84
CRR (Competitor-to-resource ration), 115
Crustaceans, 59–60
Crypsis, 236
Cryptic female choice, 134
Cuckolding, 202
Cues, 12–13, 30, 33, 44
Culture, learning and, 11
Curio, Eberhard, 33
Currency, 4, 197
Current benefits hypothesis, 282
"Customer" fish, 86
Cuttlefish, 32

Dark-eyed juncos, 59
Darwin, Charles: altruism, 327; on dogs, 63; on female choice, 152–53; on honey bees, 327–28; mating systems, 100; *On the Origin of Species*, 327, 328; secondary sexual traits, 104; sexual selection, 118–19, 145; on shared ancestry of species, 348; on social communication, 353
Darwin-Fisher theory of sexual selection, 104
Dawkins, Richard, 76–77
Deception, and communication, 76
Decision rules, 9–10
Decision to flee, 39–40
Decreasing marginal returns, 11
Deduction, 339
Deep-sea anglerfish, 32
Deer, 23–24, 70, 231
Defense against predators, 26, 41–42
Dehydroepiandrosterone (DHEA), 235
Density in aggregations, 226–27

The Descent of Man and Selection in Relation to Sex (Darwin), 100
Desert ants, 6
Desert iguanas, 229
Desertion, 201
Desert locust, 237–38, 273
Desert spiders, 302
Detection, 24, 30–33
Detection phase, 25–26
Detritus, 335
Developmental constraints, 299
Dewlaps, 55
DHEA (Dehydroepiandrosterone), 235
Diana monkeys, 54, 84
Dicots, 230
Diel vertical migration (DVM), 28, 29
Diet selection, 9
Dill, Lawrence, 24–25, 39
Dilution effect, 232, 270–71
Dimetrodon, 347–48
Dingemancse, Niels, 308
Dinosaurs, 344–48
Diploid eggs, 280
Direct observation, 339
Dispersed social monogamy, 225–26
Displacement activities, 71, 72–73, 74
Displays, 54; ritualization and origins of, 71–74; typical- and variable-intensity, 74–76
Disruptive coloration, 32
Diurnality, 29
Divergence, 186
Diversity in animal behavior, 14–15
DNA fingerprinting, 104
Dogs, 61, 62–63, 71, 75, 82
Dolphins, 11
Domenici, Paolo, 42–43
Domestic pigs, 303
Dominance, 270–71, 303–6
Dorsal sails, 347–48
Dragonflies, 41
Drosophila melanogaster. *See* Fruit flies
Ducks, 176–77
Dung flies, 185
Duration of displays, 74
Duvall, David, 122
DVM (Diel vertical migration), 28, 29
Dynamic signals, 55

Earwigs, 212
Eastern tent caterpillars, 274
Eavesdropping, 53–54, 263
Ecological consequences of aggression, 310–15
Ecological constraints hypothesis, 276
Ecological mating-system influences, 133–34
Economically defendable, 275
Ectospermalege, 178–79
Edaphosaurus, 347–48
Edinburgh University, 182
Egg clutches, 344–45
Elephants, 56, 205, 242
Elk, 146
Emancipation, 74
Emberizine finches, 43
Emerald cockroach wasp, 11
Emlen, Stephen, 101–2
Emus, 345
Encounter, 24, 28–30
Endler, John, 82–83
Environment, importance of in communication, 82–83
Environment, information about, 70
Environmental potential for polygamy, 101–2
Escape, from predators, 42–43
Estradiol, 261
Estrogen, 235, 263
Ethological isolating mechanisms, 68
Eurasian penduline tit, 200–201
Eurasian red squirrels, 231
European cuckoos, 268
Eusociality, 226, 258–59, 280–81, 332–33
Evolution, 4, 145
Evolutionary consequences of aggression, 315–16
Evolutionary ecology, 149
Evolutionary history of behavior: introduction to, 339–40; linking fossils to behavior of living species, 340–48; recovering past behaviors, 348–54; work of studying, 354–56
Evolutionary spite, 268
Evolutionary tug-of-war, 174
"The Evolution of Altruistic Behaviour" (Hamilton), 329

The Evolution of Animal Communication (Searcy and Nowicki), 82
Experience, in predator/prey encounters, 44, 45
Experimental evolution, 180
The Expression of the Emotions in Man and the Animals (Darwin), 353
Extended families, 203–4. *See also* Families, conflict and cooperation in
Extinction, 186–87, 242
Extrafloral nectaries, 10
Extra-pair copulation, 149, 202
Extra-pair paternity, 104

Facultative, 209
Families, conflict and cooperation in: care givers, 200–203; communication signals in, 210–12; extended, 203–4; family planning, 204–6; introduction to, 195–96; parental care, 196–200; parental favoritism, 206–10; parent-offspring conflict, 212–14, 215; summary, 214–16
Family planning, 204–6
Fathead minnows, 44, 45
Feedback mechanism, 166–67
Female choice: chase-away model, 156–60, 167; male-male competition and, 148–49; models based on adaptive female choices, 161–62; models based on arbitrary female choices, 153–61; in ornament evolution, 152–53; rooster plumes and, 159–60; runaway model, 153, 155, 157–58, 160–62, 167; sensory exploitation model, 155–56; species recognition hypothesis, 153–55
Female defense polygyny, 275
Fernández-Juricic, Esteban, 30–31
Ferrari, Maud, 44
Fertilization, sexual conflict in, 176
Fiddler crabs, 271
Filial cannibalism, 207
Fire ants, 261–62
Fish, 202, 207
Fisher, Ronald, 104–5, 153, 160
Fisher's fundamental theorem, 165
Fishing spiders, 296–97
Fish schools, 226–27

Fitness, 4, 27, 158–59, 197
Fitness proxies, 4
Fleeing behavior, 36, 40
Flight-initiation distance, 35
Flight-intention movements, 72
Florida scrub-jay, 276
Flower constancy, 8–9
Flycatchers, 154, 155
Font, Enrique, 81
Food sources, and spacing systems, 228–31
Foraging for food: communication and, 64–67; decision to, 5–7; efficiency, 273–74; explaining diversity in animal behavior, 14–15; extracting reward, 11–12; finding and selecting food, 7–8; food and lifestyle, 14; foods to avoid, 8–11; foraging trip, 5; individual tastes, 13–14; introduction to, 1–5; social behavior for increased efficiency, 273–74; social foraging, 12–13
Foraging gene, 261
Forced copulation, 175–76
Form of displays, 74
Fossils, linking to behavior of living species: dinosaurs with egg clutches, 344–45; introduction to, 339–40; polygyny, 345–48; sabertooth cats and mastodons, 341–44
Fowler, Kevin, 182
Francis, Richard, 303
Frederick, William, 39–40
Free-tailed bats, 264
Frequency of displays, 74
Freshwater fish, 309
Friberg, Urban, 184
Frog-eating bats, 54
Frogs, 37–39, 202–3
Fruit flies: mating behavior, 103, 158; sexual conflict in, 180, 182–85, 187, 188; social behavior, 273; tactile communication, 59
Functional spite, 268
Future benefits hypothesis, 282

Galápagos hawks, 277
Game theory, 151, 203
Generalist foragers, 6
Genes, social behavior and, 261–62

INDEX

Genetic bottlenecks, 312
Genetic covariance, 316
Genetic drift, 352
Genetic mating systems: definition of, 106; vs. social mating systems, 104, 131; terminology, 108; understanding sexual selection, 105
Genetic relatedness, 327
Genic capture model, 166
Gestational diabetes, 214
Glucocorticoids, 260
Good genes hypothesis, 165
Good parent model, 165
Gopher snakes, 36, 37, 85
Gould, James, 66, 67
Grey-sided voles, 229
Ground squirrels, 68; Belding's ground squirrels, 272, 330–32; California ground squirrels, 36, 37, 43, 85; smelly predators and, 348–51; squirrels exploited smelly predators, 265
Group augmentation, 279
Group coordination, and communication, 64
Group size, and predator/prey behavior, 43–44
Grouse, 60
Guanacos, 271
Guppies, 234, 237
Gustatory signals, 57

Habitat structure, 222
Hamadryas baboons, 275
Hamilton, William D., 329–30
Hamilton's rule, 266
Handicap principle, 77, 163
Haplodiploidy, 280–81, 333
Haploid eggs, 280
Harassment, 186
Hard-life hypothesis, 277
Harris's hawks, 273, 277
Hatching asynchrony, 209
Hawaiian honeycreepers, 229
Head bobs, 51, 52, 353
Head crests, 347–48
Hen-feathered gene, 159
Herbivores, 1–2
Herds, 226–27

Heritability, 165
Hermaphrodites, 122
Hermit crabs, 79
Hermit warblers, 313
Heterospecifics, 222
Hinde, Robert, 76, 78
Home ranges, 225, 232
Honesty in animal communication, 76–79, 150–51, 162–64, 211–12
Honey badgers, 86
Honey bees: alarm pheromone, 53, 58, 67; as central-place foragers, 3; costs and benefits of group living, 269; Darwin and, 327–28; diversity in behavior, 14–15; extended families, 204; foraging trips, 5; as generalist foragers, 6; social behavior, 274; social foraging, 12; waggle dance, 65–67, 70
Honeyeaters, 84
Honeyguides, 86
Hormones, 235–36, 260
Hornbill birds, 54
Hosken, David, 185–86
Hotshot hypothesis, 274
Hotspot hypothesis, 274
House crickets, 305
Houseflies, 181
House mice, 197–98
Howard, Henry E., 224
HPA (Hypothalamic-pituitary-adrenal) axis, 260
Hughes, William, 333
Hummingbirds, 15
Humpback whales, 205
Hymenoptera, 280–81, 332–33
Hypothalamic-pituitary-adrenal (HPA) axis, 260

Impulsivity, 9–10
Incest avoidance, 279
Inclusive fitness theory, 329–34
Index of resource monopolization, 112, 125–26
Indiana University, 59
Indicator models, 162–63
Indicator traits, 151
Indirect genetic effects, 316

Indirect reciprocity, 267
Individual identity, 68
Individual recognition, 306
Individual spacing, and aggression, 310–12
Individual tastes, 13–14
Infanticide, 175, 206–7, 225, 281–82
Inference, 339
Information, 52
Information center, 12, 270
Information content: in bird song, 53; environment, 70; individual identity, 68; metacommunication, 71; physiological condition/motivation, 69–70; sexual identity, 69; species identity, 67–68; status, 70–71
Infrasonic sounds, 56
Insects: extended families, 203–4; parental care, 200; sexual conflict in, 177–79, 185; siblicide, 208; tactile communication, 59–60
Insurance hypothesis, 209
Intention movements, 71–72
Interlocus sexual conflict, 174
Intersexual selection: chase-away model, 156–60, 167; definition of, 148; intrasexual selection as foundation of, 167–68; models based on arbitrary female choices, 153–61; runaway model, 153, 155, 157–58, 160–62, 167; sensory exploitation model, 155–56; species recognition hypothesis, 153–55
Interspecific communication, 51, 83–86
Intralocus sexual conflict, 174
Intrasexual competition, 100, 235
Intrasexual selection, 148, 150, 167–68
Intraspecific communication, 51
Intraspecific variation in animal spacing: aggregations, 226–28; aggression and individual spacing, 310–12; consequences of, 238; developmental effects on spacing behavior, 236–38; effects on conservation, 241–42; effects on mating patterns and reproductive success, 238–39; effects on population-level processes, 239–41; genetic basis for spacing behavior, 234–35; introduction to, 221–23; neurobiological basis for spacing behavior, 235–36; proximate mechanisms underlying, 228–42; random distribution, 224; summary, 242–43; type of, 223–28; uniform distribution, 224–26. *See also* Spacing behavior and systems
Invasion, and aggression, 312–15
Irrationality, 9
Isotope ratios, 343–44
Iwasa, Yoh, 122

Japanese grasshoppers, 186
Japanese serow, 225–26
Jones, Adam, 103–4, 120, 125, 127–30

Kalahari meerkats, 279
Keeley, Ernest, 234
Ketterson, Ellen, 59
Kin recognition, 68, 264–66, 334–35
Kin selection, 203, 266, 281, 328, 330
Kinship, 327–28. *See also* Altruism
Kirkpatrick, Mark, 160
Kleptoparasitism, 269
Krebs, John, 76–77

La Brea tar seeps, 341–43
Ladybird beetle, 269
Laidre, Mark, 79
Lake trout, 44, 45
Lande, Russell, 160, 162
Langerhans, R. Brian, 24–25, 27
Leafcutter ants, 204
Leal, Manuel, 84–85
Lek paradox, 165–66
Leks, 274
Leopards, 84
Lesser spotted woodpecker, 231
Life history traits, 197
Lifelong monogamy, 173
Ligon, J. David, 159
Lima, Steve, 24–25, 43
Limited resources, social behavior and access to, 275–76
Linkage disequilibrium, 160
Linnaeus, Carl, 348
Lions, 34, 173, 242
Living in groups, costs and benefits of, 268–70

INDEX

Lizards, 40, 52, 55
Lloyd, Monte, 105
Lone hunter lifestyle, 341
Long-tailed tits, 265
Long tails, 160–61
Lordosis, 69
Lorenz, Konrad, 264, 353
Lott, Dale, 222, 224, 233

Magrath, Rob, 54
Maher, Christine, 224, 233
Major histocompatibility complext (MHC), 265–66
Male-female pairs, spacing and, 225–26
Male-male competition, 148–50, 159
Mammalian herds, 226–27
Mammals, and parental care, 198–99
Mandrills, 147
Manipulation, and communication, 76
Mantids, 177–78
Mantis shrimp, 63, 68
Many-eyes effect, 44, 231–32
Marginal offspring, 208
Marler, Peter, 67, 70
Martin, Oliver, 185–86
Masquerade, 31–33
Mastodons, 341–44
Mate choice vs. competition, 149–50
Mate guarding, 175, 178, 180, 225
Mate-sharing, 277–78
Mathematical models of central-place foragers, 3–4
Mating, sexual conflict in, 176
Mating calls, reconstruction of, 351–53
Mating competition, 127
Mating patterns, 238–39, 345–48
Mating systems and measurement of sexual selection: Bateman gradient, 111, 120, 121, 122–24; Bateman's principles, 103, 119–20; commonly used terms, 106; definition of, 99–100, 106; ecological mating-systems influences, 133–34; empirical examples, 127–30; future challenges, 130–35; history of, 100–105; index of resource monopolization, 125–26; introduction to, 99–100; maximum standardized selection differential, 125; measurement of sexual selection, 108–9; molecular markers and genetic mating systems, 104–5; Morisita index, 111, 125–26; operational sex ratio, 102, 109, 113–16, 127–30, 149; opportunity measures, 120–21; polygamy, 101–2, 106–7, 346; postcopulatory sexual selection, 122, 134–35, 188; potential reproductive rate, 116–17; qualitative and quantitative methods, 110–12; qualitative measurements of sexual selection, 109; qualitative vs. quantitative methods for measurement, 110–12; quantitative measurements of sexual selection, 118–19; selection on sexually selected traits, 126–27; sex difference in sexual selection, 124–25; and sexual conflict, 180–81; social vs. genetic mating systems, 131; spatial and temporal mean crowding, 109, 110, 117–18; summary, 135; terminology, 105–8; theoretical quantitative framework, 103–4; variation, 131–33
"Mating Systems and Strategies" (Shuster and Wade), 103
Mating-system variation, 131–33
Matocq, Marjorie, 85
Matthews, Lauren, 225
Maximum standardized selection differential, 111, 125
Maynard Smith, John, 155
McCaffery, Alan, 238
McPhail, J. Donald, 234
Mean crowding, 106, 117–18
Mean spacial crowding, 109, 110, 117–18
Mean temporal crowding, 109, 110, 117–18
Mendel, Gregor, 328
Merlins, 43
Mesopredatory fish, 28
Message content, 52
Meta-analysis, 297
Metacommunication, 71
MHC (Major histocompatibility complex), 265–66
Mice, 58, 197–98, 206, 236, 307–8
Microsatellite DNA, 104
Microtine voles, 311
Mobbing, 42, 67, 348

Modeling studies, 15
Molecular markers, and genetic mating systems, 104–5
Møller, Anders, 78
Monandrous, 333
Monandry, 174, 181
Monogamy, 102, 106, 108, 174, 215–16, 346
Monogyny, 174, 181
Moose, 43
Morisita index, 111, 125–26
Morphological adaptations, 30, 41–42, 43
Morphological signals, 55
Morphs, 335
Morton, Gene, 228
Moths, 61
Motivation, 69–70, 75–76, 78
Mountain bluebirds, 311–12, 313
Mouthbrooders, 63–64, 207
Mule deer, 271
Munz, Tania, 66
Muskoxen, 271
Mutualisms, 13, 86, 266

Naked mole-rats, 226
Natal populations, 331
Natural selection: altruism, 327; antipredator adaptations, 25, 27, 44, 46; constancy in expression, 299; evolution by, 145; predatory behavior and, 36; predatory defenses, 42
Nazca boobies, 208–9
Neotropical singing mice, 58
Nest construction, 196
Neuropeptides, 235–36
New Zealand spiny lobsters, 237
Nicrophorus orbicollis, 206
Ninespine sticklebacks, 316
Nobel Prize, 67
Nocturnality, 29
Nocturnal prosimians, 225
Nomadic species, 199
Nonbehavioral messages, 53
Nonterritorial species, 238–39
Noonan, Brice, 37–39
Northern pike, 44, 45
Northern spotted owls, 313
North Sea harbor seal, 242

Norway lemmings, 312
Nowicki, Steve, 82
Nuptial gift, 196
Nurse ants, 204
Nuthatches, 54

Obligate, 208–9
Obligatory, parental care as, 198
Ocean skaters, 271
Olfactory cues, 30, 33, 44, 207
Olfactory signals, 57, 59
On the Origin of Species (Darwin), 327, 328
Ontogenetic shifts, 236
Operational sex ratio (OSR), 102, 109, 110, 113–16, 127–30, 149
Opportunity for selection, 110, 120
Opportunity for sexual selection, 111, 120–21
Opportunity measures, 120–21
Optimal foraging, 2
Optimality models for decision to flee, 40
Order-of-gamete-release hypothesis, 202
Oring, Lewis, 101–2
Ornaments and armaments: armaments, evolution of, 150–52; in dinosaurs, 346–48; female choice and male-male competition, 148–49, 159; intrasexual selection as foundation of intersexual selection, 167–68; introduction to, 145–46; lek paradox, 165–66; mate choice vs. competition, 149–50; mechanism for evolution via adaptive mate choice, 161–62, 166–67; models of intersexual selection based on arbitrary female choices, 153–61; ornamental traits in territorial animals, 152; ornaments, evolution of, 152–62; sexual ornaments, 346–48; sexual selection, 145–49; sexual selection based on adaptive female choice for ornaments, 161–62; signal honesty, 162–64; summary, 168
OSR (Operational sex ratio), 102, 109, 110, 113–16, 127–30, 149
Oviposition, 238
Owings, Don, 36

Pair bonds, 62
Paleontologists, 341, 343, 345

Paleontology, 339–40
Paradise fish, 303
Paradise flycatcher, 154, 155
Parasites, 199
Parasociality, 258–59
Parental care: aggression, 306–8; beetles, 181; birds, 198–99; care givers, 200–203; communication and, 63–64; in dinosaurs, 344–45; favoritism, 206–10; insects, 200; males, 148; provision of, 196–200
Parental favoritism, 206–10
Parental investment, 176, 197, 213
Parental investment theory, 101
Parental table, 105
Parent-offspring conflict, 213–14, 215
Parent-offspring identification, 63
Parker, Geoff, 174, 187
Partial loads, in foraging, 3
Partridge, Linda, 182
Passerine birds, 307
Passive aggregations, 270–71
Past behaviors, recovery of: caterpillar communication, 353–54; mating calls, 351–53; squirrels exploited smelly predators, 348–51
Paternity, 202
Payne, Katy, 56
Peak shift phenomenon, 8
Pearson, Scott, 313
Pedetta, Silvia, 306
Peiman, Kathryn, 316
Penguins, 264–65, 334
Peregrine falcons, 43
Permeability, 259
Personalities, 13–14
Pfennig, David, 335
Phenotype matching, 265
Pheromones, 57, 58
Phylogenetic comparative method, 340
Phylogenetic relationships, 349
Phylogenetics, 221, 339–40
Physiological condition, 69–70
Pied flycatcher, 270
Piloerection, 72
Pipefishes, 108, 123, 128, 129–30, 134–35
Piscivorous fish, 28
Plasticity, 298

Plastic learning capabilities, 33
Platyfish, 156
Play bow of dogs, 71
Pleiotrophy, 305
Plugs, 175, 180
Poison dart frogs, 37–39
Polak, Michal, 188
Poli's stellate barnacle, 224
Polyandry, 102, 106, 108, 124, 174, 181, 333
Polygamy, 101–2, 106–7, 346
Polygynandry, 106, 107, 108, 124
Polygyny, 102, 106, 124, 149, 174, 345–48
Pond skaters, 178
Population biology, 222
Population cycles, and aggression, 310–12
Population density, 133–34, 232–33, 240–41, 310–11. *See also* Spacing behavior and systems
Population-level processes, 239–41
Positive allometry, 347–48
Postcopulatory sexual selection, 122, 134–35, 188
Potential reproductive rate (PRR), 110, 116–17
Prairie voles, 230, 235, 239
Praying mantis, 178
Predation and antipredator behavior: assessment and communication, 34–39; attack and flight, 39–41; communication and, 35–36, 64–67; detection and recognition, 30–34; encounter, 24, 28–30; ground squirrels, 349; introduction to, 23–28; pursuit, escape, and defense, 41–43; social behavior as protection against, 271–73; spacing behavior and, 236–37; stress and, 260; summary, 43–46
Predation risk, and spacing systems, 231–32
Predator confusion, 231
Predator/prey encounters: four main steps, 24; frameworks for understanding, 24–25; interspecific communication, 51, 84–85; recognition, 25–26, 33–34, 43, 63, 264–66
Predators, 10, 199
Preeclampsia, 214
Prenuptial molt, 261

Prey, and spacing patterns, 231
Primates, 59, 196–97, 306
Private information, 12
Production costs, 163
Promiscuity, 174, 180, 215
Pronghorn antelope, 207–8, 229, 303
Protean behavior, 42, 43
Provisioning, 6
Proximate mechanisms: ecological and social determinants of spacing systems, 228–34; of social behavior, 261–66; in spacing systems, 221, 223; underlying spacing systems, 228–42
PRR (Potential reproductive rate), 110, 116–17
Pruett-Jones, Stephen, 333–34
Pteranodon, 346–48
Public goods games, 267
Public information, 12, 270
Puerto Rican lizards, 84–85
Pursuit and capture, 41
Pursuit-deterrence signals, 84–85, 272

Quantitative models/measurements, 3–4, 103–4, 110–12, 118–19
Quinn, John, 40

Rainbow trout, 44, 45, 233
Randall, Jan, 85
Random distribution, 223–24
Random walk, 7
Range limits, and aggression, 312–15
Rape. *See* Forced copulation
Rashed, Arash, 188
Ratchet model, 157, 167
Rats, 12, 261
Rattlesnakes, 36, 37, 85, 348–50
Reále, Denis, 308
Receiver, 52–53
Reciprocal altruism, 266–67
Recognition, 25–26, 33–34, 43, 63, 264–66
Red-cockaded woodpeckers, 276
Red deer, 205–6, 258
Red-eared slider, 313
Red foxes, 314
Red grouse, 311
Redirected behaviors, 71, 73–74

Red jungle fowl, 159–60
Redshanks, 40, 43
Red-sided garter snakes, 175, 177
Red-tail hawks, 31
Red-winged blackbirds, 209
Reliable information, 77–78
Rémy, Alice, 241
Reproduction, and social behavior, 274–75
Reproductive competition, 279
Reproductive effort, 205
Reproductive isolation, 185–86
Reproductive skew, 120, 276–80
Reproductive success, 14, 119, 197, 238–39
Resource defense hypothesis, 225
Resource-defense social mating systems, 118
Resource monopolization, index of, 112, 125–26
Resource-tracking hypothesis, 209
Rewards, extracting, 11–12
Ring-necked pheasants, 149
Ring-tailed lemurs, 60
Risk assessment, 34–35, 39
Risk aversion, 197
Risk prone, 197
Risk sensitivity, 9
Ritualization and origins of displays: ambivalent behaviors, 71, 73; autonomic behaviors, 71, 72; displacement activities, 71, 72–73, 74; evolution of communication, 353–54; intention movements, 71–72; redirected behaviors, 71, 73–74
Ritualized displays, 62
Robinson, Beren, 316
Rodent mating systems, 215–16
Rodents, 228
Rooster plumes, 159–60
Rough-skinned newts, 127–29
Rowe, Matt, 36
Rowher, Sievert, 313
Runaway model, 153, 155, 157–58, 160–62, 167
Rundus, Aaron, 36
Ryan, Michael, 80

Sabertooth cats, 341–44
SAC (Sexually antagonistic coevolution), 179

INDEX

Safe sleeping sites, 275
Salamanders, 127–29, 314, 316
Sand tiger sharks, 207, 210
Sandwich terns, 228
Satisficing, 7
Sawflies, 273
Scatter-hoarding, 231
Scent marking behaviors, 72
Scheel, David, 34
Scheuerell, Mark, 28
Schindler, Daniel, 28
Schreckstoff, 33
Scrounging, 12
Seahorses, 108, 131
Search image, 8, 30
Searcy, William, 82
Secondary sexual traits, 104, 149, 201
Sedentary species, 199
Seismic communication, 55
Selection on sexually selected traits, 126–27
Selfish genetic element (SGE), 187
Selfish herd, 231, 270–71
Selfishness, 266, 267–68
Seminal fluid proteins (SFPs), 183–84, 185
Sender, 52
Sen Sarma, Moushumi, 66, 67
Sensory bias, 80, 353
Sensory drive, 82–83
Sensory exploitation model, 155–56
Sensory traps, 80, 81
Sequential choice theory, 10
Serengeti wildebeest, 343
Serial polygamy, 200
Serotonin, 236, 300
Sex attractants, 61
Sex difference in sexual selection, 111, 118
Sex peptide (SP), 184
Sex pheromones, 57, 69
Sex ratio (SR), 187
Sex ratio meiotic drive, 187
Sex-role reversal, 109, 113, 122, 123, 201
Sexual behavior, and communication, 61–62
Sexual conflict: consequences of, 185–87; definition of, 174; in fruit flies, 180, 182–85, 187, 188; in insects, 177–79, 185; introduction to, 173–75; link with mating system, 180–81; operation of, 179–80; parental investment, 176; summary, 187–88; as a universal truth, 175–79
Sexual dimorphism, 224–25, 263
Sexual excitants, 61
Sexual identity, 69
Sexually antagonistic coevolution (SAC), 179
Sexually monomorphic, 346
Sexually selected traits, 126–27
Sexual ornaments, 346–48
Sexual selection: among males, 201; antipredator adaptations, 25, 27, 44, 46; Darwin-Fisher theory of, 104; definition of, 99; measurement of, 108–9; models based on adaptive female choice for ornaments, 161–62; opportunity for, 111, 120–21; ornaments and armaments, 145–49; postcopulatory, 122, 134–35, 188; qualitative measurements of, 109; quantitative measurements of, 118–19; sex difference in, 124–25; sexual conflict and, 179–80. *See also* Mating systems and measurement of sexual selection
Sexual selection gradient, 120, 122
Sexual size dimorphism, 345–46
Sexy sons, 160–61
Seychelles warblers, 276
Seyfarth, Robert, 80–81
SFPs (Seminal fluid proteins), 183–84, 185
SGE (Selfish genetic element), 187
Sherman, Paul, 330–32
Shuster, Stephen, 103
Siblicide, 206, 207–9
Signal detection theory, 8
Signal honesty, 162–64
Signals, 12–13, 52, 79–83, 210–12
Skewed sex ratio, 148–49
Slabbekoorn, Hans, 82
Smith, W. John, 53, 76, 78
Snake-harassment behavior, 36
Snakes: diversity in behavior, 15; gopher snakes, 36, 37, 85; ground squirrels and, 43; Puerto Rican lizards and, 84–85; rattlesnakes, 36, 37, 85, 348–50; red-sided garter snakes, 175, 177; sexual conflict in, 177
Snowshoe hares, 312

Social behavior: access to clumped or limited resources, 275–76; categorization of, 258–59; cooperative breeding and reproductive skew, 276–80; eusociality, 226, 258–59, 280–81, 332–33; foraging efficiency, 273–74; genes and behavior, 261–62; introduction to, 257–58; kin recognition, 68, 264–66, 334–35; living in groups, 268–70; passive aggregations, selfish herd, and the dilution effect, 270–71; protection against predators and intruders, 271–73; proximate mechanisms of, 261–66; proximate vs. ultimate causation, 259–61; range of, 268–81; reproduction, 274–75; study of, 281–83; ultimate causes of, 266–68; vocal learning, 262–64
Social communication, 353–54
Social foraging, 12–13
Sociality, 258, 340
Social learning, 203
Social mating systems: definition of, 106; vs. genetic mating systems, 104, 131; resource-defense, 118; spacing behavior and, 238–39. *See also* Mating systems and measurement of sexual selection
Social mediation, 150
Socioecology, 222
Sockeye salmon, 29, 30
Soldier termites, 204
Soma, Kiran, 235–36
Songbirds, 151–52, 262–64
Song sparrows, 236, 263–64
Sooty shearwater, 6
South American bush dog, 60
SP (Sex peptide), 184
Spacing behavior and systems: aggregations, 226–28, 236–37, 268–69; aggression, 239–40, 310–12; consequences of, 238; conservation effects, 241–42; developmental effects on spacing behavior, 236–38; ecological and social determinants of, 228–34; genetic basis for spacing behavior, 234–35; hormones and, 235–36; introduction to, 221–23; mating patterns and reproductive success effects, 238–39; neurobiological basis for spacing behavior, 235–36; population-level processes effects, 239–41; predation and antipredator behavior, 236–37; predation risk, 231–32; proximate mechanisms underlying, 228–42; random distribution, 224; summary, 242–43; type of, 223–28; uniform distribution, 224–26
Spadefoot toads, 210, 335
Spanish terrapins, 313
Sparrowhawks, 40, 43
Spatial mean crowding, 106, 117–18
Specialist foragers, 6
Speciation, 186
Species identity, 67–68
Species recognition hypothesis, 153–55
Sperm competition, 134, 183
Sphecid wasps, 234
Spiders, 15, 41, 59–60, 181, 240. *See also* Brown recluse spiders; Comb-footed spiders; Desert spiders; Fishing spiders; Wolf spiders
Spider webs, 41
Spiny lizard, 234
Spite, 266, 268
Sponking, 84
Spotted hyenas, 196–97
Spronking, 70
Squirrels, 348–51. *See also* Belding's ground squirrels; California ground squirrels; Eurasian red squirrels; Ground squirrels
SR (Sex ratio), 187
Stable equilibria, 161
Standardized mating differential, 112
Standardized selection differential, 112
Standardized selection gradient, 112
Stankowich, T., 39
Static signals, 55
Status, 70–71, 167–68
Steelhead trout, 234
Stereotyped behavior, 54
Sticklebacks, 73, 207, 308, 316
Stillness, 32
Stotting, 43, 70
Stress, 260
Submission, 62–63, 303
Subsociality, 258–59
Sugano, Yoshikazu, 186

INDEX

Superb fairy wren, 84, 261, 278–79, 333–34
Super-colonies, 312
Survival costs, 163
Survivorship, 27
Swordtail fish, 156, 302
Syngnathids, 108
Syracuse University, 187

Tactile communication, 59–60
Tail autotomy, 40
Tail flagging, 36, 70, 85
Taiwanese tree frogs, 210
Tammar wallabies, 43
Task specialization, 280
Teaching, and social foraging, 12
Teleost fish, 207
Temperament traits, 308–9
Temperature, 133
Templates, 262, 334–35
Temporal mean crowding, 106, 117–18
Tent caterpillars, 65
Termites, 204
Territorial animals, 151–52
Territorial behavior, 224
Territoriality: effect on population dynamics, 239–41; effects on mating patterns and reproductive success, 238–39; food abundance, 229, 233–34; hormones and, 235; uniform spacing, 224–25
Territory, 224
Territory in Bird Life (Howard), 224
Testosterone, 215, 235, 261, 301, 307–8, 311
Thomson's gazelles, 43
Threespined sticklebacks, 207, 308
Thrushes, 313
Tiger moth, 258
Time in, 113
Time out, 113
Tinbergen, Niko, 73, 260–61, 353
Townsend solitaires, 229
Townsend's warblers, 313
Toxicity, 39
Trade-offs, 163, 197, 204–5, 296, 307, 309
Transgenic fruit flies, 187–88
Transmission of parasites and pathogens, 242
Transparency, 32

Traplines, 7–8
Treehoppers, 64–65, 269
Tree shrews, 225
Tuco-tucos, 233
Túngara frogs, 54, 79–80, 83, 351–52
Typical-intensity displays, 74–76

Ultimate mechanisms: of social behavior, 259–61; in spacing systems, 221
Ultrasonic sounds, 56
Uniform distribution, 224–26
Uniform spacing, 224–26
University College London, 182
University of Cincinnati, 188
University of Michigan, 330
Unreliable information, 77–78
Urination in dogs, 57

Vampire bats, 266
Variable-intensity displays, 74–76
Variance, 201
Várzea lakes, 28
Vasopressin, 215, 235, 236
Veen, Jan, 228
Velvet swimming crabs, 302
Venner, Samuel, 41
Vervet monkeys, 272
Vibrational communication, 59–60
Visual communication, 55–56
Visual systems, 30–31
Vocal communication, in dinosaurs, 345, 351–53
Vocalization, 265
Vocal learning, 262–64
Vocal signatures, 334
Von Frisch, Karl, 66–67

Wade, Michael, 103, 124
Waggle dance, 65–67, 70
Wallace, Alfred, 152–53
Warning coloration. *See* Aposematism
Wasps, 268
Waterfowl, 175–77
Water mites, 80
Water striders, 178
Wattled lapwing, 164
Weasels, 349

Western bluebirds, 282, 302, 307, 311–12, 313, 315
White-browed scrubwren, 84
White-crowned sparrows, 31, 263
White suckers, 44, 45
White-tailed deer, 32
Whittaker, Danielle, 59
Wigby, Stuart, 184
Wild turkeys, 274
Wilson, Edward O., 259
Wingfield, John, 235–36
"Win-stay, lose-shift," 8
Wolf spiders, 158
Wolves, 43, 303
Woodpeckers, 229, 258, 277–78
Worker termites, 204

Wrens, 313
Wright, Dominic, 234
Wright, Sewall, 329

Ydenberg, Ron, 39
Yello casqued hornbills, 84
Yellow-headed blackbird, 211

Zahavi, Amotz, 77, 163
Zebra finches, 261, 263, 302
Zebrafish, 234
Zeus bugs, 178
Zig-zag dance, 73
Zooplankton, 28
Zuberbühler, Klaus, 84
Zwartjes, Patrick, 159